COVID-19: Impacts on Health and Health Systems

Rais Akhtar
Editor

COVID-19: Impacts on Health and Health Systems

Springer

Editor
Rais Akhtar
Formerly Professor of Geography
University of Kashmir
Srinagar, Jammu and Kashmir, India

ISBN 978-3-031-90254-3 ISBN 978-3-031-90255-0 (eBook)
https://doi.org/10.1007/978-3-031-90255-0

This Springer imprint is published by the registered company Springer Nature Switzerland AG
The registered company address is: Gewerbestrasse 11, 6330 Cham, Switzerland

Foreword

I have learnt a great deal from this book. The first thing I learnt is that the impacts of the great COVID epidemic showed huge inequalities. One of these inequalities was between different nations, and resulted from such factors as the state of their healthcare systems, and the demographic characteristics of the population (including age profiles). Climatic differences may also have played a role, as COVID tends to thrive more under cold conditions than warm.

Then there were differences within countries. One factor here was the age of the population, with older people being much more susceptible to COVID. Another factor was population density, for in densely populated areas social distancing was more difficult, and reduced air quality could lead to greater susceptibility to respiratory problems associated with COVID. Yet another factor was variations in wealth within a country, with the poorest members of society, as for example, in slum areas, having less access to healthcare.

It has also been apparent that COVID's effects on health not only were those caused by the virus itself, but also by knock-on effects. The concentration on treating seriously ill victims of COVID inevitably led to less ability to treat other conditions and to provide routine healthcare. Care giving in homes under lockdown could be problematic, which meant that some vulnerable people suffered. Moreover, some isolated people found access to food was rendered difficult, which could also have health implications. Secondly, there were serious impacts on mental health. Factors included isolation during social distancing, fear of the virus, economic worries caused by economic stress or loss of jobs, loss of dear ones, and the seemingly unending stress caused by long COVID.

In some countries, the economic effects of the COVID pandemic were also severe, with social distancing and long-distance transport restrictions causing great problems, inter alia, for the tourism and hospitality sectors. Inevitably, some countries suffered more than others in this respect.

There are many lessons that need to be learnt from the COVID pandemic so that we can cope all the better if another global pandemic afflicts us in the future. Many of the contributors to this book, with their experience in many different countries, address this important issue and point us to the way forward.

Prof. Andrew Goudie
Emeritus Professor
University of Oxford
Oxford, UK

Preface

The book entitled *COVID-19: Impacts on Health and Health Systems*, discusses catastrophic health consequences, and policy failure in both developed and developing countries leading to **7,010,681** deaths globally. The book comprises 24 chapters encompassing both developed and developing countries.

John Connell from the University of Sydney explained that in attempts to revive their struggling economies, Pacific island states, including **Vanuatu and Fiji**, cautiously reopened their borders in 2022 as the COVID pandemic faded. Both countries had fared well in terms of the incidence of COVID-19 by closing their borders, but the pandemic revealed basic problems in healthcare systems that proved difficult to resolve with limited fiscal space and growing skilled labor shortages. Tourism, the core of island economies, quickly revived and returned to a familiar pattern centered on urban resorts and later cruises, rather than toward a more sustainable structure supporting a blue economy. Here too, expansion was frustrated by labor shortages, fragmented and costly supply chains and cyclone damage. Loss of skilled health workers and trades people was Pacific Australia Labour Mobility (PALM) scheme attracting workers to Australia for longer periods. A brain and skill drain was intensified. Both the restructuring of the health sector and economic growth were thus constrained and diverted by global factors linked to risk, inflation, strategic issues, and regional labor supply and demand.

The chapter on Fiji by Eberhard Weber deals with the impact of COVID-19 on **Fiji**, a Small Island Developing State in the Pacific Islands region. Fiji has had to cope with the impacts of natural hazards for as long as people remember. In recent years the dangers of worsening climate change have made societal action indispensable. In early 2020 another threat was added, which pushed climate change into the background for some time.

In addition, the action to face natural hazards became fragmented and thus lost its focus and effectiveness. This was most evident in cases where hazards such as tropical cyclones or floods occurred simultaneously to the COVID-19 threat. It soon became clear that existing capacities were not sufficient to deal with various hazards satisfactorily.

The effects of the COVID-19 pandemic on the environment have been highlighted from a wide variety of perspectives. Two of these perspectives stand out: (1) COVID-19 impacts on the global economy, global exchange of goods and people. Among many negative impacts it also had positive effects on the environment. (2) At the same time several authors highlighted that environmental factors influenced the transmission of SARS-CoV-2 through various factors.

For Fiji declining economic activities, significantly reduced mobility and a considerable reduction in tourism led to lower energy consumption and therefore lower emissions of CO_2, and other Greenhouse Gases (GHG).

Mei-Hui Li on **Taiwan** asserts that because infectious diseases transmission is complex and influenced by multiple factors, the dynamics of emerging and reemerging diseases is difficult to predict. This study investigates the pattern of dengue occurrence in Taiwan by considering the relationship between disease dynamics and environmental elements. In Taiwan, the patterns of dengue outbreaks have changed over the past 120 years. First, the author, I analyze dengue fever occurrences over the last century classified into three major periods to determine which factors potentially affect dengue outbreaks. Interestingly, changes in the patterns of dengue occurrences in Taiwan coincided with changes in the form of government (i.e., from Japanese colonialism to Kuomintang authoritarian military rule to multiparty democracy). Subsequently, the author compared the incidence of dengue outbreaks before, during, and after the COVID-19 pandemic in Taiwan. Moreover, meteorological factors associated with dengue transmission were examined in Tainan city, one of the dengue hotspots in Taiwan. Finally, the author reviewed the current intervention policy for dengue outbreaks in Taiwan. The changes in the patterns of dengue incidence in Taiwan reflect the interplay among different factors on multiple levels.

The first chapter on **Japan** by Kazutaka highlights that the increase in temperature due to climate change and the heat island effect has led to significant health impacts, such as heat stroke, in Japan during the summer. The risk of heat stroke is expected to become more severe with the intensification of climate change. This chapter examines the combined impacts of heat stroke associated with high temperatures and other factors using the COVID-19 pandemic as a case study. It first summarizes scientific findings from existing studies on changes in heat stroke risk during the COVID-19 pandemic. The measures implemented by the Japanese government, local authorities, and academic groups to balance COVID-19 and heat stroke countermeasures are also introduced. Finally, this chapter discusses scientific findings and future strategies.

Another chapter on **Japan** by Ai Tashiro opined that in May 2023, Japan downgraded the legal status of coronavirus disease 2019 (COVID-19) from Class 2 to Class 5, placing it in the same class as common infectious diseases. This chapter provides an overview of Japan's medical system after reclassifying COVID-19 as Class 5, including changes in subsidies for dispatching healthcare workers and public financial support. Drawing on lessons learned from the COVID-19 pandemic, this chapter explores Japan's internal health strategy and how it aims to transform universal health coverage (UHC) into a global health strategy with two critical policy goals: establishing a global health architecture and promoting sustainable UHC in Japan and abroad in the post-COVID-19 era.

The chapter **Indonesia** by Budi Haryanto, Hanifa M. Denny, Ira Nurmala and Rico J. Sitorus, has long faced challenges with communicable diseases such as dengue fever, tuberculosis (TB), and malaria, alongside the rising threat of air pollution-related diseases. These health issues were exacerbated during the COVID-19 pandemic, disrupting prevention, diagnosis, and treatment programs. Pre-COVID-19, Indonesia recorded one of the highest rates of TB globally, with over 845,000 cases reported in 2019. Dengue fever remained endemic, peaking during the rainy season, with around 138,000 cases annually. Malaria, although largely controlled in certain regions, still impacted provinces in Eastern Indonesia like Papua, with a prevalence of 5.4 per 1000 people. Air pollution, particularly in urban areas like Jakarta, contributed significantly to respiratory diseases, driven by vehicle emissions and industrial activities. The COVID-19 pandemic posed further complications. Health services were redirected to manage the pandemic, leading to a decline in diagnosis and treatment for TB, with an estimated 30% reduction in detected TB cases during 2020. Malaria control programs were disrupted due to movement restrictions, affecting the distribution of bed nets and indoor residual spraying. The incidence of dengue fever saw fluctuations, with initial reports of lower cases, possibly due to underreporting amidst the pandemic. Additionally, air quality in major cities temporarily improved during the early lockdowns, reducing pollution-related respiratory diseases. Post-COVID-19, the healthcare system is gradually recovering, but the pandemic's long-term impact is evident. TB programs are regaining momentum, although cases remain underreported. Dengue and Malaria control initiatives are being reactivated, yet public health efforts are strained by resource allocation. Air pollution remains a significant public health issue, with pollution levels rebounding to pre-pandemic levels as economic activities resume. The post-pandemic recovery requires integrated public health strategies to tackle these endemic diseases alongside emerging global health threats.

According to authors Uma Langkulsen and Gamuchirai Munjoma, in **Thailand**, the dynamic relationship between diseases and environmental factors, highlighted by the COVID-19 pandemic, underscores the critical role of public health interventions. In Thailand, urbanization, air quality, and climate shaped COVID-19 spread, with densely populated areas bearing the brunt. The government responded with expanded healthcare access and innovative services. Meanwhile, shifts in disease patterns like diabetes and cardiovascular ailments necessitate proactive surveillance. Addressing these complexities demands integrated approaches, considering environmental, social, and medical factors. By prioritizing evidence-based strategies, public health initiatives can effectively navigate this intricate interplay to safeguard communities and enhance overall well-being.

The chapter on **Malaysia**, by Nasrin Aghamohammadi, Logaraj Ramakreshnan, Rama Krishna Supramanian, and Yin Cheng Lim assert that the COVID-19 pandemic has had a significant impact on healthcare and environmental systems with far-reaching consequences influencing the ecological well-being, economic stability, social cohesion, and institutional trust. The pandemic has worsened prevailing health disparities by disproportionately impacting vulnerable communities with limited access to healthcare resources nationwide. In addition to overwhelming healthcare systems due to escalating infections, hospitalizations, and fatalities, the pandemic has caused significant disruptions to regular healthcare services, that led to the postponement of elective surgeries, screenings, and other treatments. The pandemic has led to an increase in mental health issues linked with fear of the virus, social isolation, and economic uncertainty. Lockdowns and social distancing measures have also influenced the physical activity levels, dietary habits, and sleep patterns, potentially affecting the prevalence and management of lifestyle-related diseases. However, the pandemic has highlighted the significance of having a well-prepared public health infrastructure prompting more investments in healthcare systems to effectively address any future health emergencies. The COVID-19 pandemic has had both positive and negative impacts on environmental issues. The implementation of lockdowns and travel restrictions led to a significant decline in energy consumption patterns and air pollution due to reduced use of fossil fuels, passive mobility, and industrial emissions. Even though reduced industrial discharges had positive impacts on water bodies, the improper disposal of waste raised concerns about marine pollution and waste management. Changes in human activities benefited some wildlife from reduced human disturbance, whereas the others faced challenges due to disruptions in food supply chains and conservation efforts. The pandemic also underscored the importance of resilient ecosystems in preventing the emergence and spread of zoonotic diseases, which led to increased emphasis on sustainable land use and wildlife conservation. The full impact of COVID-19 on the healthcare and environmental systems is continuously unfolding, and long-term comprehensive and coordinated efforts are crucial to tackle these dual concerns.

The chapter on **India** and **Canada** by Hem H. Dholakia and Malav D. Jhala show that COVID-19 activity has been stable or decreasing. Vaccination efforts, awareness creation, quarantine measures, and travel advisories in both countries have contributed to this. Over time, new variants of COVID-19 have emerged in both countries. Currently, the BA.2.86 variant is dominant in Canada while BA.2.75 (Omicron) has remained the dominant variant in India. It is expected that new variants may emerge in future and surveillance efforts need to continue. Elderly or people with pre-existing diseases constitute the majority of hospital admissions suggesting that containment efforts should be directed toward vulnerable populations.

In **Nepal**, Yogendra Shah and colleagues assert that the, Government of Nepal took several measures for prevention and control of COVID-19 and provided support to vulnerable groups. The High-Level Coordination Committee for COVID-19 Prevention and Control coordinated the response and provide necessary policy directions against the COVID-19. The COVID-19 Crisis Management Centre monitored the stock of essential supplies and conducted other essential activities for prevention and control of COVID-19. Nepal started COVID-19 vaccination in early 2021. The coordinated response activities resulted relatively a smaller number of mortalities due to COVID-19 in Nepal. Taking lessons from COVID-19, there are several policy responses and activities to address pandemics in future. This chapter aims to summarize epidemiology of COVID-19 and policy response including post-COVID health situation in Nepal.

In **Kuwait** the COVID-19 pandemic profoundly affected healthcare systems, highlighting vulnerabilities with far-reaching implications beyond health disciplines. In response, Kuwait invested significantly in pandemic containment and mitigation strategies, bolstering its health system's resilience. This chapter explores the adaptational changes within Kuwait's healthcare landscape post-COVID-19. It highlights the nation's strategic initiatives, including community-based surveys for SARS-CoV-2, the implementation of an electronic vaccine registration system, and the establishment of a vaccine adverse event reporting system. These measures represent a paradigm shift in Kuwait's healthcare approach, signifying a broader impact of the pandemic on healthcare policy and practice.

With a focus on **Turkey**, the authors Nur Konyalilar, Muge Akpinar-Elci, and Hasan Bayram referred to studies which suggest a positive association between increased levels of air pollutants, such as particulate matter (PM), ozone, and nitrogen dioxide, and Coronavirus Disease 2019 (COVID-19), mortality and morbidity worldwide, including Turkey. PM has been shown to carry SARS-CoV-2 virus. Air pollutants can induce SARS-CoV-2 infection in human beings by both stimulating the expression of proteins (i.e., ACE2 and TMPRSS2) required for viral entry and impairing host immune defense. The objectives of this chapter are to discuss the recent epidemiological evidence, as well as the in vivo and in vitro mechanistic studies involving the impact of air pollutants on COVID-19 mortality and morbidity together with the Turkish perspective on COVID-19 from an epidemiological standpoint, focusing on air pollution and health infrastructure during and after the pandemic.

The chapter on **Mozambique** by Maria dos Anjos Ernesto Hauengue summarizes the impact of pandemic COVID-19 in Mozambique, particularly in the health system, the risk factors associated to the disease and the country epidemiological profile in the post-COVID-19 period. In Mozambique, the disease infected a total number of 137,413 people and caused 1690 official deaths. The negative impact caused by this disease was enormous, affecting different areas of development. However, for the interest of this document, we underline here the consequences brought to National Health System in terms of provision of services; demand of services by patients especially in the area of maternal—child healthcare. It also analyzes the implication of the disease on the epidemiological features of the country; measures implemented by the government to overcome the situation.

At the end, the reports recommend the need for more investment either in financial resources or material and specific equipment to increase the capacity of public health hospitals to prevent and minimize the prevalence and incidence of the disease, and revert the epidemiological profile of the country.

The chapter on **Russia Federation** by Svetlana M. Malkhazova, Fedor I. Korennoy, Tamara V. Vatlina, Dmitry S. Orlov present a medical-geographical analysis of SARS-CoV-2 infection course in the Russian Federation (RF) until September, 2021. The major focus on regional distribution of main epidemiological indicators based on official statistics was explored with regard to most general demographic and environmental factors. Time series analysis revealed the clusters of regions with similar dynamics of COVID incidence during the three epidemic waves. Calculation of the basic reproductive ratio was conducted for each epidemic wave. An analysis of vaccination dynamics and its effect on the epidemic course is also presented in the chapter.

Concentrating on **Rome, Italy** Cosimo Palagiano and colleagues demonstrate that the COVID-19 pandemic has highlighted the importance of dynamic spatial distribution systems, especially for non-pharmaceutical interventions and real-time monitoring of epidemiology for making predictions. A judicious use of these technologies could enable making decisions in advance (e.g., increasing the number of hospital beds and medical personnel, enhancing equipment for specific areas) and assessing the effectiveness of measures taken (e.g., border closures, vaccination). Additionally, it can be used to monitor the varied distribution of viral genetic variants of concern for epidemiological and therapeutic purposes. In this chapter, after a synthetic overview of some aspects concerning SARS-CoV-2 and COVID-19 pandemic, we provide an applied example of geocoding and data aggregation in GIS environment which may be used both in the first phase of emergencies, to identify spatial clusters and to hypothesize the evolution trend, and over the time to monitoring the disease patterns. For this example, also in the perspective of precision preparedness, we focus the GIS elaborations on a set of data regarding the period from 25 February to 26 September 2020 showing for the territory of Rome (Italy) some potentialities of geocoding and data aggregation.

The second chapter on **Italy** by Giovanni De Santis, and Annamaria Bartolini explains the concepts of universality, equality, and globality are the basis of the Italian healthcare system established in 1978. However, political intervention in 1992 forced its corporatization, the divestiture of Territorial Medicine and the closure of small hospitals. As a result, the healthcare system was unprepared for the COVID-19 pandemic, causing symptomatic patients to come to the emergency room, which accelerated the spread of the infection. In preparing the new pandemic plan, authorities in charge did not take into account the critical issues that also emerged in the private sector; they identified the necessary emergency measures, but did not prepare the steps to implement them.

Isabelle Roussel from **France** stated that the World Health Organization (WHO) raised the highest level of alert on the COVID-19 pandemic in May 2023, which does not mean that COVID-19 has disappeared. It's time to draw conclusions. In the aftermath of the pandemic (2020–2022), the way we look at illness and health has changed in France since this episode which caused the death of 116,000 people. On the one hand, it is a question of learning the lessons of this pandemic which has led people to say "that nothing will ever be the same again" and, on the other hand, the spread of this virus has left many after-effects which are necessary to deal with. The hospital crisis has not disappeared, the psychological consequences combined with solastalgia linked to climate change are numerous, mistrust against vaccines, and against the scientific approach to medicine has increased while prevention in the framework of global health is increasingly necessary.

Focusing on Health System Policy and the Pandemic in **England**; assessing its impact and shape, Michael Calnan brings out the policy response to the pandemic in England that was characterized by a mixture of non-pharmaceuticalized and pharmaceuticalized policies with the former generally being a precursor to the latter. However, the success of these policies hinges on the health system (both healthcare and public health). Thus, the role played in this policy response by the health system and the extent to which was prepared for a pandemic will be the focus of this analysis. It will consider how effective it was at managing the pandemic and why and what were the consequences for patient care, the public's health and the health of its workforce. Drawing on the evidence given at the current public inquiry in England it will aim to identify any lessons learnt for future pandemic policy.

With a focus on **USA**, Michele Kekeh, and Hedieh Sirjani asserted that COVID-19 has caused severe adverse outcomes among older adults. According to the Centers for Disease Control and Prevention, the death rate among older adults was 533.5 per 100,000 in 2020. Their risks increased with age and underlying medical conditions. In the United States, the number of people with at least one chronic disease in the older adult population is estimated to increase from 71.52 million in 2020 to 142.66 by 2050. Within that timeframe, this population is expected to grow from 137.25 to 221.13 million. The rapidly growing aging population is still at high risk of adverse health outcomes from any future pandemic. This chapter explores how healthcare access and quality were altered for older adults, highlights emerging challenges and opportunities, and discusses innovative approaches and policies to improve healthcare service delivery for the aging population. Additionally, the study investigates requirements from caregivers as they adapt to new protocols to ensure the well-being of the older adults for whom they provide care.

The second chapter on **USA** by Olaniyi Olayinka and Terence Tumenta examines the impact of the COVID-19 pandemic on mental health and substance use in the United States. The global health crisis triggered a cascade of psychological and behavioral effects, reshaping healthcare services and delivery. Drawing on a review of studies, clinical reports, and epidemiological data, the chapter highlights the connection between the pandemic and changes in mental health disorders and substance misuse patterns and care. The chapter concludes with a discussion of the consequences for healthcare systems, including the expansion of telemedicine that increased access to mental health and substance use treatment. By offering a nuanced understanding of the intricate relationship between the COVID-19 pandemic, mental health, and substance use, this chapter aims to contribute to our understanding of the enduring impact of the crisis on the well-being of individuals and communities in the United States.

The chapter on **Brazil** by Rhavena Barbosa dos Santos, Júlia Alves Menezes, and Rose Ane Pereira de Freitas explains the environmental transformations, climate change, deforestation, and urbanization that have significant impacts on human health. This chapter aims to address the complex relationship between these changes and the emergence of diseases, as occurred with the COVID-19 pandemic, in addition to understanding the relationships between environmental changes and the spread of infectious diseases, considering the Brazilian context. The pandemic highlighted human vulnerability to emerging diseases, in addition, the uncontrolled exploitation of ecosystems can facilitate the transmission of viruses from animals to humans. It is crucial to consider the interconnection between human health and the environment and encourage sustainable practices to prevent future pandemics and protect global health.

The chapter on **Argentina** by Daniel Oscar Lipp assessed the impact of the pandemic that was widespread in the country, affecting all areas of the economy, in addition, of course, to the numerous direct consequences on health. This chapter describes what happened in the socioeconomic field particularly affected by the crisis. The growing levels of poverty, lower levels of well-being, growing public spending, informality and unemployment show how extensive and deep our economic and social deterioration has been as a result of the pandemic.

Srinagar, India Rais Akhtar

Acknowledgements

In the process of writing, editing, and preparing this book, there have been many people who have encouraged, helped, and supported me with their skill, thoughtful evaluation of chapters, and constructive criticism.

First of all I am indebted to all the contributors of chapters from both developed and developing countries for providing the scholarly and innovative scientific piece of research to make this book a reality. I am also thankful to the reviewers who carefully and timely reviewed the manuscripts.

I am also grateful to Prof. Andrew Goudie, University of Oxford, for writing Foreword, which adds greatly to the book with his thoughtful insights.

I am thankful to my family—wife Dr. Nilofar Izhar, daughter Dr. Shirin Rais, Assistant Professor, AMU, and my son-in law Dr. Wasim Ahmad, Associate Professor, IIT Kanpur, who encouraged and sustained me in developing the structure of the book and editing tasks, and I am deeply grateful for their support and indulgence.

Finally and most essentially, I am deeply obliged to the Springer and the entire publishing team, without whose patience, immense competence and support this book would not have come to fruition. I specially thank Dr. Robert K. Doe whose energizing leadership and inspiration ensured that this book would indeed translate to reality. I am also thankful to Prof. A. R. Kidwai, Doris Bleier, and Ms. Saranya Kalidoss for their constant cooperation.

Srinagar, India | Rais Akhtar

Contents

About the Editor

Rais Akhtar is formerly Professor of Geography, and Dean, University of Kashmir, Srinagar, National Fellow and Emeritus Scientist (CSIR), CSRD, Jawaharlal Nehru University, New Delhi, and Visiting Professor, Department of Geology, AMU, Aligarh.

He has taught at the Jawaharlal Nehru University, New Delhi, University of Zambia, Lusaka and The University of Kashmir, Srinagar. He is recipient of a number of international fellowships, including Leverhulme Fellowship (University of Liverpool), Henry Chapman Fellowship (University of London), and Visiting Fellowship, (University of Sussex), Royal Society Fellowship, University of Oxford, and Visiting Professorship, University of Paris-10. Prof. Akhtar was elected Fellow of Royal Geographical Society, London and Royal Academy of Overseas Sciences, Brussels.

Prof. Akhtar delivered invited lectures in about 55 universities, geography departments and medical colleges, including London School of Hygiene and Tropical Medicine, Liverpool School of Tropical Medicine, Departments of Geography, University of Edinburgh, Leicester, Strathclyde, Glasgow, and Liverpool, Institute of Geography, National University, Mexico City, University of Hawaii, Honolulu, and University of Akron, and School of Public Health, Johns Hopkins University, Baltimore, Bologna Hospital, Institute of Health, Universities of Perugia and Camerino, and Institute of Public Health Pisa.

He was Lead Author (1999–2007), on the Intergovernmental Panel on Climate Change (IPCC), which is the joint winner of Nobel Peace Prize for 2007. Prof. Akhtar is the recipient of Nobel Memento.

Prof. Akhtar has to his credit 94 research papers and 22 books published from India, United Kingdom, United States, Germany, and The Netherlands. His books entitled: *Climate Change and Human Health Scenario in South and Southeast Asia*, was published in 2016 by Springer, and his other books include *Climate Change and Air Pollution* (with C. Palagiano) published by Springer, and *Geographical Aspects of Health and Disease in India* (with Andrew Learmonth) published in 2018 by Concept (New Delhi). His latest books include: *Extreme Weather Events and Human Health*, was published by Springer in early 2020, *Coronavirus (COVID-19) Outbreaks, Environment and Human Behavior*, and *Coronavirus (COVID-19) Outbreaks, Vaccination, Politics and Society*, were also published by Springer in 2021 and 2022 respectively. Prof. Akhtar's latest book entitled *Climate Change and Human Health Scenarios*, was published by Springer in 2023.

Prof. Rais Akhtar was Member of Expert Group on Climate Change and Human Health of the Ministry of Health and Family Welfare, Government of India. Prof. Akhtar has been appointed as Corresponding Member of Italian Geographical Society in October 2018. Prof. Akhtar is a Member of Belgium Journal of Geography Editorial and Advisory Boards.

Chapter 1
Introduction: Impact of COVID-19 on Health and Health Systems

Rais Akhtar

Abstract The impact of the corona crisis was a short-term devastating effect causing at the same time severe suffering all over the world. Therefore, researchers have rightly urged both developed and developing countries to invest in strategies to improve the resilience of their health systems for future emergencies. Geddes has rightly urged that "Health system resilience must become a central component of national health plans. Given the likelihood of future pandemics and other major shocks, there is an urgent need to design more resilient health systems capable of addressing a crisis while maintaining essential functions." (Geddes in How the COVID-19 pandemic has affected healthcare around the world, 2022).

Keywords Health system resilience · Pandemics waves · Dr. Anthony Fauci · Preventive and curative services · Zoonoses · Neglected Tropical Diseases (NTDs)

1.1 Introduction

The coronavirus disease (COVID-19)—the global pandemic—has caused severe acute respiratory syndrome coronavirus-2 (SARS-CoV-2), and wreaked havoc on the world socio-economically, and nearly half of the world's 3.3 billion global work-force were at the risk of losing their livelihood (Akhtar 2021). The pandemic has also widened socio-economic inequalities in both developed and developing countries and has devastated the world resulting in 704,753,890 infections and close to 7,010,681deaths till April 13, 2024. In the USA alone 1,219,487 deaths were reported until 13 April 2024. Prevalence of COVID-19 had caused a significant decline in the number of patients seeking routine health services related to communicable and cardiovascular diseases in almost all countries belonging to both developed and

R. Akhtar (✉)
Formerly Professor of Geography, University of Kashmir, Srinagar, Jammu and Kashmir, India
e-mail: raisakhtar@gmail.com

R. Akhtar (ed.), *COVID-19: Impacts on Health and Health Systems*,
https://doi.org/10.1007/978-3-031-90255-0_1

developing regions. Dr. Anthony Fauci,[1] former advisor to former US President Joe Biden, asserted that domestic policy failure and the lack of a public health infrastructure for handling an outbreak of such proportions have resulted in such huge human loss.

Besides, COVID-19 made a massive dent in the diagnostics sector with a steep fall in patient volumes since the beginning of 2020. The pandemic developed an haunting sense of fear among people, which eventually impacted hospital/clinic and lab visits. Regular check-ups became cumbersome, stressful, and time-consuming.[1]

Though COVID-19 pandemic has affected all sectors directly or indirectly, the crisis is worse on the already overburdened health systems in many developing countries. It is evident that significant level of health service delivery was impacted, especially, during the early times owing to the steady spread of the virus across all settings. Many of essential health services have been delayed, patients were also unable to attend follow-ups and acute care visits due to the fear and anxiety they experienced during the pandemic waves (Haileamlak 2021) It presented a severe challenge on health facility availability and utilization and medical supply management in both developed and developing countries., particularly in resource impoverished regions, and has exposed the existing gaps in the health system and rattled both preventive and curative services for communicable and non-communicable diseases/Neglected tropical Diseases. Many of essential services have been delayed by the healthcare facilities, patients were also unable to attend follow-ups and acute care visits due to the fear and anxiety they experienced during the pandemic waves.

In summation, "health systems in many low-and middle-income countries (LMICs) were already weak before the COVID-19 pandemic. They continued to be massively challenged to the point of collapse across the world, affecting public health programmes for both infectious diseases and non-communicable diseases" (Ehrenberg et al. 2021).

---

[1] Dr. Anthony Fauci served as the director of the National Institute of Allergy and Infectious Diseases at the National Institutes of Health, a role he held until stepping down in December 2022. He also served as the chief medical advisor to former President Joe Biden and was the face of America's public health response during the COVID-19 pandemic. His closed door, 14-hour, two-day testimony in January 2024 has served as a critical component of the Select Subcommittee's investigations into the origins of COVID-19, pandemic-era domestic policy failures, and improvements to the United States' public health system. In conjunction with the transcript, the Select Subcommittee also released a new staff memo that highlights the key takeaways from Dr. Fauci's transcribed interview. Dr. Fauci described how the evolving virus—its variants, how it was transmitted has impacted the public's trust in science and revealed a misunderstanding of the scientific approach to a biological, public health problem. Dr. Fauci laid emphasis that "Emerging infections are a perpetual challenge. The only way you address a perpetual challenge is by perpetual preparedness. Dr. Fauci asserted in July 2020, the lack of a public health infrastructure for handling an COVID-19 outbreak of such proportions has resulted such huge human loss in the USA. By the way, Dr. Anthony Fauci in 2008 while referring to pandemics, epidemics, and infectious diseases, warned in a Foreword to a book entitled *Encyclopedia of Pestilence, Pandemics and Plagues,…the next pandemic waits in the wings for some convergence of critical determinants not only imagined by any of us. How we respond may make a difference not only for ourself, but for the rest of the world as well.* It is unfortunate we have not heeded his warning."

1.2 Some Examples

In the context of USA, Cameron E. E and colleagues have rightly asserted, "While the American health system often struggled during the COVID-19 pandemic—a result of both structural issues and underinvestment in pandemic preparedness and foundational and functional public health capabilities" (Cameron et al. 2022). Newly emerging infectious diseases have had damaging impacts on healthcare systems throughout history and have been most acute in weaker health systems. With the COVID-19 pandemic, even the well-developed health systems are struggling most to cope as the crisis deepens in the regions of Europe and America (Rathnayake et al. 2020).

Public health policy paralysis in U.K. had resulted of a growing disconnect between the government's policies and the needs of the population. "The UK's initial public health response to the pandemic comprised a delayed lockdown, shortages of personal protective equipment (PPE), insufficient testing capacity, and ambivalence about mask wearing. The pandemic's first wave ravaged the health and care sectors. Subsequently, with experience and tight testing regimes, management of COVID-19 in the care sector was improved enormously. Hospitals reduced their workload to a bare minimum initially, followed by designing separate pathways to facilitate elective work, underpinned by testing and infection control. In addition to the elderly and frail, those on the fringes of society—for example, homeless, refugees, asylum seekers, and prison populations—experienced high rates of infection and mortality. Nation-wide restrictions on movement were propped by an economic support program" (Misra 2023).

Similarly, in Ethiopia, a developing country, following the first COVID-19 reported case in March 2020, the health system was challenged heavily. The ever-increasing COVID-19 cases demanded reshuffling health care workers and repurposing healthcare facilities. On the other hand, patients with various disease conditions shy away from visiting health facilities. Comparison of the pre-COVID-19 era service utilization with the COVID-19 period showed that there was a substantial disparity in service delivery practice of the health facilities. A huge decline in the patient flow for routine services was noted (Haileamlak 2021).

The COVID-19 challenge passed strong message to the world on the need of building resilient and sustainable health system. In doing so, strong investment to strengthen the health systems including the health workforce development, creating a decent working conditions, providing training and equipment, especially in relation to personal protective equipment and occupational safety is required. We must adopt holistic approach in health system development that takes care of both cardiovascular (life style) NTD, as well as climate change/environmental-based diseases. Social dialogue is essential to building resilient health systems and therefore has a critical role both in crisis response and in building a future that is prepared for health (Haileamlak 2021).

In relation to France, Roussel explained that in the aftermath of the pandemic (2020–2022), the way we look at illness and health has changed in France since this episode which caused the death of 116,000 people. On the one hand, it is a question of learning the lessons of this pandemic which has led people to say "that nothing will ever be the same again", and, on the other hand, the spread of this virus has left many after-effects which are necessary to deal with. The hospital crisis has not disappeared, the psychological consequences combined with solastalgia linked to climate change are numerous, and mistrust against vaccines and against the scientific approach to medicine has increased while prevention in the framework of global health is increasingly necessary. National difficulties combine with a chaotic global geopolitical situation abandoning multilateralism at a time when health and climate issues are global (Kickbusch and Liu 2022). The climate, zoonoses, and trade are globalized, and withdrawal, whether at the level of individuals or that of nations, is an obstacle to the good management of crises which affect all of humanity. Yet a new vision of society seems to be emerging at the end of the pandemic, calling into question values that seemed consensual such as consumerism, competition, exploitation of resources, and anthropocentrism? The time of confinement has undoubtedly contributed to considering health, life, and death in a different way while a certain distrust of all collective systems, including the health system, has tended to establish itself. The model developed by scientific ecology invites us to rethink the proper functioning of ecosystems which conditions the balance of man on the planet, the prevention of zoonoses, and numerous pathologies which have the environment as their origin. A better knowledge of the environment emphasizes prevention in a timely manner because the consequences mainly affect the most disadvantaged populations. Hence, the harmful influence of disinformation, populism, and anti-science movements because transformations in society require quality information, calm debates, and a high degree of public trust in institutions. This more systemic approach to health requires a change in outlook which must respond to a more integrative vision essential to promoting a new conception of overall health and well-being.

In Japan, on 8 May 2023, coronavirus disease 2019 (COVID-19), caused by infection with severe acute respiratory syndrome coronavirus 2 (SARS-CoV-2), became a Class 5 (the Ministry of Health, Labour, and Welfare: MHLW 2023b) instead of a Class 2 infectious disease under the Infectious Disease Control Law in Japan. One year has now passed since the transition to Class 5. Since the days of the COVID-19 pandemic, Japanese society has regained its normal lives. Infection control measures are now based on individual decision-making. The transition to a normal medical care delivery system is gradually underway. March 2024 marked the end of public financial support for medical expenses for patients with COVID-19 and the issuance of fees to medical institutions for securing hospital beds (MHLW 2023a). However, the essential themes of "infection prevention measures" and "securing the healthcare delivery system" that match the situation remain unchanged. Strict infection control measures and behavioural restrictions are no longer in place, and the free vaccination and subsidies for expensive coronary drugs ("special exception on medical fees" for SARS-CoV-2 infections), which had been provisionally in place, ended at the end of 2024. However, they will be extended if required. The shift to Class 5 does not mean COVID-19 has disappeared or is no longer dangerous (From Chapter by Ai Tashiro). The lack of global solidarity in Asia, according to Ai Tashiro problems with global health regulations, and the absence of a global response coordination mechanism are the main issues that require active involvement from the international community. Global health governance is a political arena comprising diverse actors, including major powers, smaller states, and non-state actors. If these actors can build strong solidarity in developing COVID-19 vaccines and treatments, ensuring fair distribution, supporting pandemic resolution, and reinforcing response frameworks, Japan's future in health governance in the post-COVID-19 era looks promising. The urgent task for global health governance is to establish a sustainable health cooperation framework that can withstand political changes while utilizing the individual roles of diverse actors.

In the USA as Michele Kekeh highlights COVID-19 has caused severe adverse outcomes among older adults. According to the Centers for Disease Control and Prevention, the death rate among older adults was 533.5 per 100,000 in 2020. Their risks increased with age and underlying medical conditions. In the USA, the number of people with at least one chronic disease in the older adult population is estimated to increase from 71.52 million in 2020 to 142.66 by 2050. Within that timeframe, this population is expected to grow from 137.25 to 221.13 million. The rapidly growing ageing population is still at high risk of adverse health outcomes from any future pandemic. This chapter explores how healthcare access and quality were altered for older adults, highlights emerging challenges and opportunities, and discusses innovative approaches and policies to improve healthcare service delivery for the ageing population. Additionally, the study by Michele Kekeh investigates requirements from caregivers as they adapt to new protocols to ensure the well-being of the older adults for whom they provide care. The experience gained from the COVID-19 pandemic could guide preparedness efforts for future pandemics and upcoming health challenges. The necessity for creating robust infrastructure in the healthcare system, including adequate and accessible medical supplies and hospital beds, equipped nursing homes, telehealth, and access to the required technology, attention to mental health, training providers and caregivers, and comprehensive health insurance coverage, was underscored during the pandemic.

The COVID-19 pandemic has damaged the control and elimination of neglected tropical diseases (NTDs) in the USA by disrupting the distribution and manufacturing of essential medical supplies. Some NTDs that have been impacted include: schistosomiasis, leprosy, and rabies.

"According to European Union data, there are at least 90 million people, who represent the 18.5% of the total population, over the age of 65 and the majority of them in this age group have multiple chronic NCDs.[3] The observed exponential increase in the incidence of the aforementioned diseases, combined with the increasingly comprehensive understanding of their adverse health effects—which are preventable—have rightly aroused the interest of the world scientific community and policymakers" (Stachteas et al. 2022).

This article by Katie Palmer and colleagues discusses how the COVID-19 outbreak and related infection control measures could hit the most frail individuals, worsening the condition of NCD patients, while further jeopardizing the sustainability of the healthcare systems. We suggest ways to define an integrated strategy that could involve both public institutional entities and the private sector to safeguard frail individuals and mitigate the impact of the outbreak (Palmer et al. 2020).

To avoid the intense impacts seen with COVID-19 in any future pandemics, it is imperative that we foster global collaboration. By considering the unique challenges and needs of each region, all countries can play a role in building a safer future for all populations, including older adults. This shared responsibility can instil a sense of unity and purpose in our efforts.

In developing country of Thailand, the dynamic relationship between diseases and environmental factors, highlighted by the COVID-19 pandemic, underscores the critical role of public health interventions. In Thailand, urbanization, air quality, and climate-shaped COVID-19 spread, with densely populated areas bearing the brunt. The government responded with expanded healthcare access and innovative services. Meanwhile, shifts in disease patterns like diabetes and cardiovascular ailments necessitate proactive surveillance. Addressing these complexities demands integrated approaches, considering environmental, social, and medical factors. By prioritizing evidence-based strategies, public health initiatives can effectively navigate this intricate interplay to safeguard communities and enhance overall well-being. In response to identified weaknesses, the Thai government implemented measures to fortify the healthcare system, including the recruitment of village health volunteers, migrant volunteers, and the implementation of tele healthcare services. Additionally, intriguing shifts observed in the disease patterns of NCDs are likely influenced by a combination of factors, including environmental influences, public health interventions, lifestyle modifications, and healthcare accessibility. These findings underscore the importance of holistic approaches to address emerging health challenges and enhance overall population health outcomes (Chapter on Thailand by Uma Langkulsen and Gamuchirai Munjoma).

In attempts to revive their struggling economies, Pacific island states, including Vanuatu and Fiji, cautiously reopened their borders in 2022 as the COVID pandemic faded. Both countries had fared well in terms of the incidence of COVID-19 by closing their borders, but the pandemic revealed basic problems in healthcare systems that proved difficult to resolve with limited fiscal space and growing skilled labour shortages. Tourism, the core of island economies, quickly revived and returned to a familiar pattern centred on urban resorts and later cruises, rather than towards a more sustainable structure supporting a blue economy. Here too, expansion was frustrated by labour shortages, fragmented and costly supply chains, and cyclone damage. Diversification of small island economies was even more challenging. Loss of skilled health workers and tradespeople was accompanied by a wider loss of labour, coinciding with the expansion of the Pacific Australia Labour Mobility (PALM) scheme attracting workers to Australia for longer periods. A brain and skill drain was intensified. Both the restructuring of the health sector and economic growth were thus constrained and diverted by global factors linked to risk, inflation, strategic issues, and regional labour supply and demand.

Combating pandemics requires a regional approach but in a climate of rapid political change in many PICs (most obviously in Vanuatu) and equally rapidly evolving external strategic interests, achieving the stability required for effective regional collaboration is scarcely feasible. A better understanding of the significance of global health issues and the distinctive role of regional organisation, such as WHO and SPC, was not enough to create genuine ongoing regional collaboration. One well-documented conclusion was that "adaptive, inclusive governance generated through transparent and trusting communication was—and will continue to be—essential to the COVID-19 pandemic response" (Phillips et al. 2022: 11). At a regional scale that had little evident traction. Similar invocations as those that suggested a more local and sustainable form of tourism were directed at food security, much threatened during COVID-19 (e.g. Farrell et al. 2020), and, again, a pandemic could have been the catalyst to change. Just as with tourism there was no real hint of an appetite for new directions (Chapter on Vanuatu and Fiji). Many neglected tropical diseases include malaria, dengue, Chagas, and other region specific diseases including Ebola, Zika, and other viral diseases transmitted by vectors include chikungunya fever, yellow fever, West Nile fever, Japanese encephalitis (all transmitted by mosquitoes), tick-borne encephalitis (transmitted by ticks), and Oropouche fever (transmitted by Culicoides flies).

Acknowledging the extent of health vulnerability, developing invested in health surveillance and disease monitoring systems to provide earlier detection and faster response times for outbreaks, including mosquito-borne diseases like malaria and dengue, ebola, zika, and chagas. Expanded distribution of insecticide-treated nets, environmental management strategies, and public awareness campaigns have helped curb vector populations and reduce transmission.

To conclude Andrew Goudie opined that It has been apparent that COVID's effects on health not only were those caused by the virus itself, but also by knock-on effects. The concentration on treating seriously ill victims of COVID inevitably led to less ability to treat other conditions and to provide routine health care. Care giving in homes under lockdown could be problematic, which meant that some vulnerable people suffered. Moreover, some isolated people found access to food was rendered difficult, which could also have health implications. Secondly, there were serious impacts on mental health. Factors included isolation during social distancing, fear of the virus, economic worries caused by economic stress or loss of jobs, loss of dear ones, and the seemingly unending stress caused by long COVID.

References

Akhtar R (ed) (2021) Coronavirus (Covid-19) outbreaks, environment and human behaviour. Springer, p 3

Cameron EE, Nuzzo JB, Bell JA, Nalabandian M, Lau SK, O'BrienLeague A, Ravi S, Meyer D, Snyder M, Mullen L, Warmbrod L (2022) Global health security index: building collective action and accountability, 2019 [January 7, 2022]. https://www.ghsindex.org/wp-content/uploads/2019/10/2019-Global-Health-Security-Index.pdf

Ehrenberg JP, Utzinger J, Fontes G, da Rocha EMM, Ehrenberg N, Steinmann P (2021) Efforts to mitigate the economic impact of the COVID-19 pandemic: potential entry points for neglected tropical diseases. Infect Dis Poverty 10:2. https://doi.org/10.1186/s40249-020-00790-4

Farrell P, Thow A-M, Wate JT, Nonga N et al (2020) COVID-19 and Pacific food system resilience: opportunities to build a robust response. Food Secur 12:783–791
Geddes L (2022) How the COVID-19 pandemic has affected healthcare around the world, (New insights into the disruption caused by COVID-19 could help strengthen health systems ahead of future pandemics.) Vaccineswork, gavi.org, July 25
Haileamlak A (2021) The impact of COVID-19 on health and health systems. Ethiop J Health Sci 31(6):1073–1074. https://doi.org/10.4314/ejhs.v31i6.1
Kapoor M, Kaur KN, Saeed S, Mohd Shannawaz M, Chandra A (2023) COVID-19 on health-care system in India: a systematic review. J Public Heath Res 12(3):22799036231186349, ID: 37461400
Kickbusch I, Liu A (2022) Global health diplomacy—reconstructing power and governance. Lancet 399(10341):P2156–2166
MHLW (2023a) Response to new coronavirus infections since April 2024. Accessed August 8, 2024 https://www.mhlw.go.jp/stf/covid-19/00003.html
MHLW (2023b) Novel Coronavirus (COVID-19)|Ministry of Health, Labour and Welfare|Government of Japan. Accessed August 8
Misra TN (2023) The public health response to COVID-19 in the UK: a view from the frontline. In: Pachauri S, Pachauri A (eds) Global perspectives of COVID-19 pandemic on health, education, and role of media. Springer, Singapore. https://doi.org/10.1007/978-981-99-1106-6_20
Palmer K, Monaco A,·Kivipelto M, Onder G,·Maggi S,·Michel J-P, Prieto R, Sykara G, Donde S (2020) The potential long-term impact of the COVID-19 outbreak on patients with non-communicable diseases in Europe: consequences for healthy ageing. Aging Clin Exp Res 32:1189–1194
Phillips G, Kendino M, Brolan C, Mitchell R et al (2022) Lessons from the frontline: leadership and governance experiences in the COVID-19 pandemic response across the Pacific region. Lancet Reg Health West Pac 25:100518
Rathnayake D, Clarke M, Jayasinghe VI (2020) Health system performance and health system preparedness for the post-pandemic impact of COVID-19: a review. Int J Healthcare Manage 14(1):250–254. https://doi.org/10.1080/20479700.2020.1836732
Stachteas P, Symvoulakis M, Tsapas A, Smyrnakis E (2022) The impact of the COVID-19 pandemic on the management of patients with chronic diseases in primary health care. Pop Med 4:1–13

Chapter 2
Beyond Barriers and Borders: Averting a Post-COVID Crisis in Vanuatu and Fiji?

John Connell

Abstract In attempts to revive their struggling economies, Pacific island states, including Vanuatu and Fiji, cautiously reopened their borders in 2022 as the COVID pandemic faded. Both countries had fared well in terms of the incidence of COVID-19 by closing their borders, but the pandemic revealed basic problems in healthcare systems that proved difficult to resolve with limited fiscal space and growing skilled labour shortages. Tourism, the core of island economies, quickly revived and returned to a familiar pattern centred on urban resorts and later cruises, rather than towards a more sustainable structure supporting a blue economy. Here too, expansion was frustrated by labour shortages, fragmented and costly supply chains, and cyclone damage. Diversification of small island economies was even more challenging. Loss of skilled health workers and tradespeople was accompanied by a wider loss of labour, coinciding with the expansion of the Pacific Australia Labour Mobility (PALM) scheme attracting workers to Australia for longer periods. A brain and skill drain was intensified. Both the restructuring of the health sector and economic growth were thus constrained and diverted by global factors linked to risk, inflation, strategic issues, and regional labour supply and demand.

Keywords COVID-19 · Vanuatu · Fiji · Health systems · Economies · Tourism

2.1 Introduction: Barriers and Borders

Pacific island states largely quarantined themselves against COVID-19, and, compared with more continental states, this worked to avert the worst health outcomes and delay others. That was more or less how they had managed to survive rather uneasily the ravages of the Asian flu pandemic almost exactly a century earlier. These lessons had been remembered (albeit especially in Samoa). But quarantining came at a cost, because of the immediate collapse of the tourism economy, the basis

J. Connell (✉)
University of Sydney, Sydney, Australia
e-mail: john.connell@sydney.edu.au

R. Akhtar (ed.), *COVID-19: Impacts on Health and Health Systems*,
https://doi.org/10.1007/978-3-031-90255-0_2

of most Pacific island country (PIC) economies, notably in Vanuatu and Fiji, and the resultant crash in employment. The impacts of COVID-19 were, however, multidimensional, even if centred on health (including mental health) care provision and broad economic issues, involving social and cultural issues, and of a complexity that PICs had never hitherto faced. Moreover, they were not isolated problems, occurring, in Vanuatu especially, alongside physical hazard events—notably cyclones and floods—that further challenged maintaining and regenerating livelihoods, as both countries faced the more perceptible consequences of climate change.

By 2022 most Pacific nations, including Vanuatu and Fiji, had reopened their borders to tourists. An absence of tourists for two years had caused disruption to economies that relied heavily on tourism and which made up over 40% of both countries' GDPs. Subsequently, with international tourism returning to pre-pandemic levels, with Fiji to the fore, PICs began to reclaim their share of tourists, as they more generally learned from the varied problems caused by COVID. This chapter traces how Fiji and Vanuatu coped with the pandemic, examines the extent to which they were able to restore their economies and health systems in its wake, and how that effectively meant reverting to the pre-pandemic economy rather than its being a catalyst for change and new directions.

2.2 A Brief Background to Vanuatu and Fiji

Both countries have many populated islands presenting a challenge for economic development and for the delivery of services, including health care, more so in this century with greater formal concern over inequitable development. Fiji has a population of about 950,000 with an increasing proportion resident in the Lami-Suva-Nausori metropolitan corridor, whereas Vanuatu has about 345,000 people with 65,000 resident in Port Vila. Poverty, unemployment, and a lack of food security exist, mainly in informal settlements, and a growing incidence of NCDs is a source of concern. Urbanisation has steadily increased the role and significance of the capital cities, and especially Port Vila, relative to the rest of the country, as their agricultural economies (including the sugar industry of Fiji) have slowed or contracted. Economic development is increasingly centred on tourism and hindered by dependence on relatively few commodity exports—dominated by the struggling Fijian sugar industry and a growing kava export industry in Vanuatu. Fish and the leasing of fisheries waters are valuable sources of income in both countries. Agriculture is vulnerable to natural disasters, while considerable distances between islands (with low volumes and high costs) and from main markets constrain profitability. In the post-COVID era this has been exacerbated by supply chain problems, and trade deficits have worsened.

2.3 Health Care

The COVID-19 pandemic was initially tangential to the PICs. Fiji's first positive case of COVID-19 was in March 2020 by which time Australia, New Zealand, the USA, and many Asian countries had been affected. Consequently, there was already international experience showing how countries were managing their situations, and the necessary urgency of response. Vanuatu had no direct experience of COVID-19, being one of the last countries in the world to have avoided the outbreak, recording its first case in November 2020. By then, as in Fiji, the country's borders had been closed and tourism abruptly ended early in 2020 (Naupa et al. 2021; Connell and Taulealo 2021). The effective closure of the national border, and of the internal 'border' between Efate and other islands, kept the incidence of COVID-19 low, so that, when the border was reopened in June 2022, hospitalisations and deaths due to COVID had been few, with 14 recorded deaths and 10,500 cases. Fiji, with a more international and interconnected economy, was less fortunate. Alongside implementing lock-downs, contact tracing, social distancing, and basic medical interventions, Fiji also resorted to self-isolation. It relied on vaccinations and while a high proportion of the population was eventually vaccinated, vaccines were rolled out inefficiently and did not immediately reach those most at risk (Singh et al 2022; Weber et al. 2022). Vanuatu relied on border closure, at national and regional level, to prevent the arrival and spread of the pandemic, and vaccinations were limited in their coverage.

A minor advantage was that Fiji had experienced several major health crises in the past. Measles (1875–1876) killed about 33% of the local population due to lack of vaccine but a second outbreak (2019) was quickly controlled with an intense national immunisation campaign and vaccination. Fiji also experienced Spanish Flu (1918) with a 5% fatality rate, but this was contained with active government warnings, food safety measures, and increased health care (Singh et al. 2022). This past experience, and the 'delayed' arrival of COVID-19 in the country, enabled a positive response to the exigencies of dealing with the new pandemic. Vanuatu's past history of epidemics has not been documented, but it too had the advantage of looking on from afar to problems that occurred elsewhere in the world.

Fiji emerged from the experience with 878 deaths and Vanuatu with 14 (all in the two provinces with towns) early in 2024. Fiji had a single quarantine breech in 2021 that resulted in 54,000 cases, a per capita transmission rate that was one of the highest globally, and 700 deaths (Bell et al. 2022). That outbreak overwhelmed the health system, with reports of shortages of beds, medicine, equipment (including PPE), and inevitably, of staff. COVID also challenged progress towards the control of other diseases such as hepatitis and tuberculosis (the most infectious disease on the planet) as attention was diverted. Ultimately both countries reached over 92 percent vaccine coverage (close to that of Australia and well in advance of the United States): a measure of the trust that both governments were able to command (including the support of the business communities) and of the willingness to cover remote populations (see McLennan et al. 2023). Politicians were, however, rather less likely to support senior healthcare workers. Despite deaths, near universal vaccination and access to effective emergency care was invaluable, enabling a remarkable degree of cohesion and unity (Phillips et al. 2022). Perhaps as never before, small PICs were willing to trust major countries, including Australia, and regional organisations, notably WHO and the Secretariat for the Pacific Community (SPC).

COVID-19 left an indelible impact on PIC healthcare systems. Nurses especially had experienced difficult times; some of them died or witnessed the death of relatives, fellow workers, and inevitably patients. The burden of care was enormous, where death rates were high, the pandemic lingered and access to vaccines, PPE etc. was unusually difficult. Overtime in crowded hospitals and resultant burnout occurred, as the normal functions of nursing and midwifery continued, amidst border closures, lockdowns, and curfews. When the Fijian government introduced a 'no jab, no job' policy in 2021 as many as 250 health workers and teachers were reported to have lost their jobs because they had not been vaccinated (Weber et al. 2022). That loss of workers could have been even more disastrous. Afterwards, in Fiji at least, resignations reached unprecedented levels and emigration often followed (see below). The crisis even left mountains of dangerous plastic waste.

As the impact of COVID faded, both countries returned to a focus on the rising incidence of NCDs. Comorbidity was such that as many as 70 percent of Fijians who died from COVID-19 were also suffering from NCDs (Movono 2022). A widespread assumption was that the scale of the COVID-19 health problem would result in better funding for public health systems. That was so on the part of the wider public, but the national budget was too stretched to make significant changes to a system that had previously been underfunded in comparison with other middle income countries (McLennan et al. 2023). The COVID crisis had revealed too many shortfalls for urgent remedies to be quickly implemented. Opportunities existed to reinvigorate the health system, including, for example, to ensure an adequate, committed, and appropriate workforce (Bourgeault et al. 2020) but the scale of loss meant that was virtually impossible whether with respect to public health or clinical care. COVID-19 was too much of a shock to the system.

In practice it was the devastated economies that revived first and revitalising healthcare systems lagged. Economic growth preceded welfare.

2.4 Economies in Crisis? the Tourism Trap

With closed borders, the pandemic had a massive impact on the economies of all PICs, and the direct impact was obvious in the collapse of formal and informal employment for tourism as tourists and cruise liners disappeared (Connell 2021; Connell and Taulealo 2021). The resultant decline in the GDP growth rate was mainly due to the closure of the tourism and its multiplier effects elsewhere: the lay-off of workers, the closure and downsizing of general businesses, including markets, and other services such as transport. Consequently, by the end of 2020, as many as 115,000 jobs had been lost in Fiji—about a third of the total workforce - and much the same proportion experienced job losses in Vanuatu, indicative of the dependence of PIC economies on tourism.

That dependence was exemplified in the considerable difficulty of seeking to encourage domestic tourism. Use of hotels as quarantine stations and the presence of various international organisations in Fiji gave it some hope. Thus the Outrigger resort that had 530 employees beforehand was able to retain 154 during the crisis, a possibility not open to hotels in Vanuatu despite attempts to encourage local visitors with reduced prices (Connell 2021). In due course that meant that reopening proved much easier for hotels that had been able to retain key staff. Otherwise the two-year collapse involved hotel and resort closures, selective losses of employment and incomes, with women (who made up around three quarters of the workforce in tourism-related activities, whether formal or informal), bearing the brunt. Markets, car hire, taxis, and the urban informal sector were particularly disadvantaged. Several thousand workers lost formal sector employment, and the informal sector was virtually wiped out. Handicraft makers and hair braiders too had no local market. The domestic value chain of resorts such as the Havannah (Naupa et al. 2021) was no longer relevant. That raised complex questions over the duration of the crisis, over self-reliance, diversification, sustainability, the future of tourism, the potential for an eventual 'new normal', and therefore of tourist development in small island states (Connell and Taulealo 2021, Naupa et al. 2021, Connell 2023). Such questions resurfaced during the post-crisis revival and especially after the twin cyclones of March 2023 that caused considerable damage to tourism infrastructure (and much else) in Vanuatu.

Since the majority of formal and informal employment in the tourism industry was of women, this proved a gendered problem, increasing social and economic burdens on women but nonetheless with women subsequently leading the way in revitalising village and national economies (Connell 2023). Many urban workers and their households, unable to afford remaining in cities, returned to rural 'homes' and sought to re-establish rural livelihoods. In some contexts, a return to village life was welcomed as a break from the more overt capitalist structures of international tourism (Movono and Scheyvens 2023, Kanemasu 2005), and such ties with home certainly averted an even worse crisis. Vanuatu was more easily able to cope. Significant parts of the country had retained a partly subsistence-oriented economy, with a local marketing component, and although incomes had fallen these livelihoods remained valuable and viable. Re-invigorating a subsistence-oriented economy was more difficult in the more commercialised agricultural system of Fiji, although even in the capital Suva, vacant urban land reverted to agriculture, assisted by the government providing seeds (Randin and Turagabaleti 2021).

However, outside the cities, as many returnees sought to re-establish agriculture, they were not always welcome when they been away for some time, land was scarce and already in use, while agricultural production could not simply be regenerated instantly. In any case there was pressure on rural households who now found it difficult to market any food production (Underhill et al. 2023). There was some indication from both PICs that where households returned to villages where fishing was productive a successful return was more likely. In due course, with limited alternative income-generating opportunities and some social tensions, such as where children missed urban life and missed out on education, most islanders were anxious to jump at the chance of returning to their place in the 'modern' economy, as they were in Vanuatu (Connell 2024a).

Many people were without alternative solutions to provide for their daily life during the pandemic, especially amongst the urban poor in Fiji where local borders and barriers prevented mobility. Negative social impacts ensued leading to increases in domestic violence and crime, partly a function of absent jobs and incomes. Education was less easily accessible as schools closed particularly disadvantaging rural children, where distance learning was impossible. Unanticipated shifts and pressures, and uncertainty over when the crisis might be over, brought worsened mental health.

In contrast, greater resilience was evident as people shared resources through Facebook groups, and younger generations used their increased spare time to learn traditional fishing or craft skills from seniors. Less money was in circulation so bartering took the place of cash transactions (Movono et al. 2021) but it was not easy to compensate for the extent of income losses, where a cash economy was invaluable. In both countries limited mobility and income, and commodity chain constraints on pharmaceutical goods, resulted in a parallel reversion to the use of traditional medicines.

Both countries were anxious to revive their shattered economies but were cautious of a limited healthcare system (despite relatively few fatalities from COVID-10). Moreover border closures had catastrophic impacts on national airlines. Fiji Airways had to ground 95% of its passenger flight until the border reopened. Air Vanuatu was similarly affected. Most of their workforces were terminated making it difficult to restart when the borders opened, especially since both airlines were close to bankruptcy. Air Vanuatu remained in crisis long after reopening with unpaid bills, unavailable parts, and cancelled flights disrupting operations. Lack of agreements with foreign airlines posed a significant problem (overseas airlines being less interested in prioritising Vanuatu, as a small market). Such combined problems reduced capacity on flights into Vanuatu, disrupted internal flights, and created a lack of confidence in the airline and thus the travel industry. Cruises too had been as slow to restart—only returning late in 2023 a year after they had returned to Fiji. Meanwhile flights within the two countries had severely contracted or disappeared—placing burdens of various kinds on residents of outlying islands.

In addition to this slow reopening, Vanuatu, often regarded as the most natural-disaster-prone nation in the world, endured two Category 4 tropical cyclones in a single week in March 2023, leaving 80 percent of its population without power or food, and the closure of the renascent tourism industry for a month, while the nation was still recovering from the Category 5 Tropical Cyclone Pam in 2015. That followed volcanic eruptions, which necessitated the (mainly temporary) evacuation of Ambae Island, earthquakes, and an (introduced) coconut rhinoceros beetle outbreak in 2019 that devastated copra production and resulted in a state of emergency.

Unlike Vanuatu, Fiji was one of the first PICs to reopen its borders to visitors in December 2021, removing any testing or quarantine requirements for visitors. It had been impatient to get to this stage, proposing a 'Bula Bubble' a special arrangement to allow travel from Australia and New Zealand, but neither country was enthused. Fears of a second wave initially discouraged growth in Fiji (and the main markets of Australia and New Zealand were still closed off well into 2022). Tourism was dependent on external decisions by governments, airlines, and cruise ship companies, none favourable to Fiji until much later in 2022. After reopening, Fiji experienced a remarkably rapid expansion of tourism with more visitors in late 2022 than comparable pre-pandemic months so that most of the tourism workforce had returned by 2023.

Vanuatu, without COVID-19, was much more cautious. In any case, multiple disruptions and their multiplier effects demonstrated the harsh realities of dependency on international tourism and especially on dependency on two high-volume and low profit margin sectors: discount packages and cruise ship tourists, largely paid for overseas. However, that was all on hold. When Cyclone Harold hit Vanuatu in 2020, foreign aid workers were banned from the country—to ensure that COVID-19 too was kept out—despite the considerable devastation. Travel was also banned between islands. In the interim, Vanuatu sought to devise strategies where it could add value to the tourism sector, while external advisers made similar suggestions (e.g. Wondirad 2023). A less well-prepared Vanuatu, with a less adequate healthcare system, watched and waited. Compared with Fiji, Vanuatu's tourism sector struggled even as COVID-19 faded from sight. As in Fiji, the first returnees went to the resorts, whether boutique resorts such as The Havannah or international chains such as Holiday Inn, which offered the pleasures and greater certainties of enclosure and introspection. Urban resorts led the rebirth, assisted by tourist conservatism, returnees familiar with the resorts and potentially accessible health care. The tourist industry experienced its own tribulations. Cyclone damage had reduced access to fresh foods of various kinds (notably such tree crops as mangoes, pawpaws, and bananas), and hotels were more dependent on imported foods, which were more costly and disrupted by fractured supply chains (hindered by fuel costs and spare parts), and less frequent shipping. Some resorts were damaged by the cyclones and had reduced capacity. Post-COVID return to normal was thus hampered as much by the cyclones, and by transport and labour supply factors (see below), as by COVID-19 itself.

Nonetheless despite multiple difficulties, and largely buoyed by the enthusiasm of previous tourists to return to familiar places, both countries—but especially Fiji—recorded rapid growth. More caution and more restructuring might have been warranted but the old order was re-established; resorts, international airlines, and relatively affluent tourists led the way. External encouragement to diversify tourism by targeting lower-volume, higher-yield markets, to diversity markets beyond Australia and New Zealand, and to diversify the economy beyond tourism, went unheeded. Booming numbers, especially from Australia and New Zealand, and external interests discouraged rethinking and restructuring. During the first 4 months of 2023, Australian tourism to major South Pacific destinations had already surpassed pre-pandemic levels and was still growing, but, with weaker airline connections, Vanuatu lingered. Indeed, the twin cyclones that hit Vanuatu in March 2023 disrupted flights and caused widespread physical damage. By late 2023, as transport connectivity improved and infrastructure was rebuilt, and more cautious Asian markets were beginning to return, a further boost to 'traditional' resort tourism.

Not only was restructuring in favour of more locally oriented tourism and more local social and economic linkages, as resorts (often international chains) revived, but tourism effectively contracted to the security and safety of these urban and coastal entities, where tourists could be entirely introspective (Connell 2016). That came at the expense of rural and remote places and islands. For most tourists experiencing urban 'cultural villages' and staged performances within resorts was enough to provide a brief experience of exotic scenery (lagoons and 'jungles') and culture (grass skirts, bamboo bands, painted cheeks, and bare chests), alongside watered-down kava and distinctive fruit (Connell 2024a). In so doing, the post-COVID industry meant that local people were more marginalised from the industry that at almost any time in the past. That was as true of cruise tourism as of other tourism ventures: as one small-scale Fijian tourism operator observed:

> In the year or so before COVID, we watched the massive cruise ships arrive in Savusavu Bay, dwarfing the town. Even then, a cruise ship in town created tension. As hundreds of tourists spilled onto the streets for just a few hours, small business owners almost came to blows in their efforts to capture some of the trade. Ironically, there wasn't very much money available anyway – most of the tourist dollars were siphoned off by players higher up the chain long before the visitors disembarked. We learned that our own tour, for which we charged F$25 (around A$16.50), was being sold on board for up to A$110. (Moorhead 2021)

Questions of whether tourism was really valuable, or whether it destroyed local values and invoked too much competition and inequity, were also being raised elsewhere. Cruises had little interest in advertising local ventures, offered their own entertainments, and ignored environmental impacts. It made nonsense of invocations along the lines of: 'post-pandemic tourism in Fiji should focus on net yield and sustainability instead of the number of tourist arrivals. Therefore, marketing efforts need to be well crafted to target tourists who can stay longer and benefit the economy while discouraging mass tourism' (Wondirad 2023: 3762). If anything, in both PICs, it had moved further away from that, while, ironically, any national focus was centred on tourism as the expense of more sustainable activities from agriculture to fisheries.

2.5 A Clamour for Mobility

For half a century Fiji has experienced a steady out-migration of people for social, political, and economic reasons, in early days to Canada, and most recently to New Zealand and Australia. Since the 2010s, Vanuatu also experienced significant migration, but mainly the temporary migration of guestworkers to work in horticulture in New Zealand and Australia, for periods of up to nine months at a time. Post-coup Fiji belatedly joined in the same systems. During COVID, most guestworkers were marooned in New Zealand and Australia, unable to go home. Perhaps remarkably, given the problems that many migrants faced in destination countries, and especially as agricultural workers in New Zealand and Australia, remittances largely held up both from long-term migrants and from guestworkers. While remittances had always been valuable in PICs (notably in Samoa and Tonga) they had enhanced livelihoods and boosted health and education in several places, albeit sometimes at considerable social cost, and had special value during the pandemic. That alone increased interest in migration.

Adopted in 2012, Australia's guestworker scheme, initially the Seasonal Worker Program (SWP) which sought to facilitate development outcomes in sending countries, expanded in 2018 to provide labour mobility opportunities in longer-term, non-seasonal employment in semi-skilled industries including hospitality, tourism, aged care, labour hire, and meat processing, and the scheme was renamed the Pacific Australia Labour Mobility (PALM) scheme. Contracts in these industries expanded to range from one to four years in length, resulting in increased earnings (Petrou and Connell 2023a, b). The extension of seasonal agricultural labour mobility programs into industry (especially abattoirs) and various service sectors (especially care giving) significantly increased interest in longer-term overseas employment. Not surprisingly workers who had been terminated from the tourism sector and from the airline industry were one cohort particularly interested in migration. The progressive lifting of COVID-19 travel restrictions from late 2020 onwards saw a virtual explosion of labour mobility participation as employers in host countries sought to fill urgent labour shortages, other migrant workers (especially working holiday makers) failed to reappear, and the PICs sought to recover from the economic impacts of the pandemic (after their tourism-oriented economies collapsed). Vanuatu became the largest supplier of PALM workers.

Alongside guestworker mobility, a potentially more permanent migration of others continued, made easier by growing diaspora populations that provided support and information via social media, and which emphasised a significant brain and skill drain, and skill loss. Fiji, Vanuatu, and other PICs have experienced a skill drain from public and private sectors: engineers, nurses, computer programmers, radio announcers, and teachers resigned or took leave from skilled positions to pick fruit or be caregivers in Australia and New Zealand (Petrou and Connell 2023a, b; Lowonbu 2023). This skill drain from the formal sector became more important under PALM, as contracts were longer, counteracting some of the welfare gains made by households. Losses of health workers were longstanding, linked to a complex culture of migration (Connell 2014). National shortages of workers in health (including dentistry), education, engineering, tourism, and other formal sectors have emerged—and accelerated after COVID, especially away from the health sector, ironically itself a victim of COVID-19 (Connell 2024b). That has hampered the processes of governance and taken a toll of both the private and public sectors, reducing the ability to achieve, manage, and direct national development. Critically, since demand is limited, this has meant a shift from earlier years, where the national goal of the scheme in the PICs was to provide opportunities for agricultural workers from disadvantaged rural areas and remote islands to an emerging bias in favour of relatively skilled urban workers.

A Vanuatu Chamber of Commerce and Industry survey (Orozco and Spencer 2023) found that half of all enterprises had lost at least one employee to the labour mobility schemes in the previous twelve months, which threatened their attempts to rebound after COVID-19. The most common reason for resignation was to work overseas. That was evident in the tourism industry of both counties, where wages were usually below national averages. Consequently many components of the tourism industry experienced labour shortages. Urban resorts in Vanuatu alone were losing skilled chefs and others (at a rate for some resorts in mid-2023, of one or two resignations per month) and replacing them at additional cost with migrants mainly from south Asia and the Philippines. The healthcare system was bringing nurses from Solomon Islands, and Vanuatu had created an emergency visa aimed at bringing in 1500 people (Petrou and Connell 2024). The most difficult positions to fill locally were trade occupations (e.g. builders, electricians, plumbers, and welders), retail jobs (e.g. salespeople, shopkeepers, cashiers, and retail assistants), and restaurant staff (e.g. chefs, bakers, bartenders, and waitstaff: often where the skills required took several years of training and workplace experience (Orozco and Spencer 2023) and also required mentoring, management, and continuity. Individual household benefits from migration were at some cost to national development.

2.6 Conclusion: Islands at Risk

As elsewhere in the world the pandemic exacerbated existing economic challenges and pointed to the burdens of inequality and of growing economic, social, and spatial disparities, some shaped by tourism. Both island state economies were devastated and public health systems—at least in Fiji—found it almost impossible to cope, as the entirety of state capacity was challenged. However, quick decision-making on closure, and the virtues of islands as barriers and the sea as border, enabled both Vanuatu and Fiji (and other PICs) to fare better than might have been expected when so close to the Chinese ground zero. Nonetheless low mortality rates came at the expense of economic disruption—a function of both countries' growing dependence on tourism—and reflected both a willingness to absorb external advice (and aid) in an unprecedented context, and a general willingness by both national populations to respond to a free flow of information and accept constraints and privations. To some extent, as island state residents, they had little choice. They were also spared the 'medical populism' of countries such as the United States and Brazil where aversion to 'expertise' and the role of 'establishment' authorities resulted in high death rates. Like other PICs, Fiji and Vanuatu had some capacity to revive subsistence-oriented economies. Even then such small triumphs came at a cost.

The poor and vulnerable, often in urban settlements, were the most affected by the pandemic—but there is little to suggest development now favours them. Ideally, since the various impacts of the COVID-19 pandemic were inherently multifaceted, involving a range of socio-economic and health problems such as rising domestic violence, new mental health issues, worsened access to education, unemployment, and disruption of the production and the supply of goods and services, and any strategy for recovery should 'also be multidimensional supported by strong regional and national policy measures and international collaboration' (Wondirad 2023: 3762). In the need to regenerate lives and livelihoods that was always impossible. No easy fixes existed for the health systems: both countries had limited fiscal space and both had scarce (and declining) human resources to cope. In the short term, recovering from the economic downturn caused by the pandemic took precedence—as livelihoods were at stake (especially in Vanuatu)—primarily requiring the opening of borders and the revival of airlines and route. In a sense it was fortuitous that tourism opened up at a time when residents of the two largest markets, Australia and New Zealand, felt cooped up, and anxious to travel again initially to places and resorts familiar to them, and where vaccination rates were high. Addressing health and other social issues, although equally vital, required more time and more subtlety, worsened by the loss of significant workers from the public sector and especially the loss of health workers.

Health problems are significant in the PICs, and skilled health workers are needed. NCDs are demanding of resources. The first outbreak of dengue fever was in 1997 which caused 24,000 infections and 13 deaths, and it has reemerged intermittently in the PICs (with Samoa experiencing an outbreak in 2024). Tuberculosis continues to pose a health risk to poor households with unhealthy lifestyles and poor sanitation infrastructure (Singh et al. 2022). Further pandemics are always possible (with a new avian flu hinting at such a possibility in mid-2024) and will return and will require effective workforces. Unlike 'conventional' shocks (such as coups and cyclones) that are immediate, pandemics are of indeterminate duration. By late 2023 in both countries, COVID-19 was regarded as a moment in history, something that people preferred to forget and was rarely mentioned even within the Ministries of Health, but, as the cruise ship revival occurred, there were already hints that 'quiet epidemics' were in place. Borders provided no health information or demanded vaccinations. The pandemic had gone from public life.

Combatting pandemics requires a regional approach but in a climate of rapid political change in many PICs (most obviously in Vanuatu) and equally rapidly evolving external strategic interests, achieving the stability required for effective regional collaboration is scarcely feasible. A better understanding of the significance of global health issues, and the distinctive role of regional organisation, such as WHO and SPC, was not enough to create genuine ongoing regional collaboration. One well-documented conclusion was that 'adaptive, inclusive governance generated through transparent and trusting communication was—and will continue to be—essential to the COVID-19 pandemic response' (Phillips et al. 2022: 11). At a regional scale that had little evident traction. Similar invocations as those that suggested a more local and sustainable form of tourism were directed at food security, much threatened during COVID-19 (e.g. Farrell et al. 2020), and, again, a pandemic could have been the catalyst to change. Just as with tourism there was no real hint of an appetite for new directions.

Meanwhile the tourist economies had reverted to the old order, centred on inclusive resorts and air travel, despite it being argued from Fiji that 'from an economic viewpoint, the crucial lesson we got from this challenging pandemic is the vitality of developing a diversified and resilient economy that resists external shocks and turbulences' (Wondirad 2023: 3762). That was a constant plea, oriented to a more regional and cooperative orientation, directed to a sustainable, locally oriented tourism (in terms of both supply and demand) with positive ties to a blue economy and in support of the sustainable development goals (SDGs). However, a rhetoric of sustainable and regenerative tourism was divorced from practice, and agricultural enterprises were too small even to provide regular supplies of such basic foods as eggs. Achieving any locally more beneficial 'new normal' was quite doubtful. Despite the widespread invocation to create and develop post-COVID regenerative, sustainable and resilient economies (e.g. Leach et al. 2021; Lehmann et al. 2021) that was scarcely evident in the return to tourism, where that might have mattered most.

Remittances have contributed to keeping household and national economies afloat since COVID-19 but at long-term costs to economic development (as businesses lose staff), welfare (as teachers and health workers migrate), and social wellbeing (as family structures are fragmented). Migration has thwarted both the revival of tourism and any plausible diversity of national economies. Skilled labour shortages and costly and increasing NCDs pose a risk to long-term growth. Apart from the adverse impacts of climate change, the issue of labour emigration—particularly to Australia and New Zealand—has long-term implications for the availability of domestic skilled and semi-skilled workers.

Ultimately the turn to 'tradition' and to past values and practices was short-lived as employment regenerated, most welcoming returning to jobs with which they were familiar and provided reasonable livelihoods. But, whether in the health sector itself, where migration was massive, or more generally in both countries, there was an uptick in international migration (and the demand for it) spurred on by a more conciliatory and more strategic Australian policy, while primarily bowing to the needs of many businesses and farms in regional Australia, that extended the opportunities that were available (Petrou and Connell 2023b). The resultant migration that boosted household economies at some cost further emphasised how key strands in development were tightly linked to overseas decisions, while entire national economies were affected by external forces, whether security concerns, cyclones, warfare, investments in airlines and resort chains, supply chains, and, of course, climate change and pandemics. Complex combinations of factors brought challenges in investing in health, education, skills, effective social protection, and infrastructure resilient to climate change, and thus to limited movement towards the SDGs.

Furthermore, the pandemic has pointed to the significance of collaboration at local, regional, and international levels to successfully manage unforeseeable events such as COVID-19. However, that has never been easy in Pacific especially with its growing strategic significance as China seeks multiple ties and interests, deep sea resources are alluring, and 'traditional' foreign interests react. On the volcanic rim of fire, cyclones and climate change offer more challenges to sustainable development while both countries shift uncertainly through Prime Ministers and feasible development strategies. In uncertain economic times, politics and development are more fragmented and cautious now than in the time of COVID. In so many ways and in so many arenas, it was unsurprising that there was no 'new normal' but a multitude of attempts to get back to more familiar, secure, and certain times.

References

Bell L, van Gemert C, Merilles O, Cash H et al (2022) The impact of COVID-19 on public health systems in the Pacific island countries and territories. The Lancet Regional Health Western Pacific 25:100498

Bourgeault I, Maier C, Dieleman M, Ball J et al (2020) The COVID-19 pandemic presents an opportunity to develop more sustainable health workforces. Human Resources Health 18:83

Connell J (2014) The two cultures of health worker migration. Pacific Perspective, Social Sci Med 116:73–81

Connell J (2016) Fiji: reflections in the infinity pool. In: Alexeyeff K, Taylor J (eds) Touring Pacific cultures. ANU Press, Canberra, pp 427–438

Connell J (2021) COVID-19 and tourism in Pacific SIDS: lessons from Fiji, Vanuatu and Samoa? The Round Table 110(1):149–158

Connell J (2023) COVID-19 and tourism in the Island Pacific: gender tribulations and transformations in different seas. In: Datta A, Momsen J, Oberhauser A (eds) Bridging worlds—building feminist geographies. Essays in Honour of Janice Monk. Routledge, Abingdon, pp 212–220

Connell J (2024a) Culture, COVID-19, cyclones and supply chains. The contraction of tourism in Vanuatu. In: Baldacchino G (ed) Archipelago tourism after the pandemic. Routledge, Abingdon

Connell J (2024b) The shortage of nurses in Fiji: practices, problems and policies. WHO, Suva

Connell J, Taulealo T (2021) Island tourism and COVID-19 in Vanuatu and Samoa: an unfolding crisis. Small States Territories 4(1):105–124

Farrell P, Thow A-M, Wate J, Nonga N et al (2020) COVID-19 and Pacific food system resilience: opportunities to build a robust response. Food Secur 12:783–791

Kanemasu Y (2005) Weapons of the workers. Employees in the Fiji hotel scene. In: Connell J, Rugendyke B (eds) Tourism at the grassroots. Routledge, London, pp 114–130

Leach M, MacGregor H, Scoones I, Wilkinson A (2021) Post-pandemic transformations: how and why COVID-19 requires us to rethink development. World Dev 138:105233

Lehmann P, de Brito M, Gawel E, Groß M et al (2021) Making the COVID-19 crisis a real opportunity for environmental sustainability. Sustain Sci 16:2137–2145

Lowonbu L (2023) Woka shotej bae i save afektem ekonomi. [Worker shortages are affecting the economy] Retrieved from https://www.vbtc.vu/woka-shotej-bae-i-save-afektem-ekonomi/

McLennan S, Movono A, Thomas-Maude J (2023) Nobody should talk about it. Fijian health system resilience and the COVID-19 pandemic. Massey University Institute of Development Studies Working Paper, Palmerston North

Moorhead, A (2021) Small steps for sustainable tourism in Fiji. ANUDevPolicyBlog, Canberra, 13 December

Movono L (2022) Fiji shifts focus from Covid-19 to NCDs. Radio New Zealand, 29 January

Movono A, Scheyvens R (2023) Tourism in a world of disorder: a return to the Vanua and Kinship with nature in Fiji. In: Connell J, Campbell Y (eds) COVID in the Islands: a comparative perspective on the Caribbean and the Pacific. Palgrave Macmillan, Singapore, pp 265–278

Movono A, Neihapi P, Scheyvens F, Steenbergen D (2021) The agency of local people in the Pacific: indigenous responses to the local pandemic. ANUDevPolicy Blog, Canberra, 25 November

Naupa A, Mecartney S, Pechan L, Howlett N (2021) An industry in crisis: how Vanuatu's tourism sector is seeking economic recovery. In: Connell J, Campbell Y (eds) COVID in the islands: a comparative perspective on the Caribbean and the Pacific. Palgrave Macmillan, Singapore, pp 231–252

Orozco L, Spencer J (2023) Vanuatu skills needs. Industry Survey Report. Vanuatu Chamber of Commerce and Industry, Port Vila
Petrou K, Connell J (2023a) Pacific Island guestworkers in Australia. The New Blackbirds? Palgrave Macmillan, Singapore
Petrou K, Connell J (2023b) Our 'Pacific family'. Heroes, guests workers or a precariat? Australian Geogr 54(2):125–136
Petrou K, Connell J (2024) Temporary labour mobility schemes: triple wins and social costs in Pacific Island states? International Migration Review
Phillips G, Kendino M, Brolan C, Mitchell R et al (2022) Lessons from the frontline: leadership and governance experiences in the COVID-19 pandemic response across the Pacific region. The Lancet Region Health Western Pacific 25:100518
Randin G, Turagabaleti R (2021) COVID-19 and food security in Fiji's urban areas: the rise of urban farming and home gardening as a 'new way of life.' In: Connell J, Campbell Y (eds) COVID in the islands: a comparative perspective on the Caribbean and the Pacific. Palgrave Macmillan, Singapore, pp 353–366
Singh R, Lal S, Khan M, Patel A, Chand R, Jain D (2022) The COVID-19 experience in the Fiji Islands: some lessons for crisis management for small island developing states of the Pacific region and beyond. N Z Econ Pap 56(1):67–72
Underhill S, Patolo S, Molimau-Samisoni S, Kumar S, Burkhart S (2023) Farmer and market vendor perceptions of COVID-19 impacts on horticultural fresh food systems in Tonga, Fiji and Samoa. Agric Food Secur 12(1)
Weber E, Kopf A, Vaha M (2022) COVID-19 in Fiji—from health and economic to major political crisis. In: Akhtar R (ed) Coronavirus (COVID-19) outbreaks, vaccination, politics and society. Springer, Cham, pp 45–66
Wondirad A (2023) Is Fiji's tourism sector on the right path to successfully recovering from the COVID-19 virus? A rapid assessment. Curr Issues Tourism 26(3):3757–3764

Chapter 3
COVID-19, Climate, and Environmental Change in Fiji

Eberhard Weber and Andreas Kopf

Abstract It is four years since reports of a mysterious disease from China were received first with interest, but soon with uncertainty and fear throughout the rest of the world. At the end of January 2020, the World Health Organization (WHO) declared a "public health emergency of international concern". In mid-February 2020, the first person infected with the new SARS-CoV-2 died in France, the first death outside Asia. From March 2020, measures were taken worldwide to contain the spread of the novel disease. Strict quarantine measures such as curfews, contact restrictions, school closures and travel restrictions, and associated severe restrictions on international travel were at the forefront of the measures introduced. Publications of various kinds have since appeared on the COVID-19 pandemic. The search term "COVID-19"yields more than 5.3 million hits in a Google Scholar search. This chapter deals with the impact of COVID-19 on Fiji, a Small Island Developing State in the Pacific Islands region. Fiji has had to cope with the impacts of natural hazards for as long as people remember. In recent years the dangers of worsening climate change have made societal action indispensable. In early 2020 another threat was added, which pushed climate change into the background for some time. The chapter also focuses on impacts of COVID-19 on Fiji's health system and has caused significant morbidity and mortality in Fiji. Individuals deferred routine and non-urgent treatments due to lockdowns and fear of contracting the virus, leading to chronic illnesses and late diagnoses of serious conditions. At the beginning of the pandemic hospitals had shortages of essential supplies, such as ventilators, personal protective equipment (PPE), and ICU beds. COVID-19 tests had to be sent to Australia. Despite many challenges COVID-19 led to the upgrade of health facilities,experience, and skills. Fiji might be better prepared to face future health challenges. The effects of the COVID-19 pandemic on the environment have been highlighted from a wide variety of perspectives. Two of these perspectives stand out: (1) COVID-19 impacts

E. Weber (✉)
Associate Professor of Geography, School of Agriculture, Geography, Environment, Ocean, and Natural Sciences, The University of the South Pacific, Suva, Fiji
e-mail: eberhard.weber@usp.ac.fj

A. Kopf
Lecturer of Sociology, School of Law and Social Sciences, The University of the South Pacific, Suva, Fiji

R. Akhtar (ed.), *COVID-19: Impacts on Health and Health Systems*,
https://doi.org/10.1007/978-3-031-90255-0_3

on the global economy, global exchange of goods, and people. Among many negative impacts, it also had positive effects on the environment. (2) At the same time, several authors highlighted that environmental factors influenced the transmission of SARS-CoV-2 through various factors. For Fiji declining economic activities, significantly reduced mobility and a considerable reduction in tourism led to lower energy consumption and therefore lower emissions of CO_2 and other greenhouse gases (GHG) (Mishra et al. 2022).

Keywords Climate change · Agriculture · Fisheries · Tourism · COVID-19 · Fossil fuel consumption

3.1 Introduction

It comes as little surprise that much has been published about climate change and COVID-19. Taking Google Scholar, which allows a quick (and maybe not entirely pure) overview about academic publications in the two fields, it comes as a surprise that "Climate Change" in mid-March 2024 received 3.35 million hits, while "COVID-19" 4.9 million hits. Publications on "Climate Change" have been done since decades, on "COVID-19" nothing could have been published prior to the year 2020, when the expression was coined.[1] Searching for the expressions "climate change" and "COVID-19" together in the same publication we get 2.34 million hits. In other words: it appears that reflection about "climate change" picked up considerably as the result of the COVID-19 pandemic.

To look at some titles of publications where both "events" are discussed together reveals a big range of possible connections. Some talk of "Wake-up call" for climate change (Leal-Filho et al. 2020), or "Road to green recovery" (Bogojević 2020), "lessons from the covid-19 pandemic as blueprint for the climate emergency" (Salas 2020), COVID-19 and the Climate Crisis: Challenges and Opportunities (Das Sarma 2020), "Addressing the triple burden of malnutrition in the time of COVID-19 and climate change" (Hickey and Unwin 2020); "Lessons from COVID-19 for Climate Shocks" (Sarkis et al. 2020), "Building a Social Mandate for Climate Action: Lessons from COVID-19" (Howarth et al. 2020), "COVID-19: Lessons for the climate change emergency" (Manzanedo and Manning 2020), "Early Warnings: The Lessons of COVID-19 for Public Health Climate Preparedness" (Sheehan and Fox 2020), "What can COVID-19 teach us about responding to climate change?" (Herrero and Thornton 2020).

[1] The official names COVID-19 and SARS-CoV-2 were issued by the WHO on February 11, 2020. Google Scholar gives more than 1 million hits for the expression "COVID-10" for the period prior to the year 2020. However, when one opens the individual links publications are from 2020 or later. Some publications from the end of 2019 already use the expression "COVID-19" (e.g., Kumar and Morawska 2019) or others just have a wrong publication year in Google Scholar (e.g., Barzegar et al. 2014, which indeed was published in 2021).

Looking at these (and thousands more titles of publications on "Climate Change" and "COVID-19") one can easily get the impression that reflections on "COVID-19" well nurtured the "climate change" debate. This already gives some ideas of how issues gain significance to the public: It appears to be necessary that the water needs to stand right up to the chin before most people start feeling worried.

COVID-19 impacts might have led to enhanced perception and awareness on environment and climate change. It is too early to say if this strengthens, e.g., traditional agriculture or more sustainable forms of production and consumption. Other, more direct impacts on the environment, e.g., the reduction in the consumption of fossil fuels, appear to have bounced back to pre-pandemic levels as soon as people and politicians found that the crisis is over. To hope changes in awareness and policies might be not more than "wishful thinking".

Five issues are in the spotlights of this chapter: (1) What environmental impacts did COVID-19 have through the reduction of tourism activities (2) How did COVID-19 influence energy consumption in Fiji (3) What were COVID-19 impacts in informal settlements in Suva, (4) What were the effects of COVID-19 on agriculture and fisheries in Fiji, and finally (5) the impact of COVID-19 on people's vulnerability to natural hazards.

3.2 COVID-19 and Environmental Impacts

Climate change favored the outbreak of COVID-19 as it has influenced the growth and distribution of bat populations in tropical countries (Beyer et al. 2021). Environmental change has also increased mortality from COVID-19 through the presence of cardiopulmonary comorbidities. Observations at the beginning of the COVID-19 pandemic showed that mortality from COVID-19 is particularly high in regions with high levels of air pollution (Münzel et al. 2021). This chapter will not follow such lines of thought. It also will not go deep into the discussion if the concentration on COVID-19 crowded out attention and resources needed for the fight against climate change (Bostrom et al. 2020; Rauchfleisch et al. 2023). The chapter concentrates on how COVID-19 impacted climate and other forms of environmental change.

Bashir et al. (2020) highlighted that COVID-19-related containment measures caused a sharp decline in total global economic production, transportation, and international trade and tourism, leading to lower energy consumption and unprecedented declines in CO_2 emissions (Nižetić 2020; Nguyen et al. 2021). In the first few months of 2020, virtually all countries introduced drastic measures to combat the spread of the coronavirus bringing economic activity almost to a standstill. Businesses closed, land and sea traffic fell drastically, airplanes were grounded, unemployment figures rose, and consumer spending collapsed. With the decline in transportation and production, emissions of greenhouse gases, most notably CO_2, were reduced.

CO_2 emissions fell by 1048 million tons of CO_2 or 8.6% during the first four months of 2020 compared to 2019 in the same period. China accounted for the largest share of the reduction with − 242 million tons of CO_2, followed by the USA

(− 207 million tons of CO_2), Europe (− 123 million tons of CO_2), and India (− 98 million tons of CO_2). On April 7, at the height of the global lockdown, emissions in regions responsible for 89% of CO_2 emissions were 83 million tons of CO_2, 17% below the 2019 average and on a par with 2006.

A reduction in CO_2 emissions of 4–7% was forecast for 2020 as a whole. By comparison, the global financial crisis of 2008/9 resulted in a reduction in CO_2 emissions of 1.4% in 2009. If such a reduction, as assumed for the whole of 2020, was achieved every year over the following few decades, there would be a realistic chance of complying with the global warming limit of 1.5 °C of the Paris Climate Agreement. An annual reduction in global CO_2 emissions of 2.8 billion tons of CO_2 over the next few decades would be needed (Evans 2020).

With reduced industrial activity, air quality in previously smog polluted areas has improved noticeably. In 2020, air pollution in New York dropped by 50% and coal use in China declined by 40% (Gautam 2020; Pitiranggon et al. 2022). This led to a significant improvement of air quality in many Chinese cities (Ming et al. 2020). Similar positive impacts on air quality were recorded in many countries, particularly in urban areas (Aman et al. 2020; Bernauer and Slowey 2020; Berman and Ebisu 2020; Bogdan 2020; Cheval et al. 2020; Debata et al. 2020). Satellite images from NASA have confirmed this trend throughout the globe (Dutta et al. 2021; Manne and Kantheti 2020). NO_2, SO_2, and other environmental pollutants in major epicenters declined by 20–38% compared to 2019 (Bar 2021; Eroğlu 2021; Kumar and Ayedee 2021; Liu et al. 2021; Sarkodie and Owusu 2021).

In many parts of the world interesting trends have been revealed concerning urban and rural areas: while urban areas saw an improvement in air quality, impacts on employment and livelihood generation were severe (Girdhar et al. 2021; Kang et al. 2021). Living in more densely populated urban areas, particularly in slums and squatter settlements, increased the exposure to the SARS-CoV-2 virus (Sridhar 2023). Other studies found that migrants were often severely affected, such as in Dhaka, Bangladesh (Sohel et al. 2022). People living in or migrating to rural areas have often done much better (Laborde et al. 2021). This is particularly true in regards to food security, where rural residents often, but not always, have shown greater resilience compared to urban dwellers (Endris Mekonnen and Kassegn Amede 2022) (Fig. 3.1).

Despite a considerable decline in emissions, there was little impact on the concentration of CO_2 in the atmosphere. To the contrary: CO_2 concentrations in the Earth's atmosphere increased also in the years of COVID-19. In April 2020, the CO_2 concentration at Mauna Loa stood at 416.21 ppm compared to 413.33 ppm a year earlier (Hwang et al. 2021). The increase of 2.9 ppm/year is among the highest increases of the previous 20 years (NOAA 2020).

Little research exists on how COVID-19 had direct impacts on climate change (Usman et al. 2021). Contreras et al. (2021) looked at media coverage on COVID-19 from January to June 2020. They discussed six important lessons that one can learn from the COVID-19 epidemic and that might have impacts on sustainable development, including mitigation of factors that lead to climate change: the reduction in the usage of fossil fuel leading to lesser greenhouse gas emissions, rapid response to the thread and global response suggesting that the world community can respond quickly

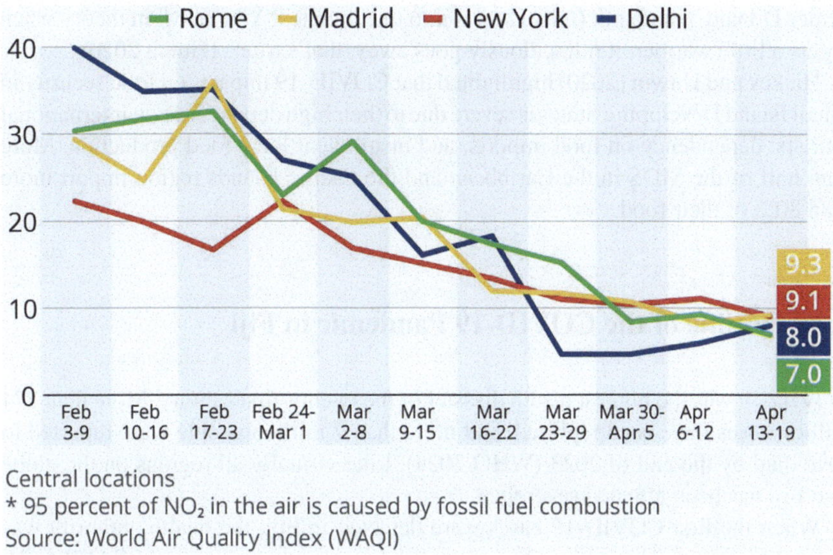

Fig. 3.1 COVID-19 improves air quality in just three months—weekly average concentration of NO_2 in the air in selected cities (Feb.–Apr. 2020). *Source* https://www.statista.com/chart/21495/no2-in-the-air-selected-cities/

when major threats emerge, the emphasis people gave to a sustainable environment, trust in science, and the limits of rugged individualism. Many observers of these trends were foreseeing an opportunity for greater, meaningful change (Contreras et al. 2021).

Fernández-Ahúja and Martínez (2021) studied COVID-19 in Spain. They looked at the spread of the SARS-CoV-2 virus taking physical factors like temperature, air pressure, rainfall, and sunlight hours into consideration. They found that there is a strong connection between COVID-19 and environmental elements. Results showed that cold regions suffered more than warm regions. In a review article, Nguyen et al. (2021) discussed how the COVID-19 pandemic has influenced climate change. Due to the pandemic situation, countries adopted lockdown strategies to save their citizens from the drastic impact of the virus. Consequently, a sharp decline was reported in CO_2 emissions. Likewise, a positive effect was also evident because of less production and limited travel between cities and regions.

Today it is evident that some initial assumptions about the physical environment and its impact on the SARS-CoV-2 virus were wrong. The virus did not share many characteristics with other viruses of the same family (respiratory syndrome viruses). Expecting that hot temperatures would kill the virus did not eventuate. Predictions that the SARS-CoV-2 virus would not last long in environments warmer than 20 °C and would not survive the summer of 2020 in the northern hemisphere or at least considerably slow down in its transmission were proven wrong (Carraturo et al. 2020; Khan et al. 2020). Even the then US President and anthropogenic climate change denier Donald Trump put (false) hopes into the climate: "You know, in theory when it gets a little warmer, it miraculously goes away, that's true". (Gunia 2020).

Hickey and Unwin (2020) highlighted that COVID-19 impacts on food security in Small Island Developing States is severe due to their high dependence on international tourists, dependency on food imports, and insufficient local food production. More than half of the SIDS in the Caribbean and the Pacific Islands region import more than 80% of their food.

3.3 Outline of the COVID-19 Pandemic in Fiji

COVID-19 was the biggest health disaster of the twenty-first century. More than 774 million cases have been reported, and more than 7 million people were reported to have died by the end of 2023 (WHO 2024). Like virtually all regions on the globe also Fiji has been affected severely.

When the first COVID-19 cases were detected in Fiji, the health authority was extremely successful in contact tracing and eliminating the spread of the SARS-CoV-2 virus in the country. A combination of preventive measures and contact tracing has been conducted in the early stages of the pandemic in Fiji. Within 52 days, from mid-March to early May 2020 almost 750,000 individuals (85%) of Fiji's population were screened for COVID-19 (FBC, May 9, 2020). By mid-June 2020, all cases in the Pacific Islands region had recovered with no deaths recorded (Diara et al. 2020). Like most Pacific Island countries, Fiji had a very low incidence of COVID-19 cases for more than a year. In the later months of 2020, the country was even declared "covid-free" (Weber et al. 2021).

In mid-April 2021, a second wave of the disease struck Fiji. Carelessness in the handling of infected people in controlled quarantine facilities allowed the SARS-CoV-2 virus to spread to public areas. Within a very short time, infection numbers exploded and could no longer be contained by contact tracing. Within a few weeks Fiji was among the countries with the highest recorded infection numbers per million population globally (Weber et al. 2022).

When tens of thousands of COVID-19 cases and hundreds of deaths were recorded, the Fiji government came under great pressure from the public and opposition parties, but by the end of 2021 Fiji had the second wave well under control. On December 1, 2021, international travel resumed, and the tourism sector slowly but steadily recovered. Throughout 2022 rare COVID-19 cases were recorded, but at a dimension authorities considered not as a threat to public health. By the end of 2023, Fiji had recorded 69,047 cases of COVID-19 and 885 deaths that were attributed to COVID-19. With these numbers, Fiji had the highest number of COVID-19 cases among the Pacific Island countries followed by Papua New Guinea (46,864 COVID-19 cases and 670 deaths attributed to COVID-19) (WHO 2024).

Fiji's vaccination campaign has achieved remarkable success, reaching higher vaccination rates than any other Pacific Island country. The measures the then government had taken were not without major criticism though from human rights activists.

Authorities perceived resistance from anti-vaccination movements as a threat. Those critical of vaccination were persecuted by state authorities; those not willing to get vaccinated lost their jobs and livelihoods ("no jab, no job"). With 1.56 million vaccinations administered 72% of Fiji's population have received a complete primary series of a COVID-19 vaccine, usually two jabs. At least one booster shot was administered to 19% of Fiji's population (WHO 2024).

3.4 COVID-19 Impacts in Fiji—Tourism

The tourism sector in Fiji has suffered most from COVID-19. The Fiji Hotel and Tourism Association reported that more than 93% of hotels and resorts were closed (FBC News, March 30, 2020). Close to 40,000 workers were laid off or worked on reduced hours or rotational shifts in hotels and resorts (Fijivillage, April 27, 2020). Fiji Airways laid off 758 workers and the Airport Terminal Services 400 workers. This meant that in Fiji almost a third of the labor force either lost their jobs or had a much lower income due to reduced working hours (Finau and Kant 2022).

In contrast to other sectors, global tourism did not only bear much of the consequences of the COVID-19 pandemic (Prideaux et al. 2020). It also contributed significantly to the rapid global spread of the disease (Guan et al. 2020; Mansi 2021). In Fiji tourism was growing well prior to the global COVID-19 outbreak. Earnings increased by 1.5 times between 2012 and 2019 (Shiiba et al. 2023) (Fig. 3.2).

The effects of extensive lockdowns to contain the pandemic on the tourism industry were massive at different levels of scale. At a global level, intercontinental travel came largely to a standstill. Almost all countries in the world (96%) had imposed far-reaching entry restrictions. At a national level, travel restrictions affected economies like that of Fiji which is heavily dependent on tourism.

The sheer impact alone can be seen from the country's international arrival figures, which can be translated into empty hotels, no work for travel agencies and event organizers, and no income for villages which provided cultural experiences to tourists. As a result, unemployment rose tremendously in a society where no significant state unemployment benefits exist. After March 2020 international visitor arrivals went from earlier 50,000 to 100,000 to usually less than 1000 per month.

Those who still came were somehow connected to the management of the COVID crisis. They arrived on special flights but provided at least some visitors to some hotels. Most hotels, however, closed soon after March 2020. When on December 1, 2021, international flights to Fiji were permitted again visitor arrivals slowly increased, although uncertainties due to newly reported COVID-19 cases dampened arrivals in the first three months of 2022. After that arrival numbers steadily increased, and 2023 has been the year with the highest arrival number ever (929,740 compared to 894,389 in 2019, which earlier had been the year with the highest visitor arrivals).

Tourism earns the country the by far biggest amount of foreign exchange, and it provides livelihoods to many people and foreign exchange to Fiji's economy. In 2018 foreign exchange earnings from the tourism sector crossed for the first time

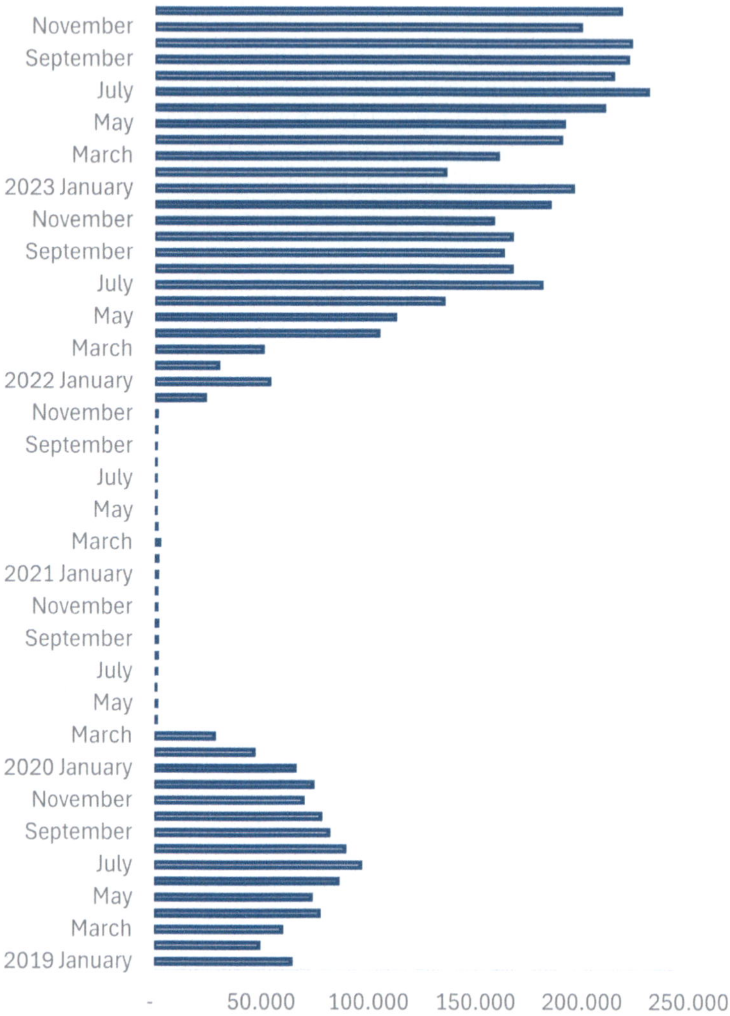

Fig. 3.2 Development of international visitor arrivals to Fiji (2019–2023) FBoS 2024

the FJD 2 billion mark (USD $900 million). In this year it exceeded the country's combined foreign exchange earnings from its top five merchandise exports (fish, water, sugar, gold, and garments). With this tourism in Fiji contributes around 40% to the country's GDP. Tourism provides direct employment to some 40,000 people and to around 116,500 people indirectly.

3.5 Fiji as a South Sea Paradise

Scientists have identified four core aspects of climate change impacts on tourism: (i) direct impacts of a changing climate (too hot, more extreme events, more rain than before), (ii) indirect impacts of environmental changes (e.g., loss of natural attractions, e.g., through coral bleaching), (iii) climate policy and global tourism mobility (e.g., through measures that can lead to an increase in travel costs), and (iv) social change in connection with reduced economic growth, changes in consumer behavior, and sociopolitical instability). If we extend such picture and include COVID-19 then on top of the list would be the decline of international tourist arrivals, but also other aspects of reduced mobility and closure of resorts and related impacts on people's livelihoods would figure prominently.

Considered a tropical paradise, a large part of tourism in Fiji is closely associated with outdoor activities amidst pristine nature and favorable climate. For that reason, the sector is highly sensitive to changes in the climate—especially in terms of precipitation and temperatures. Climate change has a major impact on demand from tourists and on the range of nature-based attractions on offer in Fiji. In addition, natural hazards that are driven by climate factors also have severe impacts on tourism. Beach resorts are inevitably connected to the shoreline, which is the major region where forces from tropical cyclones are the biggest.

Already in the past, there has been much discussion of environmental impacts of mass tourism. Mass tourism has significant environmental impacts on natural ecosystems (Miralles et al. 2023). Environmental degradation of natural resources, entire ecosystems, overconsumption, and the generation of waste stand at the center of unsustainable tourism (Holden 2005). Alone from environmental pollution/degradation perspectives, the following issues have been addressed: Loss of biodiversity caused by habitat destruction and fragmentation (Ren et al. 2022), damage to coral reefs biodiversity and reef structures through activities like snorkeling, diving, and boating (De et al. 2020; Tkachenko 2023), as well as air and water pollution (Baloch et al. 2023). Emissions from increased air travel and transportation cause higher levels of greenhouse gas emissions, air pollution, and noise pollution (Mateus et al. 2023). Inadequate treatment of wastewater and disposal from hotels, resorts, and cruise ships lead to pollution of coastal waters and marine ecosystems (Pásková et al. 2024). Tourism generates large amounts of solid waste, including plastic bottles, packaging, and food waste, which can lead to littering, pollution, and harm to wildlife (Nutsugbodo et al. 2024; Suman and Kumar 2023). Tourism facilities usually have an increased demand for water leading to water scarcity and even groundwater depletion (Aliyev and Suleymanov 2023; Mammadova and Negri 2024). The activities of tourists can cause soil erosion, lead to vegetation loss, and degradation of hiking trails and natural landscapes (Fang and Ng 2024; Tomczyk and Ewertowski 2023). The conversion of natural landscapes and agricultural land into tourist resorts and infrastructure can lead to land degradation and loss of productive land (Lagarias 2023; Ramaano 2023).

A gross reduction in tourist arrivals had a positive aspect on the environment when one considers those publications that highlight negative impacts of tourism on the environment and on climate change. Studies that look at environmental and climate change impacts **on** the tourism sector, however, are far more frequent than studies that take an opposite view (Agrawala et al. 2003; Becken 2005; Belias et al. 2022; Jiang et al. 2012; Moreno and Becken 2009; Singh et al. 2021; Wolf et al. 2021). At the same time one can certainly not argue that the sharp decline of tourism to Fiji was an overall benefit as it reduced the negative impacts of tourism.

3.6 Impacts on the Consumption of Fossil Fuels

Fiji relies heavily on imported fossil fuels, which provide 88% of the country's primary energy supply (MPWTMS 2024). Figure 3.3 indicates that Fiji imported 918,193 m^3 of fossil fuel in 2019. The following year imports declined to 705,482 m^3, and in 2021, it further went down to 586,941 m^3, or 64% of the volume of the last pre-pandemic year of 2019. In 2022 imports of fossil fuels "recovered" to 787,374 m^3. Exact figures for 2023 are not available yet, but monthly figures until September 2023 show that the highest importation of fossil fuels (2019) most likely will be reached, if not even exceeded instantly with the hot phase of COVID-19. It appears that very little has been learned from the pandemic that is useful for climate change.

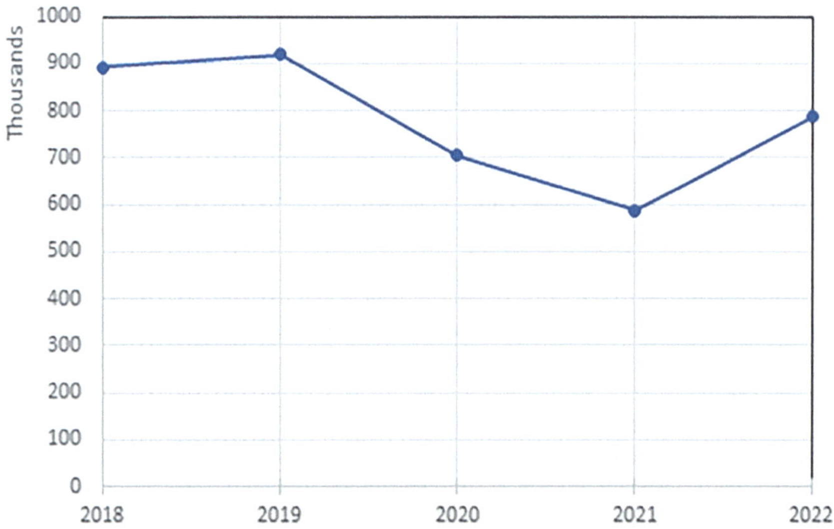

Fig. 3.3 Import of fossil fuels to Fiji (2018–2022, Petroleum oils and oils obtained from bituminous minerals (excl. crude) in Cubic meters). *Source* Trademap (2024)

3.7 COVID-19 and Environmental Security in Informal Settlements

Environments such as those often found in informal settlements favor the spread of contagious diseases including COVID-19 (Drummond et al. 2023). COVID-19 affected people living in informal settlements more than people living elsewhere. Reviews of existing literature (Salama 2023), conceptual studies (Kamalipour and Peimani 2021), and case studies from various African cities (Alonge and Wadinga 2021; Gibson and Rush 2020; Keil et al. 2023; Makumbe 2022; Mvuyana et al. 2020; Osuteye et al. 2020) and Latin American cities (Basile 2023; Duque Franco et al. 2020; Salamanca and Vargas 2020) show that people living in informal settlements have been particularly exposed to risks of infection relating to the coronavirus.

The matter of health and informal settlements is considered crucial enough to publish a series of the Lancet on the health of people who live in informal settlements (Ezeh et al. 2017; Lilford et al. 2017). Through the experience with COVID-19, this particular exposure was certainly confirmed. Again it became obvious that space is a significant variable in relation to epidemiology in the context of informal settlements. Such considerations played an important role already in colonial times when the segregation between native populations and colonial masters led to so-called cordon sanitaires (buffer zones) for "public" health considerations (and to be able to better control and dominate native people) (Colombijn and Barwegen 2009).

Simple aspects in the precaution against COVID-19 are difficult to achieve in many informal settlements. When regular handwashing becomes an essential protection against COVID-19, access to sufficient, safe, and affordable water becomes even more challenging than it anyway is. Protection against COVID-19 and universal and equitable access to water, sanitation, and hygiene highlight how important water is to achieve positive health outcomes.

From such a perspective it is very worrying that more than 50% of the global population do not have access to adequate sanitation, and 75% of households in low-income and middle-income countries are unable to wash with soap and water (Armitage and Nellums 2020).

Water supply is one of the costliest infrastructure provisions. Governments hesitate to expand the distribution networks, and particularly informal settlements are regularly left out when improvements in water supply are undertaken. In informal settlements, water infrastructure is even more underfunded denying people access to safe and affordable water. Such mechanisms have been discussed for a long time looking at climate change (Giri et al. 2021; Heath et al. 2012; Pandey et al. 2023). Without access to safe water, COVID-19 disproportionately affected individuals living in these settings (Fig. 3.4).

In Fiji's capital Suva there is evidence that COVID-19 has disproportionately affected informal settlements. Demographic, socioeconomic as well as environmental conditions are the reasons. Empirical studies that allow us to get a more precise picture hardly exist.

Fig. 3.4 Availability of water and COVID-19 adapted from World Bank Calendar 2019

In informal settlements, people live very close to each other. Overcrowded conditions paired with inadequate infrastructure, and limited access to basic services create an environment in which the SARS-CoV-2 virus can easily spread. Often the lack of clean water and safe sanitation facilities contribute to the spread of diseases. Once infected, people have limited access to health care. Higher infection rates were the result of limited physical distancing. "Even the most basic COVID-19 prevention measures, such as hand washing (limited water supply), social distancing (high population density), and working from home (informal sector employment), are not possible for informal settlement residents" (French et al. 2021).

Preliminary research in some of Suva's informal settlements revealed that low health status results not only from spatial and environmental conditions but inequalities existing in these settlements. The most vulnerable are the elderly, women, children, and people with disabilities. Such residents fall sick more easily, often suffer from various sicknesses at the same time, and are highly susceptive to COVID-19 (French et al. 2021).

3.8 COVID-19 and Its Impact on Agriculture and Fisheries

For Fiji it appears that a significant amount of suddenly unemployed workers of the tourism sector, and urban businesses went to their villages to cultivate food for subsistence (Wairiu et al. 2020). "These new arrivals in rural areas have also increased the challenges of access to limited resources such as water and food, leading to disputes among relatives and other community members. Such disputes arise mainly around newcomers' access to land and clearing of forests unsustainably to increase cultivation of root crops and vegetables" (Iese et al. 2021, 4).

Through COVID-19 agriculture and fisheries increased in importance and status and with it those who had access to land and fishing grounds usually benefitted a lot (Ferguson et al. 2022). The intensification of the use of land and fishing grounds led to unsustainable intensive cultivation with the potential to cause long-term soil infertility, reducing agricultural production and fish catches in the future (Iese et al. 2021).

Another feature that became important was bartering. Widespread unemployment led to the scarcity of cash, and people needed alternative means to sustain and support themselves and their families. It is even recorded that doctors accepted barter in exchange of medical services (Finau and Kant 2022). Early in the pandemic the Facebook site *Barter for Better Fiji (BFBF)* has been established which allowed to manage all sort of barter deals. It still exists today and has a membership of almost 180,000 (https://www.facebook.com/groups/2964591663604507/).

When people returned to their villages to engage in agriculture and fisheries, it was to gain food for their own consumption, but also as a source of livelihood at a time when labor markets were stagnating. Agriculture and fisheries attained an importance they rarely had before.

1. When multiple measures were announced to prevent the spread of COVID-19 in Fiji a sharp reduction in fishing effort was the result. Although throughout 2020 and early 2021 small-scale fishermen and women could obtain special permits to go fishing in the night, fish catch, and the sale of fish declined in many cases. Livelihoods from fishing went down considerably. Research looking at Indo-Fijians in the commercial coastal fisheries sector showed that in 2020 almost three quarters of participants interviewed recorded a decline in the sale of fish due to lockdown-imposed movement restrictions, the breakdown of international and local tourism, and a diminishing domestic purchasing power as unemployment sharply increased, and therefore, people had no money to buy fish (Mangubhai et al. 2021). The situation got exacerbated when Tropical Cyclone Harold hit the archipelago in 2020 (see above). In some areas, the sale of fish declined by half putting pressure on small-scale fishermen to provide enough food for themselves and their families. Lower commercial demand for fish likely led to a reduction in the volume of fish caught (WCS 2020).
2. Similar experiences are reported from native (iTaukei) fisher people from communities in Ba, Cakaudrove, Lau, Macuata, Nadroga, Navosa, Ra, Rewa, and Serua Provinces. Also here the impacts of multiple hazards played a crucial role, again

COVID-19 and Tropical Cyclone Harold occurring at the same time. It appears that impacts on food security were not as severe as expressed by some observers. From the 50 fishermen interviewed, an overwhelming 77% of respondents (77%) felt that there was enough (50%), more than enough (16%), or lots of food available for villagers (10%). Only a quarter (23%) expressed not having enough food available (LMMA Network 2020).

3. Different experiences are reported from agriculture. When many businesses had to close, many dismissed workers went back to their villages to cultivate land that previously was not used. A study looking into people looking for employment in the primary sector revealed a considerable increase in employment in these sectors. "Hotel workers who were laid off didn't have farms and just started planting. Their cassava crops are not mature to eat yet. People don't have money to buy food from shops like before." (LMMA Network, CI, FLMMA, Reef Explorer, and WiFN-Fiji 2020, p. 3).
4. Traditional systems of support were reactivated such as *solesolevaki*—a combination of traditional work assistance and support for those in urgent need. "We are doing *solesolevaki* in the village. The men are planting cassava, taro, yams, and the women are collectively weaving pandanus. There are also groups of villagers who are going out fishing." (LMMA Network 2020, p. 3). The concept of sharing is a major part of *solesolevaki* and is entwined in the traditional way of communal life of most iTaukei villages. Some sixty percent of people interviewed noted that such form of sharing still occurs, but now it most likely happens within family units, targeted toward those who needed assistance (LMMA Network 2020). It has to be assumed that this feature of communal life was a major factor in attracting unemployed family members from urban areas to move back to the village. In some cases, people sent produce from the farm to relatives in urban areas.
5. Strategies in dealing with job loss caused by COVID-19 had also negative impacts on the environment. While earlier restrictions on fish size have been observed now undersized fishes and banned fish species were caught as well. Restrictions to avoid fishing in Marine Protected Areas (MPA) were often not observed. People also more often went fishing in other groups' traditional fishing areas. The use of illegal fishing methods, e.g., through poison and fishing and selling fish without a license did also happen more often than before the COVID-19 pandemic started.
6. Some people ascribed the missing respect to environmental rule and considerations to a lack of leadership. Village meetings happened less frequently and discussions about environmental conservation moved to the background amidst the crisis (LMMA Network, CI, FLMMA, Reef Explorer, and WiFN-Fiji 2020).
7. Whether COVID-19 has also intensified urban agriculture is not entirely clear. Already before the pandemic hit Fiji much of urban land suitable for farming had been used, particularly land in and around informal settlements. Already in the 1970s, Thaman (1975, 1995) stressed the importance of urban agriculture for Fiji's food security. Residents of informal settlements in the eastern outskirts of Suva said that already before COVID-19 they used whatever spaces were idle and suitable to grow crops. In an attempt to support urban gardening, the Ministry

of Agriculture was rather successful in distributing seeds to urban residents. To strengthen its Home Gardening Program for urban and semi-urban residents, the Ministry distributed packages with 12 varieties of seeds for short-life cycle crops. "Demand was high, and in less than a month the MoA had distributed more than 11,600 seed packages to 5000 households".

Among the positive impacts of COVID-19 are that many communities all over the world prioritized natural and pure foods. Avoiding canned and processed foods often has been a necessity born out of interrupted supply chains. Di Renzo et al. (2020) highlighted that such behavior can create new trends for the future. A survey conducted by FMCG Gurus shows that 73% of people are more conscious about healthy foods and drinks. Almost 60% increased their daily fruit intake, close to 50 brought down their sugar intake, and some 40% expressed that they intended to increase their protein intake (Hughes 2020). Similar positive outcomes on nutritional behavior were reported by Webster et al. (2023). Whether such trends played a role in Fiji is not certain.

3.9 COVID-19 and Other Hazards Occurring Simultaneously

Vulnerabilities to hazards relate closely to people's social and economic circumstances and the position they have in their societies. Certain groups such as the poor, the elderly, women-headed households, and others are at great(er) risk of being adversely affected by hazards. Simultaneous exposure to multiple hazards enhances social vulnerability (Weber 2022). Multiple hazards of differing categories (natural, health, social, economic, and political) and differing time dimensions (sudden, long-term, etc.) need to be considered when assessing overall vulnerabilities. There is a need to conduct more research on multiple hazards (Leichenko and O'Brien 2002; O'Brien and Leichenko 2000; Pagliacci and Russo 2020).

Multiple hazard events exist when the same populations are exposed to events that exceed their capacities to cope. The repetition of the same hazard types in quick succession like in 2012 when two major (century) floods hit the city of Nadi, Fiji, within three months are such examples.

During the COVID-19 pandemic, two major events have hit Fiji in the first half of 2020. This period marked the beginning of the COVID-19 crisis. In early April 2020 Tropical Cyclone (TC) Harold, one of the strongest cyclones ever recorded in Fiji, caused severe damage throughout the country. Also TC Cody in January 2022 made it difficult for people as well as authorities to adequately respond to the cyclone hazard and COVID-19.

Soon after Fiji had announced its first measures to mitigate COVID-19 impacts, Tropical Cyclone Harold damaged more than 2000 houses in Fiji. Damages to agricultural production and infrastructure were more than USD 22.6 million (Mahul and Signer 2020). Challenges from COVID-19 were also that more than 6000 people had to seek refuge in almost 200 evacuation centers making social distancing impossible. Bangladesh made similar experience when Tropical Cyclone Amphan happened in May 2020 during the COVID-19 pandemic (Rahman et al. 2021). The volcanic eruption in Tonga in January 2022 caused damages to some 80% of the population from a tsunami and ashfall. Tonga recorded only one case prior to January 2022. The person was taken to a quarantine facility where he recovered. Following the volcanic eruption in January 2022, the virus spread in the country through people who brought air and emergency supplies to Tonga (Burki 2022). As of January 31, 2023, there have been 16,734 recorded COVID-19 cases and 13 fatalities from COVID-19.

Pacific Islands frequently experience natural hazards that affect a big share of the population of a country in every case. Damages are far beyond what governments can finance. From such a perspective COVID-19 caused an additional burden on people and governments in Fiji and other Pacific Island countries.

From a development perspective, multiple hazards are among the major challenges that hinder people to develop. Progress that has been achieved is severely disrupted; people are thrown back in their personal development, not only once but several times. They can hardly take a breath and reach levels they already had accomplished. Similar processes can be seen with states: their expenses for disaster rehabilitation consume crucial resources that otherwise could be utilized more constructively.

Tropical Cyclones (TCs) that were recorded in Fiji during the period COVID-19 restrictions were in place were the strongest ever recorded in the country (TC Herold 2020 and TC Yasa 2020). The most severe challenges are when multiple hazards happen at the same time. Between 2020 and 2022 five tropical cyclones (TC) happened in Fiji and a major flooding event. The worst were two category 5 tropical cyclones (TC Harold and TC Yasa) and two category 3 tropical cyclones (TC Cody, and TC Ana). When these hazards happened the SARS-CoV-2-virus made rescue, relief and rehabilitation activities difficult. Tens of thousands of people seeking refuge in emergency centers cannot observe social distancing as it would be required to protect from COVID-19 infections (Weber et al. 2021, 2022).

COVID-19 restrictions in Fiji were not even a month old when TC Harold (April 2020), a category 5 cyclone, struck the country, one of the strongest TC that made ever landfall in Fiji. TC Harold caused considerable damage, particularly to islands in the south-east of the country. As Suva was under lockdown relief efforts were operated from Natovi jetty, some 60 km north of the capital. Eight months later (December 2020) TC Yasa was the second category 5 TC that hit Fiji in 2020. Maximum wind speeds (260 km/h) were even higher than during TC Harold. More than 2000 homes were completely destroyed and around 6200 partially damaged. Around 23,000 people spent days in 456 evacuation centers. (Weber et al. 2021, 2022).

It did not take even two months before TC Ana caused severe damage end of January, early February 2021. The fourth TC happening subsequently with the COVID-19 pandemic was TC Cody, which struck Fiji in early January 2022. The western and central divisions experienced widespread flooding with 4630 people who had to seek refuge in 162 evacuation centers (RNZ, 13.01.2022).

COVID-19 and natural hazards at the same time were severe challenges as it became necessary to provide people protection against cyclones and floods in evacuation centers and at the same time ensure that people were not exposed to the highly contagious Delta and later Omicron variant of SARS-CoV-2 virus. Another logistical challenge was to provide food for a large number of people in emergency centers and areas where COVID-19 lockdown existed (Weber 2024).

3.10 Conclusion

When comparing COVID-19 and climate change a meaningful aspect was that media coverage about climate change significantly dropped in early 2020, when COVID-19 started. COVID-19 did not only dominate title pages, but crowded many important topics out, including climate change. "Muted by a Crisis?" (Lyytimäki et al. 2020) suggested that it will not take too long before climate change will return to where discourses and media attention had been before 2020. This might be true, but whoever thinks that COVID-19 led to a fundamental change in awareness and perception about global challenges might be disappointed. In early 2024 there is little that would suggest that the world has learnt much from COVID-19, and there is even less to assume that COVID-19 has strengthened awareness about climate change like many voices had expressed that it would.

It appears that there is no new paradigm arising from COVID-19 that can be of any use for climate change mitigation. Small(er) is more beautiful is just not happening. It is not only "business as usual", but the prevailing motif appears to be "more than ever". In Fiji efforts are in place to get even more tourists into the country than before 2020. Reflections on the negative impacts of tourism are lip services, not followed by decisive action. The Deputy Prime Minister is happy to announce that the government is pleased to achieve a record-breaking visitor arrival of 923,740 for 2023. He added that the Fiji tourism sector has even reached capacity constraints. The industry has identified that we need an additional 5000 rooms, so hopefully, it is not going to happen overnight, but given the confidence that investors have and those who see the potential of the industry in the future, we will be attracted toward investing in the hotel infrastructure industry in Fiji." (FBC New, January 24, 2024).

The positive impact the corona crisis was a short-term effect causing at the same time severe suffering all over the world. Long-term improvements can only be achieved with targeted climate and environmental policies that sustainably and permanently change production structures, infrastructures, and consumption and mobility patterns. The goal is not to return to pre-industrial societies, but to invest in resilient economies and communities that can prosper without damaging their foundations (Newell and Dale 2021). The short time after COVID-19 hot phase does not indicate that any meaningful change is happening.

Some might have hoped that the corona crisis not only slowed down production, movement, and reduced the consumption of fossil fuels, but it also slowed down the pace societies are "progressing" leaving more space and time to reflect, to build awareness where Planet Earth is heading toward (Rousseau and Deschacht 2020). Some authors even expressed that "COVID-19 (and its associated disarray of events) is a dress rehearsal for dealing with climate change in the future" (Somani et al. 2020). Other authors saw a distinct difference between COVID-19 and climate change. The perceived fear from both showed different results: people did fear COVID-19 different than climate change. It obviously made them worry more although impacts arising from climate change are not less serious (Lidskog et al. 2020).

Jankó (2020) highlighted that there needs to be a top-down initiative for a "climate friendly transition. We need the state as the main creator and maintainer of fear to operate the fear regime and to stimulate fundamental changes in the economy and society. Yet, the exhaustion of draconian lockdown measures shows that fear regimes cannot last in the long run, at least not in liberal societies. Still, the maintenance of a certain level of fear appears crucial to many in keeping proper anti-pandemic measures at the personal level. However, fear regimes should be replaced with regimes of hope in the long run" Jankó (2020, p. 310).

Ecker et al. (2020) found that COVID-19 does not fit as a "trial run" for future climate action. To the contrary participants in their study highlighted that the impacts of the pandemic justify at least "temporarily dismissing climate change as a secondary concern, while prioritizing economic recovery" (Ecker et al. 2020, 101,464). Similar Botzen et al. (2021, 1) who considered COVID-19 "as a rapid learning experiment about how to cope more effectively with climate change and develop actions for reducing its impacts before it is too late". Preventative and anticipatory measures are more cost-effective than coping actions. This is true for COVID-19 as well as climate change, both in terms of lives and resources saved (Heyd 2021).

Zambrano-Monserrate et al. (2020) highlighted that "It is essential to mention that although the emissions of some GHGs have decreased as a result of the pandemic, this reduction could have little impact on the total concentrations of GHGs that have accumulated in the atmosphere for decades. For a significant decline, there should be a long-term structural change in the countries' economies. This result can be achieved through the ratification of the environmental commitments made. Furthermore, the decrease in GHG emissions currently observed in some countries is only temporary. Since once the pandemic ends, countries will most likely revive their economies, and GHG emissions will skyrocket again".

To continue with business as usual will bring Planet Earth in no time deeper into a changing climate and its many impacts. Economic stimulus/investment packages that are crucial to get economies back on track need to be combined with sustainability goals. Climate protection, emission reductions, and resource conservation are crucial measures to continue. Whether overcoming COVID-19 encouraged structural changes toward climate protection and sustainability (Manzanedo and Manning 2020) is still to be seen. It might be that people (incl. politicians) only respond and take decisive action until the water is up to one's neck.

References

Agrawala S, Ota T, Risbey J, Hagenstad M, Smith J, van Aalst M, Prasad B et al (2003) Development and climate change in Fiji: focus on coastal mangroves. OECD, Paris, France, pp 1–56

Aliyev V, Suleymanov F (2023) The protection of water resources for sustainable tourism under climate change in South Caucasus: in the context of Azerbaijan. GeoJournal Tour Geosites 47(2):515–522

Alonge SK, Wadinga A (2021) Rethinking urban renewal programmes for pandemic control in Nigeria. African J Health, Saf Environ 2(1):14–24

Aman MA, Salman MS, Yunus AP (2020) COVID-19 and its impact on environment: improved pollution levels during the lockdown period–a case from Ahmedabad, India. Remote Sens Appl: Soc Environ 20:100382

Armitage R, Nellums LB (2020) Water, climate change, and COVID-19: prioritising those in water-stressed settings. The Lancet Planet Health 4(5):e175

Baloch QB, Shah SN, Iqbal N, Sheeraz M, Asadullah M, Mahar S, Khan AU (2023) Impact of tourism development upon environmental sustainability: a suggested framework for sustainable ecotourism. Environ Sci Pollut Res 30(3):5917–5930

Bar H (2021) COVID-19 lockdown: animal life, ecosystem and atmospheric environment. Environ Dev Sustain 23(6):8161–8178

Barzegar A, Ghadipasha M, Rezaei N, Forouzesh M, Valizadeh R (2014) New hope for treatment of respiratory involvement following COVID-19 by bromhexine. Pharmacol Ther 28(2):98–108

Bashir MF, Ma B, Shahzad L (2020) A brief review of socio-economic and environmental impact of Covid-19. Air Qual Atmos Health 13:1403–1409

Basile P (2023) Vulnerability, neglect, and collectivity in Brazilian favelas: surviving the threats of the COVID-19 pandemic and the state's necropolitics. Urban Studies 60(9):1690–1706

Becken S (2005) Harmonising climate change adaptation and mitigation: the case of tourist resorts in Fiji. Glob Environ Chang 15(4):381–393

Belias D, Rossidis I, Valeri M (2022) Tourism in crisis: the impact of climate change on the tourism industry. In: Tourism risk: crisis and recovery management. Emerald Publishing Limited, pp 163–179

Berman JD, Ebisu K (2020) Changes in US air pollution during the COVID-19 pandemic. Sci Total Environ 739:139864

Bernauer W, Slowey G (2020) COVID-19, extractive industries, and indigenous communities in Canada: notes towards a political economy research agenda. Extr Ind Soc 7(3):844–846

Beyer RM, Manica A, Mora C (2021) Shifts in global bat diversity suggest a possible role of climate change in the emergence of SARS-CoV-1 and SARS-CoV-2. Sci Total Environ 767:145413

Bogdan EA (2020) How a deadly pandemic cleared the air: narratives and practices linking COVID-19 with air pollution and climate change. Space Cult 23(3):293–300

Bogojević S (2020) Covid-19, climate change action and the road to green recovery. J Environ Law 32(3):355–359

Bostrom A, Böhm G, Hayes AL, O'Connor RE (2020) Credible threat: perceptions of pandemic coronavirus, climate change and the morality and management of global risks. Front Psychol 11:578562

Botzen W, Duijndam S, van Beukering P (2021) Lessons for climate policy from behavioral biases towards COVID-19 and climate change risks. World Dev 137:105214

Burki TK (2022) Volcanic eruption in Tonga and effects on respiratory health. Lancet Respir Med 10(3):e31–e32

Carraturo F, Del Giudice C, Morelli M, Cerullo V, Libralato G, Galdiero E, Guida M (2020) Persistence of SARS-CoV-2 in the environment and COVID-19 transmission risk from environmental matrices and surfaces. Environ Pollut 265:115010

Cheval S, Mihai Adamescu C, Georgiadis T, Herrnegger M, Piticar A, Legates DR (2020) Observed and potential impacts of the COVID-19 pandemic on the environment. Int J Environ Res Public Health 17(11):4140

Colombijn F, Barwegen M (2009) Racial Segregation in the (Post) Colonial City: the case of Indonesia. Urban Geogr 30(8):838–856
Contreras GW, Burcescu B, Dang T, Freeman J, Gilbreth N, Jacobson J, ..., Markenson DS (2022) Drawing parallels among past public health crises and COVID-19. Disaster Med Public Health Prep 16(6):2634–2640
Das Sarma D (2020) COVID-19 and the climate crisis: challenges and opportunities ("The Times They Are A-Changin'"). ACS Energy Lett 5(9):2916–2918
Das A, Ghosh S, Das K, Basu T, Dutta I, Das M (2020) (In) visible impact of inadequate WASH Provision on COVID-19 incidences can be not be ignored in large and megacities of India. Public Health 185:34–36. https://doi.org/10.1016/2Fj.puhe.2020.05.035
De K, Nanajkar M, Mote S, Ingole B (2020) Coral damage by recreational diving activities in a Marine Protected Area of India: unaccountability leading to 'tragedy of the not so commons. Mar Pollut Bull 155:111190
Debata B, Patnaik P, Mishra A (2020) COVID-19 pandemic! It's impact on people, economy, and environment. J Public Aff 20(4):e2372
Di Renzo L, Gualtieri P, Pivari F, Soldati L, Attinà A, Cinelli G, ..., De Lorenzo A (2020) Eating habits and lifestyle changes during COVID-19 lockdown: an Italian survey. J Transl Med 18:1–15
Diarra I, Muna L, Diarra U (2020) How the Islands of the South Pacific have remained relatively unscathed in the midst of the COVID-19 pandemic. J Microbiol Immunol Infect 54(1):131
Drummond JA, Malamud BD, Mulligan J, Bukachi V, Talib M, Wandera A, Taylor FE et al (2023) COVID-19 Interventions in an informal settlement: a spatial analysis of accessibility in Kibera, Kenya. J Transport Geography 113:103704
Duque Franco I, Ortiz C, Samper J, Millan G (2020) Mapping repertoires of collective action facing the COVID-19 pandemic in informal settlements in Latin American cities. Environ Urban 32(2):523–546
Dutta V, Kumar S, Dubey D (2021) Recent advances in satellite mapping of global air quality: evidences during COVID-19 pandemic. Environ Sustain 4:469–487
Ecker UK, Butler LH, Cook J, Hurlstone MJ, Kurz T, Lewandowsky S (2020) Using the COVID-19 economic crisis to frame climate change as a secondary issue reduces mitigation support. J Environ Psychol 70:101464
Endris Mekonnen E, Kassegn Amede A (2022) Food insecurity and unemployment crisis under COVID-19: Evidence from sub-Saharan Africa. Cogent Social Sciences 8(1):2045721
Eroğlu H (2021) Effects of Covid-19 outbreak on environment and renewable energy sector. Environ Dev Sustain 23(4):4782–4790
Evans S (May 19, 2020) Daily global CO_2 emissions 'cut to 2006 levels' during height of coronavirus crisis. CarbonBrief. https://www.carbonbrief.org/daily-global-co2-emissions-cut-to-2006-levels-during-height-of-coronavirus-crisis
Ezeh A, Oyebode O, Satterthwaite D, Chen YF, Ndugwa R, Sartori J, Lilford RJ et al (2017) The history, geography, and sociology of slums and the health problems of people who live in slums. The Lancet 389(10068):547–558. https://doi.org/10.1016/S0140-6736(16)31650-6
Fang W, Ng SL (2024) Trail degradation caused by mountain biking and hiking: a multi-dimensional analysis. J Environ Manage 351:119801
Ferguson CE, Tuxson T, Mangubhai S, Jupiter S, Govan H, Bonito V, Waide M et al (2022) Local practices and production confer resilience to rural Pacific food systems during the COVID-19 pandemic. Mar Policy 137:104954
Fernández-Ahúja JML, Martínez JLF (2021) Effects of climate variables on the COVID-19 outbreak in Spain. Int J Hyg Environ Health 234:113723
Finau G, Kant R (2022) Bartering as a form of resilience during COVID-19: case study of barter for better Fiji Facebook page. In: COVID-19 and social protection: a study in human resilience and social solidarity. Springer Nature Singapore, Singapore, pp 193–204
French M, Ramirez-Lovering D, Sinharoy SS, Turagabeci A, Latif I, Leder K, Brown R (2021) Informal settlements in a COVID-19 world: moving beyond upgrading and envisioning revitalisation. Cities & Health 5(sup1):S52–S55

Gautam S (2020) COVID-19: air pollution remains low as people stay at home. Air Qual Atmos Health 13(7):853–857

Gibson L, Rush D (2020) Novel coronavirus in Cape Town informal settlements: feasibility of using informal dwelling outlines to identify high risk areas for COVID-19 transmission from a social distancing perspective. JMIR Public Health Surveill 6(2):e18844

Girdhar A, Kapur H, Kumar V, Kaur M, Singh D, Damasevicius R (2021) Effect of COVID-19 outbreak on urban health and environment. Air Qual Atmos Health 14:389–397

Giri M, Bista G, Singh PK, Pandey R (2021) Climate change vulnerability assessment of urban informal settlers in Nepal, a least developed country. J Clean Prod 307:127213

Guan D, Wang D, Hallegatte S, Davis SJ, Huo J, Li S, Gong P et al (2020) Global supply-chain effects of COVID-19 control measures. Nat Hum Behav 4(6):577–587

Gunia A (February 28, 2020) Will warmer weather stop the spread of the Coronavirus? Don't count on it, say experts. https://time.com/5790880/coronavirus-warm#weather-summer/

Heath TT, Parker AH, Weatherhead EK (2012) Testing a rapid climate change adaptation assessment for water and sanitation providers in informal settlements in three cities in sub-Saharan Africa. Environ Urban 24(2):619–637

Herrero M, Thornton P (2020) What can COVID-19 teach us about responding to climate change? The Lancet Planet Health 4(5):e174

Heyd T (2021) Covid-19 and climate change in the times of the Anthropocene. The Anthropocene Rev 8(1):21–36

Hickey GM, Unwin N (2020) Addressing the triple burden of malnutrition in the time of COVID-19 and climate change in Small Island Developing States: what role for improved local food production? Food Security 12(4):831–835

Holden A (2005) Achieving a sustainable relationship between common pool resources and tourism: the role of environmental ethics. J Sustain Tour 13(4):339–352

Howarth C, Bryant P, Corner A, Fankhauser S, Gouldson A, Whitmarsh L, Willis R (2020) Building a social mandate for climate action: lessons from COVID-19. Environ Resource Econ 76:1107–1115

Hughes M (2020) Evolving eating habits as a result of COVID-19. Retrieved from https://www.newfoodmagazine.com/article/109890/evolving-eating-habits-as-a-result-of-covid-19/

Hwang Y, Roh JW, Suh D, Otto MO, Schlueter S, Choudhury T, Um JS et al (2021) No evidence for global decrease in CO_2 concentration during the first wave of COVID-19 pandemic. Environ Monit Assess 193:1–15

Iese V, Wairiu M, Hickey GM, Ugalde D, Salili DH, Walenenea J Jr, Ward AC et al (2021) Impacts of COVID-19 on agriculture and food systems in Pacific Island countries (PICs): evidence from communities in Fiji and Solomon Islands. Agric Syst 190:103099

Jankó F (2020) Fear regimes: comparing climate change and the Covid-19 pandemic. Geoforum 117:308–310

Jiang M, Wong E, Klint LM, DeLacy T, Dominey-Howes D (2012) Tourism adaptation to climate change–analysing the policy environment of Fiji. Int J Tourism Policy 4(3):238–260

Kamalipour H, Peimani N (2021) Informal urbanism in the state of uncertainty: forms of informality and urban health emergencies. Urban Design Int 26:122–134

Kang Y, Baidya A, Aaron A, Wang J, Chan C, Wetzler E (2021) Differences in the early impact of COVID-19 on food security and livelihoods in rural and urban areas in the Asia Pacific Region. Glob Food Sec 31:100580

Keil R, Ali SH, Treffers S (2023) Extending the boundaries of 'urban society': the urban political ecologies and pathologies of Ebola virus disease in West Africa. In: Turning up the heat. Manchester University Press, pp 207–221

Khan N, Fahad S, Naushad M (2020) Climate impact on corona virus in the world. Retrieved from https://www.researchgate.net/publication/340488255

Kumar A, Ayedee N (2021) An interconnection between COVID-19 and climate change problem. J Stat Manag Syst 24:1–20. https://doi.org/10.1080/09720510.2021.1875568

Kumar P, Morawska L (2019) Could fighting airborne transmission be the next line of defence against COVID-19 spread? City Environ Interactions 4:100033
Laborde D, Martin W, Vos R (2021) Impacts of COVID-19 on global poverty, food security, and diets: Insights from global model scenario analysis. Agric Econ 52(3):375–390
Lagarias A (2023) Impervious land expansion as a control parameter for climate-resilient planning on the Mediterranean Coast: evidence from Greece. Land 12(10):1844
Leal-Filho W, Nagy GJ, Ayal DY (2020) Climate change, health and pandemics–a wake-up call from COVID-19. Int J Climate Change Strat Manage 12(4):533–535
Leichenko RM, O'Brien KL (2002) The dynamics of rural vulnerability to global change: the case of southern Africa. Mitig Adapt Strat Glob Change 7(1):1–18. https://doi.org/10.1023/A:1015860421954
Lidskog R, Elander I, Standring A (2020) COVID-19, the climate, and transformative change: comparing the social anatomies of crises and their regulatory responses. Sustainability 12(16):6337
Lilford RJ, Oyebode O, Satterthwaite D, Melendez-Torres GJ, Chen YF, Mberu B, Ezeh A et al (2017) Improving the health and welfare of people who live in slums. The Lancet 389(10068):559–570
Liu F, Wang M, Zheng M (2021) Effects of COVID-19 lockdown on global air quality and health. Sci Total Environ 755:142533
LMMA Network, CI, FLMMA, Reef Explorer, and WiFN-Fiji (2020) COVID19 Update #6: Fiji. 20 August 2020. LMMA, Network. Suva. https://lmmanetwork.org/wp-content/uploads/2021/08/Fiji-iTaukei-Coastal-Communities-1.pdf
Lyytimäki J, Kangas HL, Mervaala E, Vikström S (2020) Muted by a crisis? COVID-19 and the long-term evolution of climate change newspaper coverage. Sustainability 12(20):8575
Mahul O, Signer B (2020) The perfect storm: How to prepare against climate risk and disaster shocks in the time of COVID-19. One Earth 2(6):500–502
Makumbe D (2022) Efficacy of lockdowns in Africa: a participant perspective. J BRICS Studies 1(1):41–54
Mammadova L, Negri S (2024) Understanding the impacts of overexploitation on the Salento Aquifer: a comprehensive review through well data analysis. Sustain Futures:100188
Mangubhai S, Nand Y, Reddy C, Jagadish A (2021) Politics of vulnerability: impacts of COVID-19 and Cyclone Harold on Indo-Fijians engaged in small-scale fisheries. Environ Sci Policy 120:195–203
Manne R, Kantheti S (2020) COVID-19 and its impact on air pollution. Int J Res Appl Sci Eng Technol 8(11):344–346
Mansi EW (2021) Environmental health aspect of the Novel Corona Virus disease and its global impact. J Appl Sci Eng Technol Educ 3(2):151–159
Manzanedo RD, Manning P (2020) COVID-19: lessons for the climate change emergency. Sci Total Environ 742:140563
Mateus C, Flor D, Guerrero CA, Córdova X, Benitez FL, Parra R, Ochoa-Herrera V (2023) Anthropogenic emission inventory and spatial analysis of greenhouse gases and primary pollutants for the Galapagos Islands. Environ Sci Pollut Res 30(26):68900–68918
Ming W, Zhou Z, Ai H, Bi H, Zhong Y (2020) COVID-19 and air quality: evidence from China. Emerg Mark Financ Trade 56(10):2422–2442
Ministry of Public Works, Transport and Meteorological Services (MPWTMS, 2024). Ministry launches 2023–2030 National Energy Policy. https://www.mims.gov.fj/ministry-launches-2023-2030-national-energy-policy/
Miralles CC, Barioni D, Mancini MS, Jordà JC, Roura MB, Salas SP, Galli A et al (2023) The Footprint of tourism: a review of water, carbon, and ecological footprint applications to the tourism sector. J Clean Prod:138568
Mishra HG, Pandita S, Bhat AA, Mishra RK, Sharma S (2022) Tourism and carbon emissions: A bibliometric review of the last three decades: 1990–2021. Tour Rev 77(2):636–658

Moreno A, Becken S (2009) A climate change vulnerability assessment methodology for coastal tourism. J Sustain Tour 17(4):473–488
Münzel T, Hahad O, Pozzer A, Lelieveld J (2021) Luftverschmutzung als wichtiger Kofaktor bei COVID-19-Sterbefällen. Kardiologe 15:645–651. https://doi.org/10.1007/s12181-021-00508-2
Mvuyana BY, Mbambo SB, Ntshangase BA (2020) Government interventions and human settlement emergency solutions for disaster control during and post COVID-19. J Public Admin 55(4):662–674
Newell R, Dale A (2021) COVID-19 and climate change: an integrated perspective. Cities & Health 5(sup1):S100–S104
Nguyen XP, Hoang AT, Ölçer AI, Huynh TT (2021) Record decline in global CO_2 emissions prompted by COVID-19 pandemic and its implications on future climate change policies. Energy Sources, Part A: Recov Util Environ Effects. https://doi.org/10.1080/15567036.2021.1879969
Nižetić S (2020) Impact of coronavirus (COVID-19) pandemic on air transport mobility, energy, and environment: a case study. Int J Energy Res 44(13):10953–10961
NOAA Earth System Research Laboratory (2020) Trends in atmospheric carbon dioxide. https://gml.noaa.gov/ccgg/trends/
Nutsugbodo RY, Anaafo D, Wireko-Gyebi RS, Wireko-Gyebi S, Agyeman YB, Afful BEB, Adams J et al (2024) Wastes in the wild: types, effects, and management of plastic waste at Mole National Park, Ghana. Cogent Social Sci 10(1):2316046
O'Brien KL, Leichenko RM (2000) Double exposure: assessing the impacts of climate change within the context of economic globalization. Glob Environ Chang 10(3):221–232. https://doi.org/10.1016/S0959-3780(00)00021-2
Osuteye E, Koroma B, Macarthy JM, Kamara SF, Conteh A (2020) Fighting COVID-19 in Freetown, Sierra Leone: the critical role of community organisations in a growing pandemic. Open Health 1(1):51–63
Pagliacci F, Russo M (2020) Be (and have) good neighbours! Factors of vulnerability in the case of multiple hazards. Ecol Indicators 111. https://doi.org/10.1016/j.ecolind.2019.105969
Pandey R, Yangchen C, Thiyaharajan M, Kishwan J (2023) Attributable factors for climate change adaptation among urban informal settlers of a least developed country, Bhutan. Habitat Int 136:102817
Pásková M, Štekerová K, Zanker M, Lasisi T, Zelenka J (2024) Water pollution generated by tourism: review of system dynamics models. Heliyon 10:e23824
Pitiranggon M, Johnson S, Huskey C, Eisl H, Ito K (2022) Effects of the COVID-19 shutdown on spatial and temporal patterns of air pollution in New York City. Environ Adv 7:100171
Prideaux B, Thompson M, Pabel A (2020) Lessons from COVID-19 can prepare global tourism for the economic transformation needed to combat climate change. Tour Geogr 22(3):667–678
Rahman MM, Bodrud-Doza M, Shammi M, Islam ARMT, Khan ASM (2021) COVID-19 pandemic, dengue epidemic, and climate change vulnerability in Bangladesh: scenario assessment for strategic management and policy implications. Environ Res 192:110303
Ramaano AI (2023) Environmental change impacts and inclusive rural tourism development on the livelihoods of native societies: evidence from Musina Municipality, South Africa. Int J Ethics Syst
Rauchfleisch A, Siegen D, Vogler D (2023) How COVID-19 displaced climate change: mediated climate change activism and issue attention in the Swiss media and online sphere. Environ Commun 17(3):313–321
Ren B, Park K, Shrestha A, Yang J, McHale M, Bai W, Wang G (2022) Impact of human disturbances on the spatial heterogeneity of landscape fragmentation in Qilian Mountain National Park. China Land 11(11):2087
Rousseau S, Deschacht N (2020) Public awareness of nature and the environment during the COVID-19 crisis. Environ Resource Econ 76:1149–1159
Salamanca JDG, Vargas G (2020) Quarantine and informality: reflections on the Colombian case. Space Cult 23(3):307–314

Salama AM (2023) Coronavirus questions that will not go away: interrogating urban and socio-spatial implications of COVID-19 measures. Emerald Open Res 1(5). https://doi.org/10.1108/EOR-05-2023-0006
Salas RN (2020) Lessons from the covid-19 pandemic provide a blueprint for the climate emergency. BMJ 370:m3067
Sarkis J, Dewick P, Hofstetter JS, Schröder P (2020) Overcoming the arrogance of ignorance: supply-chain lessons from COVID-19 for climate shocks. One Earth 3(1):9–12
Sarkodie SA, Owusu PA (2021) Global assessment of environment, health and economic impact of the novel coronavirus (COVID-19). Environ Dev Sustain 23(4):5005–5015
Sheehan MC, Fox MA (2020) Early warnings: the lessons of COVID-19 for public health climate preparedness. Int J Health Serv 50(3):264–270
Shiiba N, Singh P, Charan D, Raj K, Stuart J, Pratap A, Maekawa M (2023) Climate change and coastal resiliency of Suva, Fiji: a holistic approach for measuring climate risk using the climate and ocean risk vulnerability index (CORVI). Mitig Adapt Strat Glob Change 28(2):9
Sridhar KS (2023) Urbanization and COVID-19 prevalence in India. Reg Sci Policy Pract 15(3):493–506
Singh S, Bhat JA, Shah S, Pala NA (2021) Coastal resource management and tourism development in Fiji Islands: a conservation challenge. Environ Dev Sustain 23:3009–3027
Sohel MS, Hossain B, Sarker MNI, Horaira GA, Sifullah MK, Rahman MA (2022) Impacts of COVID-19 induced food insecurity among informal migrants: insight from Dhaka, Bangladesh. J Public Affairs 22:e2770
Somani M, Srivastava AN, Gummadivalli SK, Sharma A (2020) Indirect implications of COVID-19 towards sustainable environment: an investigation in Indian context. Bioresource Technol Reports 11:100491
Suman RK, Kumar V (2023) Litter free beach tourism in India: challenges and solutions. J Tourism 24(1):83
Thaman R (1975) Urban gardeningin Papua New Guinea and Fiji: present status and implications for urban land use planning. University of the South Pacific
Thaman RR (1995) Urban food gardening in the Pacific Islands: a basis for food security in rapidly urbanising small-island states. Habitat Int 19(2):209–224
Tkachenko KS (2023) Degradation of coral reefs under complex impact of natural and anthropogenic factors with Nha Trang Bay (Vietnam) as an example. Biol Bull Rev 13(5):442–459
Tomczyk AM, Ewertowski MW (2023) Landscape degradation and development as a result of touristic activity in the fragile, high-mountain environment of Vinicunca (Rainbow Mountain), Andes, Peru. Land Degrad Dev 34(13):3953–3972
Trademap (2024). https://www.trademap.org/Product_SelCountry_TS.aspx?nvpm=1%7c242%7c%7c%7c%7c27%7c%7c%7c4%7c1%7c1%7c1%7c2%7c1%7c1%7c2%7c1%7c1
Usman M, Husnain M, Riaz A, Riaz A, Ali Y (2021) Climate change during the COVID-19 outbreak: scoping future perspectives. Environ Sci Pollut Res 28(35):49302–49313
Wairiu M, Iese V, Navunicagi O, Fesaitu J, Salili DH, Tabe T, Ward CA et al (2020) Assessing nutrition and socio-economic impact of COVID-19 on Rural and Urban Communities in Fiji. Suva
WCS (2020) COVID-19 Impact on Indo-Fijians in the coastal fisheries sector: FIJI. Wildlife Conservation Society, Suva. https://lmmanetwork.org/wp-content/uploads/2021/08/Fiji-indo-Fijians-Coastal-Fisheries-Sector.pdf
Weber E (2022) People's management of risks from extreme weather events in the Pacific Island region. In: Ongoma V, Tabari H (eds) Climate impacts on extreme weather: current to future changes on a local to global scale. Elsevier, pp 253–270
Weber E (2024) Climate change and human health in Fiji: policies and equity. In: Climate change and human health scenarios: international case studies. Springer Nature Switzerland, Cham, pp 33–47

Weber E, Kopf A, Vaha M (2021) COVID-19 pandemics in the Pacific island countries and territories. In: Coronavirus (COVID-19) outbreaks, environment and human behaviour: international case studies. Springer International Publishing, Cham, pp 25–47

Weber E, Kopf A, Vaha M (2022) COVID-19 in Fiji—from health and economic to major political crisis. Coronavirus (COVID-19) outbreaks, vaccination, politics and society: the continuing challenge. Springer International Publishing, Cham, pp 45–66

Webster J, Moala A, McKenzie B, et al (2023) Food Insecurity, COVID-19 and diets in Fiji—a cross-sectional survey of over 500 adults. Glob Health 19(1):99. https://doi.org/10.1186/s12992-023-01004-w

Wolf F, Filho WL, Singh P, Scherle N, Reiser D, Telesford J, Kovaleva M et al (2021) Influences of climate change on tourism development in small pacific island states. Sustainability 13(8):4223

World Health Organization (2024) WHO COVID-19 dashboard. https://data.who.int/dashboards/covid19/cases

Zambrano-Monserrate MA, Ruano MA, Sanchez-Alcalde L (2020) Indirect effects of COVID-19 on the environment. Sci Total Environ 728:138813

Chapter 4
Patterns of Dengue Incidence Before, During, and After the Coronavirus Disease-2019 Pandemic in Taiwan

Mei-Hui Li

Abstract Because infectious diseases transmission is complex and influenced by multiple factors, the dynamics of emerging and reemerging diseases is difficult to predict. This study investigates the pattern of dengue occurrence in Taiwan by considering the relationship between disease dynamics and environmental elements. In Taiwan, the patterns of dengue outbreaks have changed over the past 120 years. First, I analyze dengue fever occurrences over the last century classified into three major periods to determine which factors potentially affect dengue outbreaks. Interestingly, changes in the patterns of dengue occurrences in Taiwan coincided with changes in the form of government (i.e., from Japanese colonialism to Kuomintang authoritarian military rule to multiparty democracy). Subsequently, I compared the incidence of dengue outbreaks before, during, and after the coronavirus disease-2019 pandemic in Taiwan. Moreover, meteorological factors associated with dengue transmission were examined in Tainan city, one of the dengue hot spots in Taiwan. Finally, I reviewed the current intervention policy for dengue outbreaks in Taiwan. The changes in the patterns of dengue incidence in Taiwan reflect the interplay among different factors on multiple levels.

Keywords Disease transmission · Human mobility · COVID-19 pandemic · Mosquito density · Border control · Southeast Asia

4.1 Introduction

The outbreak of infectious diseases stems from multiple factors acting in concert and is thus difficult to predict. Changes in disease dynamics are associated with changes in several environmental and human factors, including the climate, urbanization, conflict, ecology, human mobility, economic activity, and political institutions. Dengue poses a big challenge to global public health because it is one of the

M.-H. Li (✉)
Department of Geography, National Taiwan University, Taipei, Taiwan
e-mail: meihuili@ntu.edu.tw

R. Akhtar (ed.), *COVID-19: Impacts on Health and Health Systems*,
https://doi.org/10.1007/978-3-031-90255-0_4

fastest growing arboviral diseases in the tropics and subtropics. In 2023, the World Health Organization estimated a record high 6.5 million dengue cases at least, worldwide (WHO 2024a). Therefore, dengue dynamics must be analyzed through multiple perspectives to obtain new insights into the prevention of dengue epidemics.

Results on the history of dengue in Taiwan over the past century can elucidate the dynamics of this disease from multiple perspectives. This study investigated the pattern of dengue occurrences in Taiwan over the past century by considering the relationship between disease dynamics and environmental change. The years between 1900 and 2023 in Taiwan can be divided into three periods: the colonial era until 1945, the no-dengue-outbreak period from 1945 to 1987, and the period of intense human mobility and urbanization after 1987. In this study, I first presented the history of dengue fever occurrence across these three periods from both the local and regional perspective. The change in pattern of dengue occurrences in Taiwan coincided with changes in Taiwan's form of government (from colonial to authoritarian to democratic). Subsequently, I compared the incidence of dengue outbreaks before, during, and after the coronavirus disease-2019 (COVID-19) pandemic in Taiwan. Next, the incidences of dengue outbreaks were examined with respect to climate change. I explored the association between meteorological factors influencing dengue transmission and the incidence of dengue since 2011 in Tainan, a dengue hot spot in southern Taiwan; crucially, I analyzed the relationship between meteorological factors and weekly dengue cases in Tainan in 2023. Finally, I discussed the current policy against dengue outbreaks in Taiwan.

4.2 History of Dengue in Taiwan

4.2.1 Dengue Epidemics from 1900 to 1945

In the first half of twentieth century, many outbreaks of dengue fever were recorded in Taiwan, which was then under Japanese colonial rule (Fig. 4.1). The earliest documentation of dengue in Taiwan dates back to 1902. Subsequently, three island-wide epidemics were recorded in 1915–1916, 1931, and 1942–1943. Throughout this period, Japanese leaders used Taiwan as a regional hub through which they expanded politically, economically, and militarily into Southeast Asia (Shirane, 2022). This geopolitical situation elevated the risk of dengue transmission from endemic to nonendemic Southeast Asian countries. The first island-wide dengue outbreak occurred in 1915–1916, when a cargo ship from Southeast Asia carried the disease to Kaohsiung in southern Taiwan. The disease subsequently spread via railroad to northern Taiwan (Chen, 2018; Hsieh, 2018). This particular outbreak had an attack rate of 25%–50%, infecting 1.7 million people. Another island-wide dengue outbreak with an 80% attack rate occurred in 1942–1943, when Japanese troops used Taiwan as a main transport hub for moving into Southeast Asia during World War II (Ko, 1989; Hsieh, 2018). Overall, both dengue viruses and mosquito vector *Aedes*

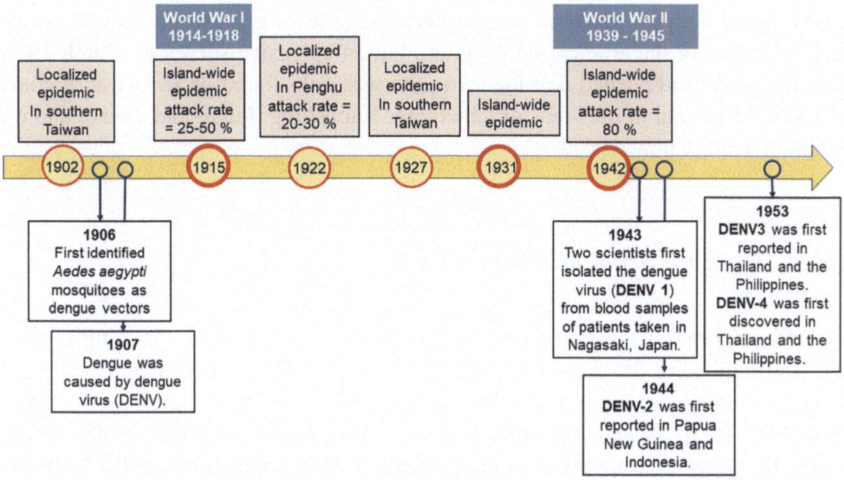

Fig. 4.1 Timeline of dengue outbreaks in Taiwan, 1900–1945 (Data sources: adapted from Ashburn and Craig, 1907; Ko, 1989 and Khan et al., 2023)

aegypti were spread throughout Southeast Asia during World War II. Consequently, the geographic distribution of dengue virus expanded, and the incidence of dengue epidemics in Taiwan increased during this period of war.

4.2.2 No Dengue Outbreak Between 1945 and 1987

Except for a local outbreak on Liuqiu Island in Pingtung County in 1981 (Ko, 1989), no dengue outbreak was recorded on the main island of Taiwan for more than 40 years from 1945 to 1987. Dengue virus serotype-2 was reintroduced from the Philippines into Liuqiu Island by local fishermen and it infected 80% of this small island's population (Hsieh, 2018). Several explanations can be offered for the absence of dengue cases on the main island of Taiwan for more than 40 years. First, the island-wide epidemic of dengue in 1942–1943 infected 5 million residents out of a total population of 6 million, resulting in the development of herd immunity against dengue in Taiwan (Ko, 1989; Chen, 2018). Second, martial law, which was implemented in Taiwan from 1949–1987, limited overseas travel, and this prevented the reintroduction of dengue virus from other countries (Hung et al., 2005). Third, the WHO and other international aids supported a malaria eradication program in Taiwan during the 1950s, which involved indoor residual spraying of dichlorodiphenyltrichloroethane (DDT) to eliminate Anopheles mosquitoes. The WHO officially certified Taiwan as a malaria-free area in 1965 (Hung et al., 2005). In addition to effectively reducing the population of Anopheles mosquitoes transmitting malaria in Taiwan, this program may have decreased the mosquito-assisted transmission of dengue virus (Hung et al.,

2005). Aedes aegypti, the main vector of dengue virus, was found in many localities in Taiwan before the eradication program, but an extensive survey in March 1965 after the program revealed that the mosquito was only found on the western coast of southern Taiwan in the areas of Chiehting, Kaohsiung, Putai, and Tainan and on Penghu Island (Chow, 1967).

4.2.3 Dengue Epidemic Since 1987

In 1987 and 1988, dengue outbreaks occurred in southern Taiwan, Pingtung county, and Kaohsiung city. Since then, indigenous dengue cases have been recorded almost every year (Fig. 4.2). Less than 100 imported cases have been reported every year from 1989 to 2004, and this annual count has exceeded 100 since 2005, except during the years overlapping with the period of border controls due to the COVID-19 pandemic (Fig. 4.2). The increase in the number of imported dengue cases was attributed to Taiwanese tourists traveling to Southeast Asia and Taiwan's labor market opening up to Southeast Asian migrant workers after martial law was abrogated in 1987 (Hung et al., 2005). The imported cases generally peaked months earlier than the indigenous dengue cases (Kuan et al., 2012). Pan et al. (2020) discovered that the high incidence of indigenous dengue fever in Kaohsiung city between 2013 and 2018 was related to the likelihood of those patients living near patients with imported dengue fever within local communities. Dengue outbreaks are usually initiated by imported dengue cases, which leads to an increase in the number of indigenous dengue cases in Taiwan under suitable climatic conditions in the presence of dengue vectors (Shang et al., 2010; Kuan et al., 2012). Southern Taiwan is especially considered as a hot spot for dengue infections. In 2015, Taiwan faced one of the most severe dengue outbreaks since 1987, with over 43,000 cases and 228 deaths. The second largest outbreak occurred in 2023, with over 26,000 dengue cases. Both these outbreaks were mainly concentrated in southern Taiwan, including Tainan city, Kaohsiung city, and Pingtung county.

4.3 Dengue Outbreaks Before, During, and After COVID-19 Pandemic

Between 2015 and 2019, the number of dengue cases significantly increased by 46% in Southeast Asia (WHO 2024b), with Taiwan recording its highest number of imported cases at 540 in 2019 (Fig. 4.2). However, only 100 indigenous cases were reported in that year, which indicates that numerous suitable conditions and factors are required in addition to imported cases to increase dengue transmission during an outbreak (Hsieh 2018). The concurrent outbreak of dengue and COVID-19 imposed a double burden on several Southeast Asian countries (Wiyono et al. 2021). In Taiwan,

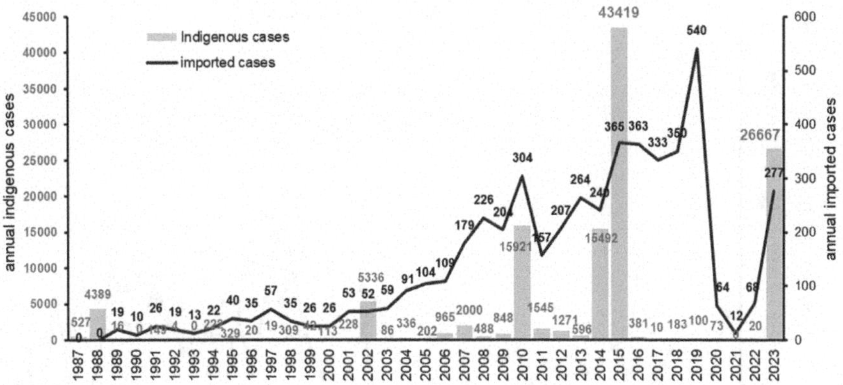

Fig. 4.2 Indigenous and imported dengue cases in Taiwan between 1987 and 2023. *Data source* Taiwan Centers for Disease Control

however, the number of dengue cases decreased during the pandemic, compared with the prepandemic phase (Fig. 4.2). Border controls were implemented on March 19, 2020, in response to the COVID-19 pandemic. Foreigners were prohibited from entering Taiwan, and travelers entering the country were subjected to mandatory quarantine for 14 days. During 2020, the first year of the pandemic, Taiwan recorded 64 imported cases and 73 indigenous cases; during 2021, only 12 imported cases and no indigenous cases; and during 2022, 68 imported cases and 20 indigenous cases, which can be explained by the fact that Taiwan gradually opened its borders and relaxed travel restrictions since 2022 (Fig. 4.2).

A major dengue outbreak occurred after COVID-19-related restrictions had been eased on March 20, 2023, with 277 imported cases and 26,429 indigenous cases (Fig. 4.2). The incidence of dengue in the prepandemic phase in 2019 differed from that in the postpandemic phase in 2023 (Fig. 4.3). One of the most severe dengue outbreaks in Taiwan happened in 2015, with 365 imported cases and 43,419 indigenous cases (Fig. 4.3). In both the 2015 and 2023 epidemics, Tainan city in southern Taiwan reported the highest annual number of indigenous cases at 22,760 and 21,513, respectively. Therefore, we examined other factors that resulted in dengue outbreak in Tainan.

4.4 Climatic Factors and Dengue Cases in Tainan

4.4.1 Meteorological Factors and Dengue Transmission

Most studies have claimed that climatic conditions can affect dengue transmission in Taiwan by altering the ecology of the mosquito vector. Chen et al. (2010) studied the relationship between meteorological factors and Aedes aegypti abundance in Taiwan

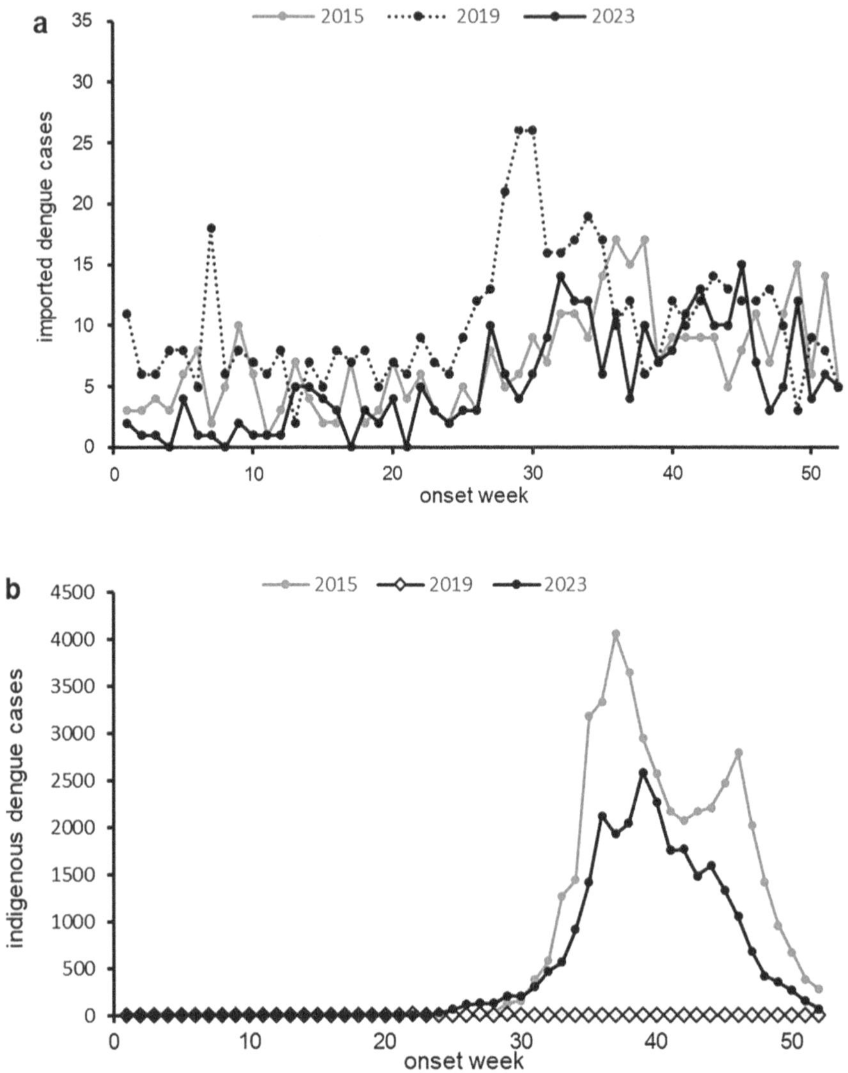

Fig. 4.3 Weekly numbers of **a** indigenous dengue cases and **b** imported dengue cases in 2015, 2019, and 2023. *Data source* Taiwan Centers for Disease Control

between 2001 and 2008 and suggested that dengue transmission could be higher in southern Taiwan because of warmer temperatures with a 3-month lag and elevated humidity with high mosquito density. Wang et al. (2016) demonstrated that the number of dengue cases was significantly correlated with the amount of precipitation and the number of consecutive days with rain during the 2014–2015 period in Tainan and Kaohsiung. You et al. (2022) analyzed the overall correlation between weekly dengue incidence rates and meteorological variables in southern Taiwan between

2007 and 2017. During the dengue epidemics (2014–2015), Tainan had the maximum temperature with a 2-month lag, and Kaohsiung had the minimum temperature with a 5-month lag. Based on the literature, temperature and precipitation were considered to influence dengue outbreaks by affecting the abundance and distribution of the mosquito vector.

4.4.2 Climatic Variables and Dengue Occurrences Between 2011 and 2023

Ordinary least squares regression revealed that only annual mean temperature exhibited a positive trend (slope = 0.06 °C/year; P-value = 0.014); no other meteorological factor changed significantly between 2011 and 2023 in Tainan (Fig. 4.4a). The annual rainfall during the years of severe dengue outbreaks was lower than the average rainfall during this entire period (1481 mm in 2015 and 1310 mm in 2023 vs. 1747 mm average; Fig. 4.4b). Tainan recorded 35 imported cases but only 31 indigenous cases in 2019. Furthermore, the annual rainfall and precipitation hour in 2019 were higher than the corresponding average values for the entire period (Fig. 4.4b), but no large outbreak occurred in Tainan in this year. In general, rainfall combined with higher temperatures can provide breeding sites for Aedes aegypti larvae, leading to an increase in dengue virus transmission (Wang et al. 2016). However, Lai (2018) reported that the incidence of dengue fever decreased as rainfall increased in Kaohsiung city between January, 2005 and March, 2014, thereby speculating that mosquito eggs and larvae may be washed out by the heavy rain. The annual number of dengue cases recorded in Kaohsiung city between 2005 and 2016 was negatively correlated with the maximum 24-h rainfall predictors during the early phases of the epidemic (Yuan et al. 2019), which may explain the absence of a large dengue outbreak in 2019 despite a high amount of rainfall and a high number of imported cases being recorded that year.

4.4.3 Dengue Cases and Meteorological Factors in 2023

I examined the relationships between meteorological factors and weekly dengue cases in Tainan as well as the time-lag effects of meteorological factors on weekly dengue cases. All meteorological data were derived from Tainan and Yongkang stations of the Central Weather Administration. The weekly meteorological data included the weekly average of the daily maximum, mean, and minimum temperatures (°C) and the daily relative humidity (%). The weekly amount of rainfall or precipitation hours was defined as the sum of the daily rainfall or precipitation hours recorded during a week. The number of weekly indigenous dengue cases and the weekly meteorological data for Tainan city in 2023 were plotted together and shown

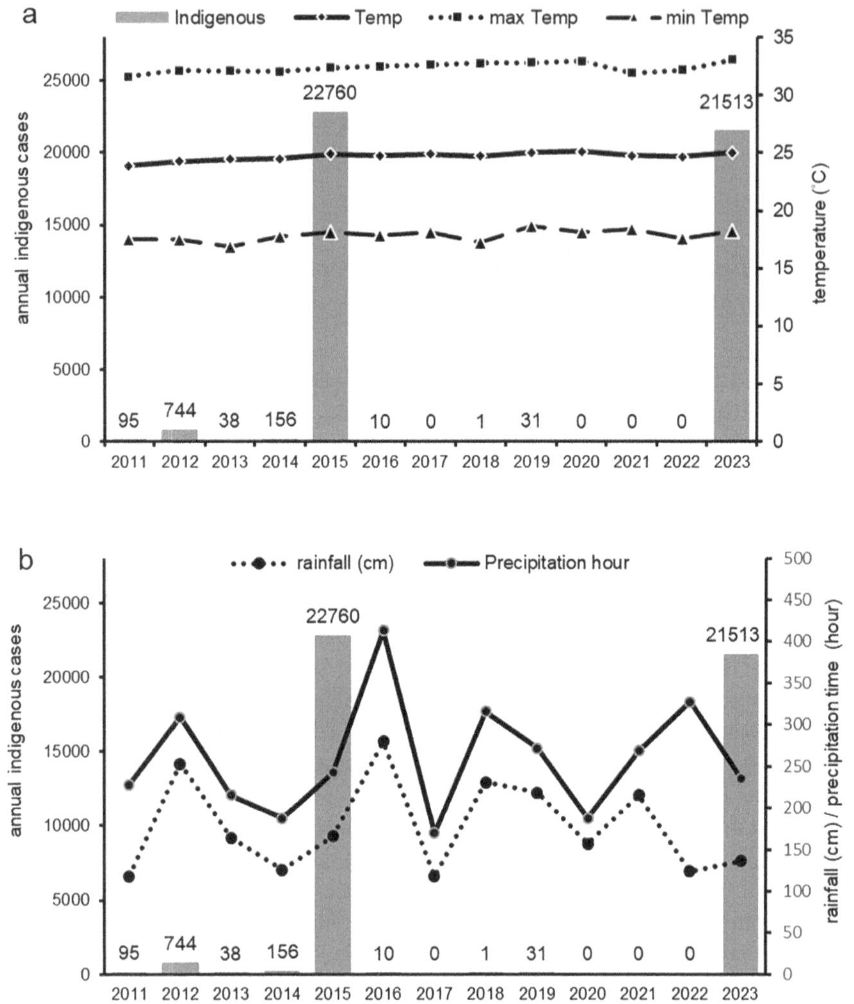

Fig. 4.4 Number of indigenous dengue cases every year between 2011 and 2023 in relation with **a** the annual maximum, mean, and minimum temperatures in this period; **b** the average annual rainfall and precipitation hour in this period. Meteorological data were obtained from Tainan and Yongkang weather stations. *Data source* Taiwan Centers for Disease Control and Central Weather Administration

in Fig. 4.5. Table 4.1 presents Spearman's rank correlation coefficients for the time-lag effects of the weekly meteorological variables on the number of weekly indigenous dengue cases. All the weekly meteorological variables displayed positive and significant relationships with the number of weekly dengue cases within 16-week lag periods. Nevertheless, the mechanisms underlying the 2023 dengue outbreak in Tainan city still remain unclear.

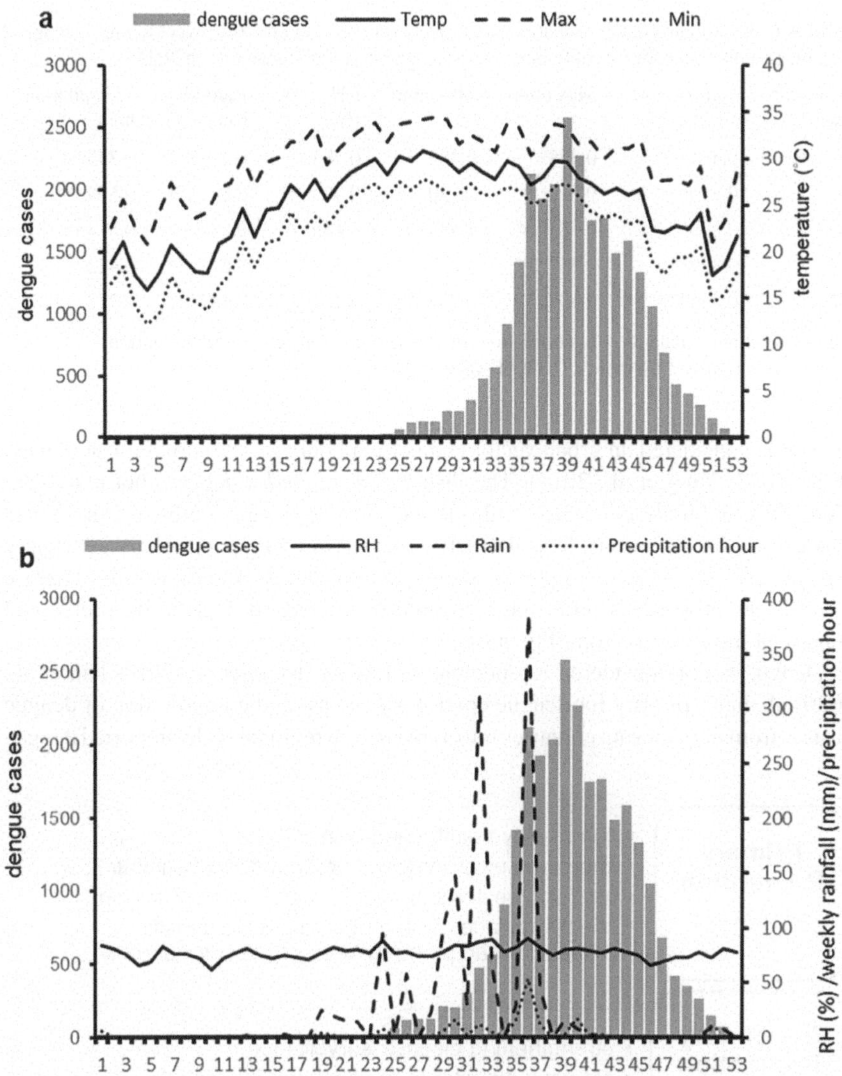

Fig. 4.5 Number of weekly indigenous dengue cases in Tainan city in 2023 in relation to the weekly **a** mean, maximum, and minimum temperatures (°C); **b** relative humidity (RH; %), rainfall (mm), and precipitation hour. All meteorological data were obtained from the Tainan and Yongkang weather stations. *Data source* Taiwan Centers for Disease Control and Central Weather Administration

4.5 Prevention of Dengue Outbreaks

In Taiwan, dengue viruses are spread by two mosquito vectors: *Aedes aegypti* and *Aedes albopictus*. *Aedes albopictus* prefers natural environments and is distributed throughout the island at elevations below 1500 m (Chen 2018). By contrast, *Aedes*

Table 4.1 Spearman's rank correlation coefficients for time-lag effects of weekly meteorological variables on the number of weekly indigenous dengue cases in Tainan city in 2023

Time-lag (week)	Temperature (°C)	Max temp	Min temp	RH (%)	Rainfall (mm)	Precipitation hour
0	0.461***	0.398**	0.483***	0.283*	0.295*	0.355**
4	0.710***	0.612***	0.730***	0.473***	0.535***	0.529***
8	0.869***	0.781***	0.885***	0.459**	0.687***	0.574***
12	0.898***	0.851***	0.877***	0.354*	0.616***	0.422**
16	0.821***	0.808***	0.783***	0.212	0.442**	0.272

All meteorological data were obtained from the Tainan and Yongkang weather stations
***P-value < 0.001; **P-value < 0.01; *P-value < 0.05

aegypti is considered an urban vector and is found only in southern Taiwan (Chang et al., 2007; Yang et al., 2014). The distribution of *Aedes aegypt*i, but not *Aedes albopictus*, is highly correlated with the locations of dengue outbreaks in Taiwan (Yang et al., 2014; Tsai and Teng 2016). Elimination of vector populations is currently the only effective strategy to prevent dengue outbreaks. Taiwan Centers for Disease Control has proposed a three-stage prevention strategy to deplete the source and density of mosquito vectors (Fig. 4.6).

Dengue is not considered as endemic in Taiwan (Kuan et al., 2012; Pan et al., 2020). A major priority for dengue control is decreasing the importation of dengue viruses from neighboring countries into Taiwan. At a regional scale, imported dengue

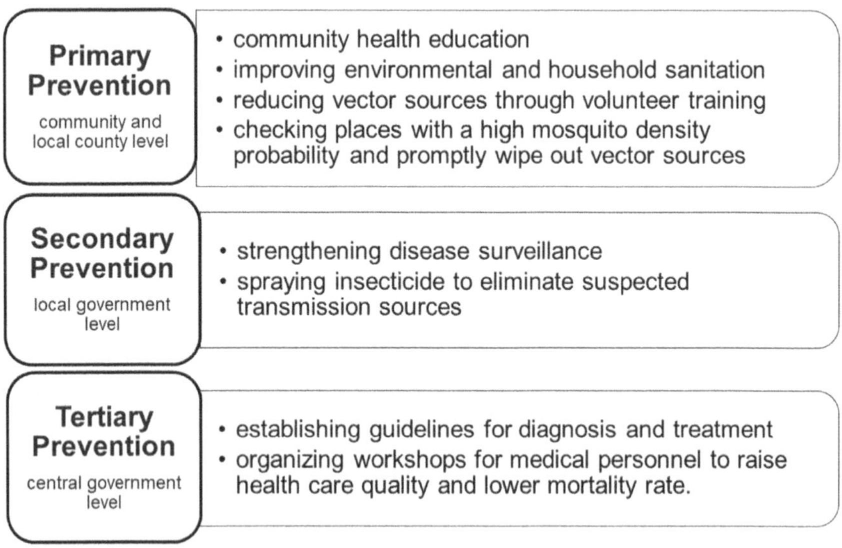

Fig. 4.6 Three-stage prevention strategy to deplete the source and density of mosquito vectors for controlling the dengue epidemic (Taiwan Centers for Disease Control, 2023)

cases are predominantly from Southeast Asia and potentially lead to an increase in indigenous cases in Taiwan. Since July 2003, after the severe acute respiratory syndrome outbreak, the Taiwan Centers for Disease Control implemented a fever screening program for symptomatic passengers at international airports and seaports to strengthen border quarantine on the importation of communicable diseases. International airports were equipped with the dengue nonstructural protein 1 antigen test to rapidly detect positive cases and promptly initiate preventive measures (Taiwan Centers for Disease Control 2023). During the COVID-19 pandemic, Taiwan Centers for Disease Control implemented tight border control with a low incidence of imported and indigenous dengue cases between 2020 and 2022.

4.6 Conclusion

Dengue transmission is influenced by multiple factors, such as temperature, rainfall, population immunity, urbanization, population mobility, and health systems with multiple scales. The year 2023 saw Taiwan experience its second most severe dengue outbreak, with 277 imported cases and approximately 26,000 indigenous cases. An increased risk of exposure to the imported cases is not adequate to explain an outbreak of such a large scale. Similarly, climatic factors alone cannot explain the low ingenious incidence of dengue in 2019 or the second most severe outbreak in 2023. The changes in the pattern of dengue outbreaks in Taiwan reflect the complex interplay among social, political, and environmental factors on multiple levels.

References

Ashburn PM, Craig CF (1907) Experimental investigations regarding etiology of dengue fever. J Infect Dis 4(3):440–475

Chang LH, Hsu EL, Teng HJ, Ho CM (2007) Differential survival of Aedes aegypti and *Aedes albopictus* (Diptera: Culicidae) larvae exposed to low temperatures in Taiwan. J Med Entomol 44(2):205–210

Chen WJ (2018) Dengue outbreaks and the geographic distribution of dengue vectors in Taiwan: a 20-year epidemiological analysis. Biomed J 41:283–289

Chen SC, Liao CM, Chio CP, Chou H, You SH, Cheng YH (2010) Lagged temperature effect with mosquito transmission potential explains dengue variability in southern Taiwan: insights from a statistical analysis. Sci Total Environ 408(19):4069–4075

Chow CY (1967) *Aedes aegypti* in the Western Pacific Region. Bull World Health Organization 36(4):544–546

Hsieh YH (2018) Dengue outbreaks in Taiwan, 1998–2017: importation, serotype and temporal pattern. Asian Pac J Trop Med 11(8):460–466

Hung CJ, Li TC, Chiou HY, Wu BH (2005) Prevention of dengue fever in Taiwan-a year 2004 report. Epidemiol Bull 21(8):241–268

Khan MB, Yang ZS, Lin CY, Hsu MC, Urbina AN, Assavalapsakul W, Wang WH, Chen YH, Wang SF (2023) Dengue overview: an updated systemic review. J Infect Public Health 16(10):1625–1642

Ko YC (1989) Epidemiology of dengue fever in Taiwan. Kaohsiung J Med Sci 5(1):1–11 (in Chinese with English abstract)
Kuan MM, Wu HS, Shu PY (2012) Enforcement of mosquito control and dengue fever prevention. Epidemiol Bull 28(14):211–222
Lai YH (2018) The climatic factors affecting dengue fever outbreaks in southern Taiwan: an application of symbolic data analysis. Biomed Eng Online 17:1–14
Pan CY, Liu WL, Su MP, Chang TP, Ho HP, Shu PY, Huang JJ, Lin LJ, Chen CH (2020) Epidemiological analysis of the Kaohsiung city strategy for dengue fever quarantine and epidemic prevention. BMC Infect Dis 20:1–9
Shang CS, Fang CT, Liu CM, Wen TH, Tsai KH, King CC (2010) The role of imported cases and favorable meteorological conditions in the onset of dengue epidemics. PLoS Negl Trop Dis 4(8):e775
Shirane S (2022) Imperial gateway: colonial Taiwan and Japan's expansion in South China and Southeast Asia, 1895–1945. Cornell University Press, p 288
Taiwan CDC (2023) Centers guidelines for Dengue/Chikungunya control. 16 edition. (in Chinese)
Tsai PJ, Teng HJ (2016) Role of Aedes aegypti (Linnaeus) and *Aedes albopictus* (Skuse) in local dengue epidemics in Taiwan. BMC Infect Dis 16:1–20
Wang SF, Chang K, Loh EW, Wang WH, Tseng SP, Lu PL, Chen YH, Chen YMA (2016) Consecutive large dengue outbreaks in Taiwan in 2014–2015
Wiyono L, Rocha ICN, Cedeño TDD, Miranda AV, Lucero-Prisno DE III (2021) Dengue and COVID-19 infections in the ASEAN region: a concurrent outbreak of viral diseases. Epidemiol Health 43:e2021070
WHO (2024a) Dengue and severe dengue. https://www.who.int/news-room/fact-sheets/detail/dengue-and-severe-dengue. Last accessed on 27 April 2024
WHO (2024b) Dengue. Dengue in the South-East Asia; 2024. Available from: https://www.who.int/southeastasia/health-topics/dengue-and-severe-dengue. Last accessed on April 18 2024
Yang CF, Hou JN, Chen TH, Chen WJ (2014) Discriminable roles of *Aedes aegypti* and *Aedes albopictus* in establishment of dengue outbreaks in Taiwan. Acta Trop 130:17–23
You SH, Chen SC, Huang YH, Tsai HC (2022) Dengue meteorological determinants during epidemic and non-epidemic periods in Taiwan. Trop Med Infectious Dis 7(12):408
Yuan HY, Wen TH, Kung YH, Tsou HH, Chen CH, Chen LW, Lin PS (2019) Prediction of annual dengue incidence by hydro-climatic extremes for southern Taiwan. Int J Biometeorol 63:259–268

Chapter 5
Japan's Internal–External Health Strategy in the Post-COVID-19 Era

Ai Tashiro

Abstract In May 2023, Japan downgraded the legal status of coronavirus disease 2019 (COVID-19) from Class 2 to Class 5, placing it in the same class as common infectious diseases. This chapter provides an overview of Japan's medical system after reclassifying COVID-19 as Class 5, including changes in subsidies for dispatching healthcare workers and public financial support. Drawing on lessons learned from the COVID-19 pandemic, this chapter explores Japan's internal health strategy and how it aims to transform universal health coverage (UHC) into a global health strategy with two critical policy goals: establishing a global health architecture and promoting sustainable UHC in Japan and abroad in the post-COVID-19 era.

Keywords Post-COVID-19 · Class 5 · UHC · Global health architecture · Global health strategy

5.1 Introduction

On May 8, 2023, coronavirus disease 2019 (COVID-19), caused by infection with severe acute respiratory syndrome coronavirus 2 (SARS-CoV-2), became a Class 5 (the Ministry of Health, Labour, and Welfare: MLHW 2023e) instead of a Class 2 infectious disease under the Infectious Disease Control Law in Japan. One year has now passed since the transition to Class 5. Since the days of the COVID-19 pandemic, Japanese society has regained its normal lives. Infection control measures are now based on individual decision-making. The transition to a normal medical care delivery system is gradually underway (Katori 2024). March 2024 marked the end of public financial support for medical expenses for patients with COVID-19 and the issuance of fees to medical institutions for securing hospital beds (MHLW 2023d). However, the essential themes of "infection prevention measures" and "securing the healthcare delivery system" that match the situation remain unchanged. Strict infection control measures and behavioral restrictions are no longer in place, and the free vaccination

A. Tashiro (✉)
Graduate School of Human and Environmental Studies, Kyoto University, Kyoto, Japan
e-mail: tashiro.ai.7m@kyoto-u.ac.jp

R. Akhtar (ed.), *COVID-19: Impacts on Health and Health Systems*,
https://doi.org/10.1007/978-3-031-90255-0_5

and subsidies for expensive coronary drugs ("special exception on medical fees" for SARS-CoV-2 infections), which had been provisionally in place, ended at the end of March 2024 (MHLW 2023c). However, they will be extended if required. The shift to Class 5 does not mean COVID-19 has disappeared or is no longer dangerous.

However, there are still numerous cases of COVID-19. Since the transition to Class 5, comprehensive surveys of all COVID-19 cases are no longer conducted in Japan. As a result, the approximate number of infected individuals can only be estimated based on the fixed-point ascertainment of patients seen at certain government-designated medical institutions, primarily by pediatricians (MHLW, n.d.). Data from this medical sentinel site surveillance shows that COVID-19 cases have steadily increased since May 2023, when the disease transitioned to Class 5. The ninth wave of the COVID-19 pandemic occurred through August (Fig. 5.1). While the number of COVID-19 cases is decreasing, there was a significant wave of COVID-19 cases from January to February 2024. As of 2024, while the number of COVID-19 cases is on a downward trend, there was a substantial wave of COVID-19 cases in January and February. Many individuals have acquired immunity against SARS-CoV-2 through vaccination or prior infection, so the rate of severe cases decreases even when the number of infections increases.

SARS-CoV-2 is also continually mutating. COVID-19 exhibits specific mutational characteristics, including the emergence of the JN.1 strain from the Omicron strain BA.2.86. The KP.2 strain has recently been rapidly spreading in the United States, Canada, and the United Kingdom, raising concerns about its potential dominance in Japan (Kaku et al. 2024). With increased foreign visitors to Japan, partly due to the weak yen, there is concern that the COVID-19 trends observed in other countries may also impact Japan. The concerning trend of SARS-CoV-2's ability to evade the immune system with each mutation persists. The latest study on the KP.2 strain reveals an enhanced ability to bypass neutralizing antibodies produced by infection with the JN.1 strain and vaccination with XBB strain-compatible vaccines

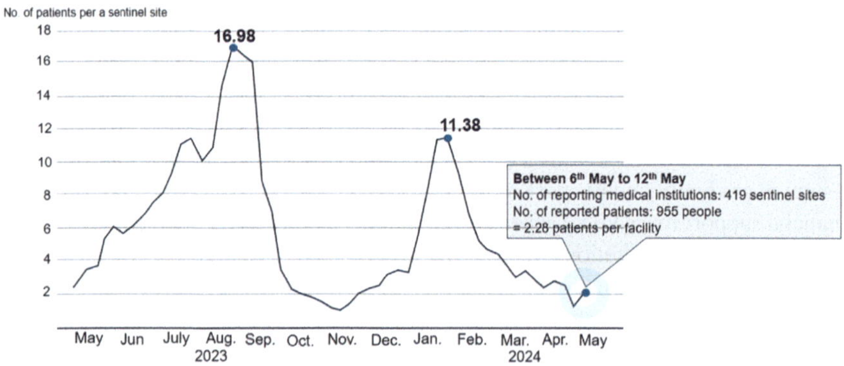

Fig. 5.1 Reported number of COVID-19 cases per sentinel medical institution in Tokyo after the transition to Class 5. *Source* Tokyo Government New Coronavirus Information No. 7, n.d., p. 1

(Kaku et al. 2024). Consequently, the protective effect against infection may be compromised.

5.2 Changes in Medical Policy

5.2.1 New Lifestyle Policy Change Before and After COVID-19

The members of the Expert Group of the MHLW summarized their views on countermeasures against infectious diseases that will be required even after revision, such as moving to Class 5, and presented them at the March 8, 2023 meeting. The opinion lists the following points as the five basic infection control measures (Okabe et al. 2023):

(1) If you feel uneasy about your physical condition or have symptoms, you should either recover at home or visit a medical institution.
(2) Wear a mask and practice cough etiquette according to the situation.
(3) Avoid crowded spaces and ensure proper ventilation.
(4) Hand washing.
(5) Engage in moderate exercise and maintain a balanced diet.

In Fig. 5.2, regarding social distancing, the 2020 policy called for maintaining a two-meter separation where possible, but at least one meter otherwise; avoiding face-to-face conversations as much as possible; and avoiding the so-called Three C of closed spaces, crowded places, and close-contact settings (Tashiro and Shaw 2020; Tashiro 2022). It also recommended avoiding crowds, choosing a quiet time for shopping, and avoid using public transportation. In contrast, the 2024 policy no longer includes specific distances for spacing between people. There are also no provisions for specific situations such as shopping, entertainment, or leisure. The policy states that avoiding the Three Cs remains an effective measure along with ventilation and that the risk of infection can be lowered by maintaining social distancing and choosing quiet times and places, especially in situations with many unknown people.

Regarding masks, the 2020 policy called for wearing a mask when going out or talking indoors, even if people have no symptoms or cannot maintain social distancing. The decision to wear a mask is based on the infection situation in the area, the surrounding congestion, the size of the space, the time spent there, and the risk of serious illness to the person in front of others. In the 2024 policy, when going out, people are expected to carry a mask with them and wear it in situations where it is encouraged.

Regarding transportation, the 2020 policy asked people to refrain from traveling to and from COVID-19 pandemic areas and to note who and where they met in case of an outbreak. The 2024 policy contains no transportation restrictions.

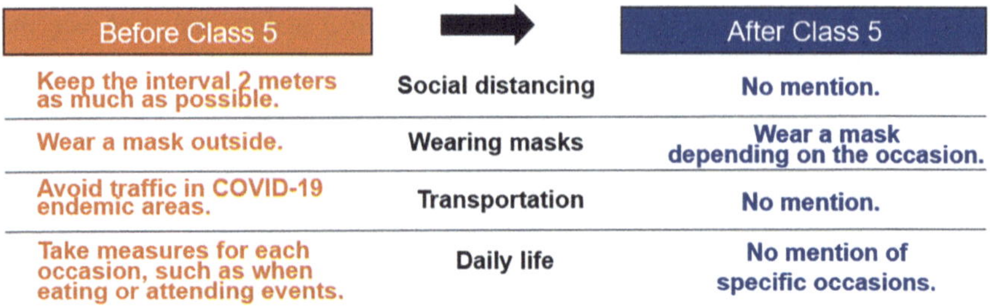

Fig. 5.2 The new lifestyle of 2020 and that of 2024. *Source* Okabe et al. (2023)

5.2.2 *Differences Between New Influenza (Equivalent to Class 2) and Class 5 Infectious Diseases*

Following the reclassification of SARS-CoV-2/COVID-19 from new-type influenza and other infectious diseases (known as Class 2 equivalent) to Class 5 infectious diseases, the government's response has shifted from a system of various government requests and involvement based on the law to one based on voluntary efforts by the public, respecting individual choices. The changes are outlined in Fig. 5.3 (MHLW 2024). The main points of change are as follows:

1. The government will no longer uniformly require basic infection control measures daily.
2. The Infectious Disease Control Law no longer mandates that individuals who test positive for SARS-CoV-2 and those in close contact with them must refrain from leaving their homes.

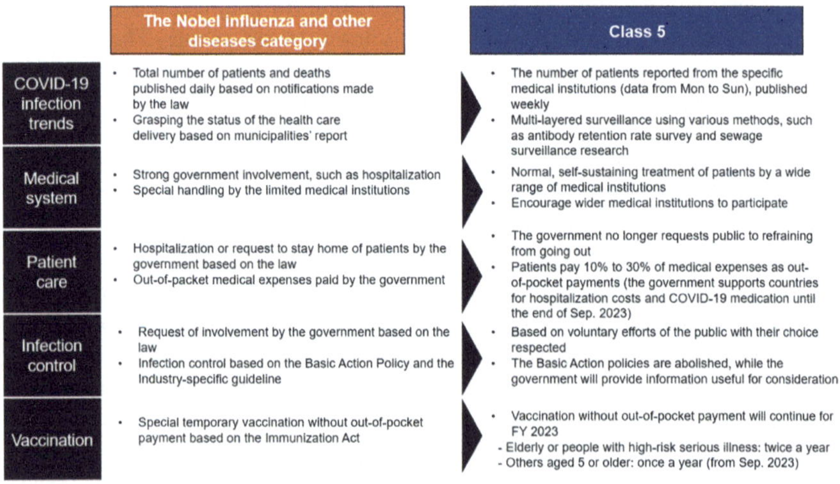

Fig. 5.3 Japan's major infection control measure changes from Class 2 to Class 5 (MHLW 2024)

3. Medical examinations will be available at a broader range of medical institutions instead of being limited to a few.
4. Health insurance will cover medical and other expenses, with patients responsible for 10%–30%. However, public financial support will still be available for a certain period.

5.2.3 Transition of Healthcare Delivery System

At the time of the government's decision in March 2023, special measures such as fees for securing hospital beds would continue until the end of September 2023, even after the transition to Class 5. These measures will be reviewed as necessary based on the results of verification of the response to the spread of infection and the status of the medical care delivery system. The detailed changes are summarized in Table 5.1.

After October 2023, the primary medical care delivery system will be further strengthened to be compatible with usual medical care. Through focused and intensive support, the system will gradually shift to a regular medical care delivery system while responding to the spread of winter infections (MHLW 2023a).

Table 5.1 Healthcare delivery changes

	Class 2	Oct. 2023–Mar. 2024	After Apr. 2024
Outpatient	Approximately 42,000 medical facilities (about 23,000 that do not limit their patients)	Approximately 50,000 medical facilities (about 39,000 that do not limit their patients)	Widespread transition to general medical care response
Inpatient	Approximately 3000 medical facilities in a wide area Maximum hospitalization of 53,000 (including approx. 31,000 secured beds and approx. 22,000 unsecured beds)	Approximately 7300 medical institutions Maximum of 65,000 inpatients (including about 19,000 secured beds and about 56,000 unsecured beds) [Transition plan]	Transition to hospitalization without secured beds ⇒ No charge for secured beds
Hospitalization coordination	State/provincial cities with public health centers are formed into special wards	In principle, the hospitalization destination is determined among medical institutions	Continue to assess hospitalization sites among medical institutions ⇒ G-MIS can continue to be used to share bed status

G-MIS Gathering Medical Information System

Source Transition of healthcare delivery system (outpatient, inpatient, hospitalization coordination) (MHLW 2023a)

5.2.4 Healthcare Public Expenditure Support

Regarding public expense support for patients with COVID-19, the special measures for the transition of COVID-19 to Class 5 ended in March 2023. Beginning in April 2024, the high-cost medical care reimbursement system in health insurance was applied, similar to other diseases. Patients will not have to pay more than certain copayments based on income (MHLW 2023a). Hospitalization measures will be shifted from a unique response by a limited number of medical institutions to a regular response by a broader range of medical institutions. Like seasonal influenza, health insurance will be applied to medical expenses, with the patient bearing 10% to 30% of the cost (MHLW 2023b). To prevent a sudden increase in the burden, a certain amount of public financial support for co-paying inpatient and outpatient medical expenses will continue for a set period.

- Response to home care patients: The health follow-up centers and overnight care facilities established by local governments will no longer need to report patient occurrences or request voluntary furloughs. As a result, these centers and facilities will be discontinued. However, consultation services for medical advice and sudden changes in physical condition will continue to ensure minimal impact on emergency, outpatient, and hospital beds. Additionally, overnight care facilities for the elderly and pregnant women will remain open for a limited time.
- Regarding inpatient and outpatient medical expensing, health insurance will cover the costs of new SARS-CoV-2 infections, similar to seasonal influenza. Patients will be responsible for paying 10%–30% of their medical expenses. To prevent a sudden increase in the financial burden, a certain amount of public financial support for co-paying inpatient and outpatient medical costs will continue for a limited period.

Tables 5.2 and 5.3 show the detailed changes in public expense support for patients with COVID-19 after it transitions to Class 5.

5.3 Estimates of Excess Mortality and Its Relation to COVID-19

Have Japan's new countermeasures against the COVID-19 pandemic been successful? Globally, the number of deaths from the COVID-19 pandemic is lower in Japan than in Europe and the USA. However, an estimated 270,000 excess deaths occurred in Japan during the COVID-19 pandemic, the most crucial indicator of the damage caused by it (Schellekens 2024). These excess deaths suggest that a significant number of citizens have lost their lives during the four-year pandemic. In many countries, the number of excess deaths is several times higher than the number of confirmed deaths. Another concerning issue is that the number of excess deaths in Japan has been increasing significantly since 2022, a fact that is not well-known to

Table 5.2 Public expense support for patients with COVID-19 after its transition to Class 5

	Until Sep. 2023	Oct. 2023–Mar. 2024	After Apr. 2024
COVID-19 treatments	All costs of COVID-19 treatments are covered by public funds (outpatient and inpatient)	• Continuation of public expenditure while requiring a certain level of copayment • The maximum copayment amount will be set in stages according to the percentage of medical expenses to be borne: 10%-payer: 3000 yen, 20%-payer: 6000 yen, and 30%-payer: 9000 yen	• Public expense sharing ends, and the usual copayment is made according to the medical insurance contribution ratio • As with other illnesses, the high-cost medical care system will be applied, and patients will be treated so that they will not have to pay more than a certain copayment depending on their income
Hospital expenses	20,000 yen reduction from the copayment limit of the high-cost medical care benefits system	• Given fairness with other illnesses, continue public funding by revising the high-cost medical care system to a reduction of 10,000 yen from the maximum copayment amount	

Source MHLW (2023c)

the public (Fig. 5.4). Western countries have transitioned to the "with COVID-19" approach after observing a decrease in hospitalizations and deaths following the emergence of the Omicron strain. Similarly, Japan has transitioned COVID-19 from Class 2 to Class 5, equivalent to seasonal influenza. However, the number of deaths is still rising.

5.4 National Strategy for Global Health

5.4.1 Global Health Architecture

The Government of Japan has prioritized global health as one of its key areas for international cooperation. It was evident at the G7 Hiroshima Summit held in May 2023, where global health was a major theme (Cabinet Secretariat, Office of Healthcare Policy 2023). In May 2022, the Government of Japan developed the "(Japan Government 2022)" in anticipation of the 2030 Sustainable Development Goals (SDG) (Japan Government 2022). It aims to contribute to the establishment of a global health architecture (GHA) that supports health security and strengthens preparedness and

Table 5.3 Support for elderly facilities

Before Sep. 2023	Oct. 2023–Mar. 2024	After Apr. 2024
Support for additional expenses in the event of COVID-19 outbreak	• The maximum subsidy amount for work allowance for responding to individuals with COVID-19 is 4000 yen per day, per person	• Various forms of support have been implemented in response to the new coronavirus infection • In the 2024 revision of long-term care compensation, the following measures have been permanently implemented in elderly care facilities to better prepare for future outbreaks of emerging infectious diseases: – Mandatory collaboration with medical institutions that specialize in treating emerging infectious diseases – Introduce an additional fee (10 units/month) to incentivize participation in training sessions on nosocomial infection control conducted by medical institutions working in partnership with those specializing in emerging infectious diseases – Introduction of an additional fee to assess the provision of on-site guidance on infection control, among other requirements, by medical institutions that meet specific infection control criteria (5 units/month) – Introduction of an additional fee to evaluate elderly care facilities that provide in-facility care during outbreaks of emerging infectious diseases (240 credits/day)
Support for in-facility medical treatment (regular assistance: 10,000 yen/day, additional assistance: 10,000 yen/day)	• Normal assistance 5000 yen/day • Additional assistance of 5000 yen/day	
Additional fee for accepting patients recovering from COVID-19 from medical institutions (An additional fee for pre-discharge coordination [500 units/day], which can be calculated for a maximum of 30 days)	• Additional fee for pre-discharge coordination (500 units/day) can be calculated for up to 14 days	

Source MHLW (2023c)

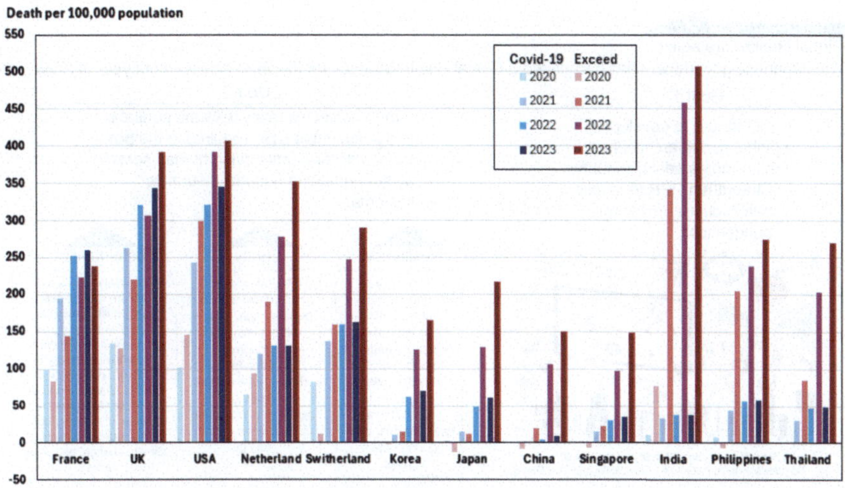

Fig. 5.4 COVID-19-related deaths and excess deaths between 2020 and 2023. *Source* Our World in Data, https://ourworldindata.org/covid-deaths#cumulative-confirmed-deaths-per-million-people

response to public health crises, including pandemics. It will implement prevention, preparedness, and response (PPR) measures, establish international norms, develop health information detection systems, and advance medicine/vaccine development, production, procurement, and distribution to achieve this (MOFA 2022a). The primary goals of the GHA are as follows (Japan International Cooperation Agency 2022):

- Contribute to building and strengthening the GHA to enhance prevention, preparedness, and response to public health crises, such as infectious disease pandemics.
- Achieve the robust, equitable, and sustainable universal health coverage (UHC) required in the post-COVID-19 pandemic era by strengthening health systems in each country.

The Government of Japan announced its Global Health Strategy in May 2022 based on lessons learned from the COVID-19 pandemic (the Headquarters for Healthcare Policy of Japan 2022). Figure 5.5 illustrates Japan's global health strategy. This strategy aims to strengthen the international community's system to better prevent, prepare for, and respond to future pandemics. It focuses on three key areas: GHA, UHC, and health innovation (JICA 2022; Kishida 2023). The G7 has expressed its commitment to leading global health efforts and contributing to the world. Specific details are provided below.

- Committed to increasing political momentum for coordinated and sustainable summit-level governance, with the World Health Organization's

Fig. 5.5 Basic global health policy of Japan

(WHO) leading role, in anticipation of the United Nations (UN) PPR high-level meeting in September 2023

- Under the "G7 Common Understanding on Strengthening Financial and Insurance Partnerships and PPR Financing," the G7 will promote financial initiatives such as strengthening partnerships, the Pandemic Fund, and surge financing
- Guided by impartiality, the G7 expressed its commitment to contribute to preparing the Pandemic Convention and the negotiations to revise the Global Health Regulations
- Affirmed the importance of strengthening and maintaining adequate and quality health human resources, including consideration of a Global Health Emergency Command (GHEC)
- Contribute to achieving resilient, equitable, and sustainable UHC
- Affirmed the promotion of addressing traditional health challenges set back by the new COVID-19 response, including infectious diseases (e.g., human immunodeficiency virus [HIV]/acquired immunodeficiency syndrome [AIDS], tuberculosis [TB], malaria, neglected tropical diseases [NTDs], and polio), non-communicable and other diseases (e.g., mental health symptoms; nutrition; and water, sanitation, and hygiene [WASH]), life stage-specific health issues (e.g., maternal and child health, sexual and reproductive health and rights [SRHR], and healthy aging), and drug resistance (e.g., antimicrobial resistance [AMR])
- Contribute to global health, increase domestic financial mobilization, and increase private sector investment in global health by making financial contributions at the summit. The G7 has contributed over $48 billion

in public and private funds (Japan has contributed $200 million to the Government Global Health Technology Fund [GHIT]
- Triple I for Global Health, an initiative to encourage impact investments in global health, was endorsed at the Hiroshima Summit
- Specific actions to achieve UHC in the G7 were issued as the "G7 Global Plan for UHC Action Agenda" at the Nagasaki Health Ministers' meeting and endorsed at the Hiroshima Summit. It was agreed that a global hub function on UHC, including finance, knowledge management, and human resources, should be noted
- Promote health innovation to address various health challenges
- Further promote health innovation, including research and development (R&D) to respond to pandemics and AMR. Regarding AMR, the need to introduce market incentives for R&D was confirmed in anticipation of 2024 UN high-level meeting on AMR
- The G7 Hiroshima Vision for Equitable Access to Medicines for Communicable Disease (MCM) was announced at the summit. It launched the "Delivery Partnership on MCMs (MCDP)," laying out principles for an end-to-end ecosystem covering the entire value chain from R&D and manufacturing to delivery

Source Kishida (2023).

Since the 2000s, the Government of Japan has also advocated for UHC as a concrete realization of human security in the health sector, which is one of the targets of the UN SDGs (Brolan 2022). Japan has achieved UHC through its universal health insurance system and the development of its healthcare system, which is considered one of the reasons Japan has become one of the world's leading countries in terms of longevity (Tashiro and Kohsaka 2019). The Government of Japan made promoting UHC one of the main themes at the G7 Ise-Shima Summit in May 2016 to assist more countries in achieving UHC. One of the outcome documents, the G7 Ise-Shima Vision for Global Health, clearly stated the need to strengthen healthcare systems toward UHC (WHO 2023). Additionally, at the Sixth Tokyo International Conference on African Development (TICAD) in August of the same year, Japan, together with the World Bank, WHO, and others, released "UHC in Africa," a framework to be used as a reference for African countries in formulating specific national strategies (Kishida 2023). At the G20 Osaka Summit in June 2019, Japan was the first to discuss strengthening UHC financing as one of the main agenda items, and it became the first G20 country to hold a joint session of finance and health ministers. These efforts to mainstream UHC also led to the UN General Assembly high-level meeting on UHC in Africa in September of the same year. Japan's achievements in the G7 health strategy are detailed below:

- Under the "G7 Common Understanding on Strengthening Financial and Insurance Partnerships and PPR Financing" endorsed at the Joint Meeting of Finance and Health Ministers, the G7 will further promote financial initiatives such as strengthening financial and insurance partnerships, the Pandemic Fund, and surge financing.

- Guided by impartiality, the G7 expressed its commitment to contribute to all stakeholders in preparing the so-called Pandemic Convention and in the negotiations for revising the Global Health Regulations.
- Affirmed the importance of strengthening and maintaining adequate and quality health human resources, including consideration of a GHEC.
- Contributing to achieving a more resilient, equitable, and sustainable UHC.
- Affirms the promotion of addressing traditional health challenges set back by the new COVID-19 response, including infectious diseases (e.g., HIV/AIDS, TB, malaria, NTDs, and polio), non-communicable and other diseases (e.g., mental health symptoms, nutrition, and WASH), life stage-specific health issues (e.g., maternal and child health, SRHR, and healthy aging), and drug resistance (e.g., AMR)
- Contribute to global health, increase domestic financial mobilization, and increase private sector investment in global health by making financial contributions at the summit. The G7 has contributed over $48 billion in public and private funds (Japan has contributed $200 million to the GHIT).
- Triple I for Global Health is an initiative to encourage impact investments in global health.
- For Global Health, an initiative to encourage impact investments in global health was endorsed at the Hiroshima Summit.
- Specific actions to achieve UHC in the G7 were issued as the "G7 Global Plan for UHC Action Agenda" at the Nagasaki Health Ministers' meeting and endorsed at the Hiroshima Summit. It was agreed that a global hub function, including finance, knowledge management, and human resources, is important for UHC.
- Promote health innovation to address various health challenges.
- Further promote health innovation, including promotion of R&D to respond to pandemics and AMR. Regarding AMR, the need to introduce market incentives for R&D was confirmed in anticipation of the 2024 UN high-level meeting on AMR.
- The G7 Hiroshima Vision for Equitable Access to Medicines for Communicable Disease (MCM) was announced at the summit.
- Launched the "Delivery Partnership on MCMs (MCDP)," laying out principles for an end-to-end ecosystem covering the entire value chain from R&D and manufacturing to delivery.

Source Ito (2023).

The Government of Japan has been working to build international momentum for global health through various conferences and has supported human resource development and financial assistance (JICA 2023). This support includes contributions to international organizations and technical cooperation in the health field in developing countries by the JICA and others.

For example, JICA has supported developing and expanding hospitals in 22 countries, which has benefited approximately 200 million people (JICA 2023). They have also provided tele-training in intensive care to more than 2500 people under the Corona Disaster Relief Program for FY2020 and beyond.

Furthermore, the Government of Japan has provided approximately US $5 billion in assistance to developing countries as part of COVID-19 measures (Japan Government 2022). In particular, Japan has provided financial assistance of up to US $1.5 billion and in-kind provision of vaccines to the COVAX program (Japan Government 2022). It has also assisted in 78 countries and regions, totaling approximately 18.5 billion yen, to develop cold chain systems and strengthen vaccination capacity among medical personnel (Japan Government 2022).

Additionally, 10.8 billion yen is being provided to assist in revitalizing developing countries' economies and societies and resuming human traffic (MOFA 2023). At the eighth TICAD in August 2022, the Government of Japan announced a contribution of up to $1.08 billion over the next three years to the Global Fund to combat the three major infectious diseases and strengthen health systems (MOFA 2022b).

Since the 2000s, the government of Japan has advocated for UHC as a concrete realization of human security in the health sector; UHC is also one of the targets of the UN SDGs (Tashiro and Kohsaka 2019). Japan has achieved UHC through its universal health insurance and healthcare system, which is considered one of the reasons why Japan has one of the longest life expectancies in the world.

5.5 Conclusion

People have faced and survived new infectious diseases in the past. After nearly four years of battling COVID-19, it seems to have become a part of history. However, the threat of new infectious diseases remains, and people must consider how to prepare for the next pandemic.

Global health governance is a framework that addresses global health issues using various methods involving states and non-state actors. With the progression of globalization and the transformation of infectious diseases into global threats, health governance has become more complex and intertwined with politics. Unfortunately, cooperation is currently hindered by inter-state conflicts and nationalistic tendencies during the COVID-19 pandemic.

The lack of global solidarity in Asia, problems with global health regulations, and the absence of a global response coordination mechanism are the main issues that require active involvement from the international community. Global health governance is a political arena comprising diverse actors, including major powers, smaller states, and non-state actors. Suppose these actors can build strong solidarity in developing COVID-19 vaccines and treatments, ensuring fair distribution, supporting pandemic resolution, and reinforcing response frameworks. In that case, Japan's future in health governance in the post-COVID-19 era looks promising. The urgent task for global health governance is to establish a sustainable health cooperation framework that can withstand political changes while utilizing the individual roles of diverse actors.

Acknowledgements This research was supported by the JSPS KAKENHI Grant Number 23K17098 and the Center for Transdisciplinary Education and Research at the Graduate School of Human and Environmental Studies, Kyoto University.

References

Brolan CE (2022) Public health and the UN sustainable development goals, Oxford Research Encyclopedia of Global Public Health. https://doi.org/10.1093/acrefore/9780190632366.001.0001/acrefore-9780190632366-e-327

Cabinet Secretariat, Office of Healthcare Policy (2023) 2023 Global health strategy initiatives, accessed August 8, 2024. https://www.kantei.go.jp/jp/singi/kenkouiryou/global_health/dai8/siryou1.pdf

Ito N (2023) Global health strategy's efforts to address climate change and health challenges and prospects, accessed August 8, 2024. https://hgpi.org/wp-content/uploads/Materials_Keynote-Speech-2_The-New-Global-Health-Strategys-Approach-to-Climate-Change-and-Health-Challenges-and-Prospects_JPN.pdf (in Japanese)

Japan Government (2022) Japan's three promises for realizing the vision of the UN charter | The Government of Japan—JapanGov, accessed August 8, 2024. https://www.japan.go.jp/kizuna/2022/10/pm_address_at_the_77th_unga.html

JICA (2022) [G7 Special 2] Global health in the post COVID-19 world, accessed August 8, 2024. https://www.jica.go.jp/english/information/topics/2023/20230518_11.html

JICA (2023) A future of protecting lives from the pandemic: the role of JICA's initiative for global health and medicine | News & Media—JICA, accessed August 8, 2024. https://www.jica.go.jp/english/information/topics/2022/20221129_12.html

Kaku Y, Uriu K, Kosugi Y, Okumura K, Yamasoba D, Uwamino Y, Kuramochi J, Sadamasu K, Yoshimura K, Asakura H, Nagashima M, Ito J, Sato K (2024) Virological characteristics of the SARS-CoV-2 KP.2 variant. The Lancet Infect Dis 24(7):e416. https://doi.org/10.1016/s1473-3099(24)00298-6

Katori T (2024) Japan's healthcare delivery system: from its historical evolution to the challenges of a super-aged society. Glob Health Med 6(1):6–12. https://doi.org/10.35772/ghm.2023.01121

Kishida F (2023) Human security and universal health coverage: Japan's vision for the G7 Hiroshima Summit. The Lancet 401(10373):246–247. https://doi.org/10.1016/s0140-6736(23)00014-4

MHLW (n.d.) Fixed-point monitoring of outbreak trends related to COVID-19, accessed August 8, 2024. https://www.mhlw.go.jp/content/001065724.pdf (in Japanese)

MHLW (2023a) Revision of the new coronavirus infections in October and thereafter. 1. Basic Policy of Revisions, accessed August 8, 2024. https://www.mhlw.go.jp/content/001147042.pdf

MHLW (2023b) Status of transmission of COVID-19 and efforts to combat it, accessed August 8, 2024. https://www.mhlw.go.jp/content/10906000/001146454.pdf

MHLW (2023c) Special measures concerning new coronavirus infections, 1. Basic concept accessed August 8. https://www.mhlw.go.jp/content/001219109.pdf (in Japanese)

MHLW (2023d) Response to new coronavirus infections since April 2024, accessed August 8, 2024. https://www.mhlw.go.jp/stf/covid-19/00003.html

MHLW (2023e) Novel Coronavirus (COVID-19)|Ministry of Health, Labour and Welfare|Government of Japan, accessed August 8. https://www.mhlw.go.jp/stf/seisakunitsuite/bunya/0000164708_00079.html

MHLW (2024) Response to COVID 19 (Novel coronavirus) after the classification change |Ministry of Health, Labour and Welfare of Japan, accessed August 8. https://www.mhlw.go.jp/stf/covid-19/kenkou-iryousoudan_00006.html

MOFA (2022a) Minister for Foreign Affairs KAMIKAWA Yoko's participation in the UN general assembly high-level meeting on pandemic prevention, preparedness and response (PPR) | Ministry of Foreign Affairs of Japan. https://www.mofa.go.jp/ic/ghs/page7e_000043.html
MOFA (2022b) White paper on development cooperation 2022 | Part III 8 Africa | Ministry of Foreign Affairs of Japan. https://www.mofa.go.jp/policy/oda/white/2022/html/honbun/b3/s8.html
MOFA (2023) Japan's international cooperation. https://www.mofa.go.jp/policy/other/bluebook/2023/pdf/pdfs/3-2.pdf
Okabe et al (2023) On considering infection control measures around us from now on (third report)—views on "new health habits". The 118th advisory board on countermeasures to combat new coronavirus infections, accessed August 8, 2024. https://www.mhlw.go.jp/content/10900000/001069238.pdf (in Japanese)
Our World in Data, accessed August 8, 2024. https://ourworldindata.org/covid-deaths#cumulative-confirmed-deaths-per-million-people
Schellekens P (2024) Japan and US are worlds apart on pandemic mortality, accessed August 8. https://pandem-ic.com/japan-and-us-are-worlds-apart-on-pandemic-mortality/
Tashiro A, Kohsaka R (2019) Universal health coverage: healthcare system for universal health coverage under partnerships. In: Leal Filho W, Marisa Azul A, Brandli L, Lange Salvia A, Wall T (eds) Partnerships for the goals. Encyclopedia of the UN sustainable development goals. Springer, Cham, pp 1319–1330
Tashiro A, Shaw R (2020) COVID-19 pandemic response in Japan: what is behind the initial flattening of the curve? Sustainability 12(13):5250. https://doi.org/10.3390/su12135250
Tashiro A (2022) COVID-19 vaccination trends and public views and responses in Japan. In: Akhtar R (ed) Coronavirus (COVID-19) outbreaks, vaccination, politics and society. Springer, Cham, pp 67–81
The Headquarters for Healthcare Policy of Japan (2022) Global health strategy of Japan, accessed August 8, 2024. https://www.kantei.go.jp/jp/singi/kenkouiryou/en/pdf/final_GHS.pdf
Tokyo Government New Coronavirus Information No. 7, p. 1. (n.d.), accessed August 8, 2024. https://www.hokeniryo.metro.tokyo.lg.jp/kansen/corona_portal/info/monitoring.files/2024No7.pdf (in Japanese)
WHO (2023) G7 Ise-Shima vision for global health—strengthening of prevention and preparedness against public health emergencies, accessed August 8, 2024. https://extranet.who.int/sph/g7-ise-shima-vision-global-health-strengthening-prevention-and-preparedness-against-public-health

Chapter 6
Compound Impacts of COVID-19 Pandemic and Heat Stroke and Its Countermeasures: Examples from Japan

Kazutaka Oka

Abstract The increase in temperature due to climate change and the heat island effect has led to significant health impacts, such as heat stroke, in Japan during the summer. The risk of heat stroke is expected to become more severe with the intensification of climate change. This chapter examines the compound impacts of heat stroke associated with high temperatures and other factors using the COVID-19 pandemic as a case study. First, it summarizes scientific findings from existing studies on changes in heat stroke risk during the COVID-19 pandemic. The measures implemented by the Japanese government, local authorities, and academic groups to balance COVID-19 and heat stroke countermeasures are also introduced. Finally, this chapter discusses scientific findings and future strategies.

Keywords Heat stroke · COVID-19 · Temperature · Ambulance transport · Countermeasures

6.1 Introduction

The average summer (June–August) temperatures in Japan have been increasing at a rate of 1.25 °C per century (JMA 2023). The year 2023 was recorded as the hottest year globally (WMO 2024), and Japan was no exception (JMA 2024). The rise in temperature in Japan has caused significant health impacts, such as heat stroke. The risk of heat stroke is expected to become more severe with the intensification of climate change (Oka et al. 2023).

The compound impacts of heat stroke associated with high temperatures and other factors have already been observed in Japan. One example is the compound effect of natural disasters and heatstroke. On September 9, 2019, one of the strongest typhoons ever recorded made landfall in the Kanto region (Cabinet office 2020).

K. Oka (✉)
Center for Climate Change Adaptation, National Institute for Environmental Studies, Tsukuba, Japan
e-mail: oka.kazutaka@nies.go.jp

R. Akhtar (ed.), *COVID-19: Impacts on Health and Health Systems*,
https://doi.org/10.1007/978-3-031-90255-0_6

The typhoon caused extensive damage, with 9 fatalities and 160 injuries in Chiba Prefecture (FDMA 2019). Power outages caused by the typhoon resulted in 498 people being transported by ambulance due to heat stroke during September 9–15, three times higher than that in the previous week (September 2–8), due to the inability to use air conditioning (The Asahi Shimbun 2019). As climate change leads to more severe natural disasters, the occurrence of such events during the summer and the accompanying power outages will increase the risk of heat stroke.

Another example is the compound impact of the coronavirus disease (COVID-19) pandemic and heat stroke (Ministry of Health, Labour and Welfare [MHLW] 2020). Government policies restricting movement and other measures to control the COVID-19 pandemic have drastically changed people's behavior (Cabinet office 2023). These policies, especially during the summer, have altered the risk of heat stroke. This chapter focuses on the compound impacts of the COVID-19 pandemic and heat stroke and presents a review of existing studies that have analyzed changes in heat stroke risk during the pandemic. It also summarizes measures implemented by the Japanese government, local authorities, and recommendations from academic groups to balance COVID-19 and heat stroke countermeasures, bridging scientific findings and practical measures.

The remainder of this paper is organized as follows: Sect. 6.2 reviews the status of COVID-19 and heat stroke in Japan. Section 6.3 summarizes scientific findings from studies analyzing changes in the heat stroke risk during the COVID-19 pandemic. Section 6.4 discusses implemented measures to balance COVID-19 and heatstroke countermeasures. Section 6.5 discusses bridges between the scientific findings and measures. Finally, Sect. 6.6 presents concluding remarks.

6.2 COVID-19 and Heat Stroke

6.2.1 Covid-19

6.2.1.1 Numbers of New Positive, Severe, and Mortality Cases for COVID-19

The first COVID-19 patient in Japan was identified on January 15, 2020 (MHLW 2021).Figure 6.1a–c shows the numbers of new positive, severe, and mortality cases of COVID-19 from January 1, 2020, to May 8, 2023 (MHLW 2024a). The peak number of new positive cases was approximately 260,000 per day in August 2022, with severe and mortality cases peaking at approximately 1800 in September 2021 and 500 in January 2023, respectively.

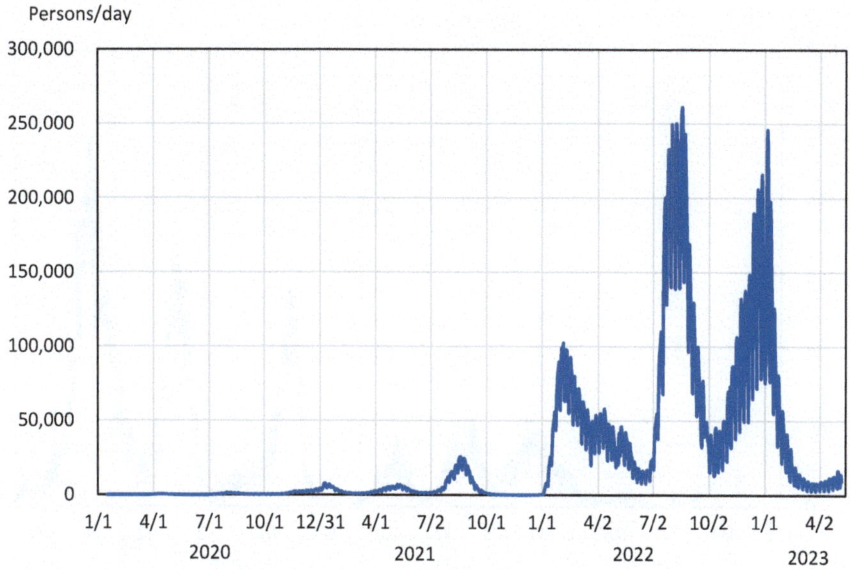

(a) New positive cases

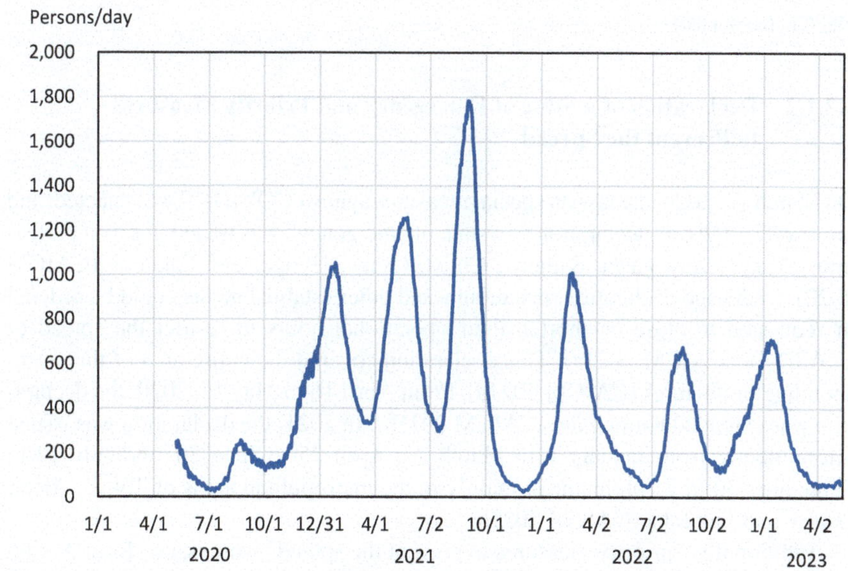

(b) Severe cases

Fig. 6.1 Numbers of **a** new positive, **b** severe, and **c** mortality cases for COVID-19

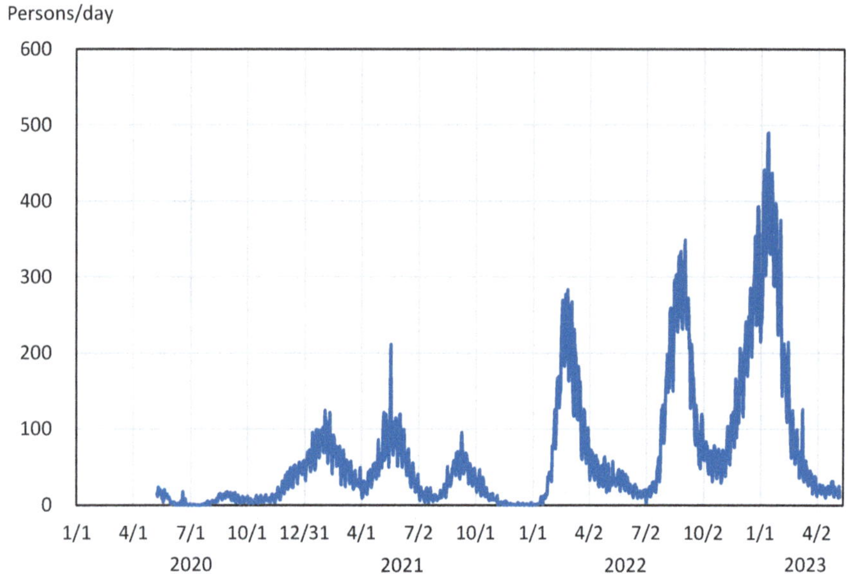

(c) Mortality cases

Fig. 6.1 (continued)

6.2.1.2 Declaration of a State of Emergency and Priority Measures to Prevent the Spread

On March 13, 2020, the law on special measures against COVID-19 was enacted, and on April 7, 2020, a "declaration of a state of emergency" was issued in seven prefectures: Tokyo, Kanagawa, Saitama, Chiba, Osaka, Hyogo, and Fukuoka (CAICM 2020). Under the declaration, restaurants and other establishments could be ordered or requested to close or shorten their operational hours to restrict the spread of COVID-19 (CAICM 2021a). This declaration covered the entire area of the corresponding prefectures (CAICM 2021a). From April 16 to May 14, 2020, the declaration was extended nationwide (CAICM 2021b). In 2021, the declaration was issued intermittently from January 8 to March 21, April 25 to June 20, and July 12 to September 30 in 21 prefectures, mainly in the metropolitan areas of Tokyo, Aichi, Osaka, and Fukuoka (CAICM 2021b).

Additionally, "priority measures to prevent the spread" were issued from April 5 to September 30, 2021, in 33 prefectures, and from January 9 to March 21, 2022, in 36 prefectures (CAICM 2022). Unlike the declaration, priority measures do not include orders or requests for closure but only reduce business hours (CAICM 2021a). The purpose of priority measures is to control the spread of COVID-19 within a specific area, and only areas designated by the governor are subject to such measures (CAICM

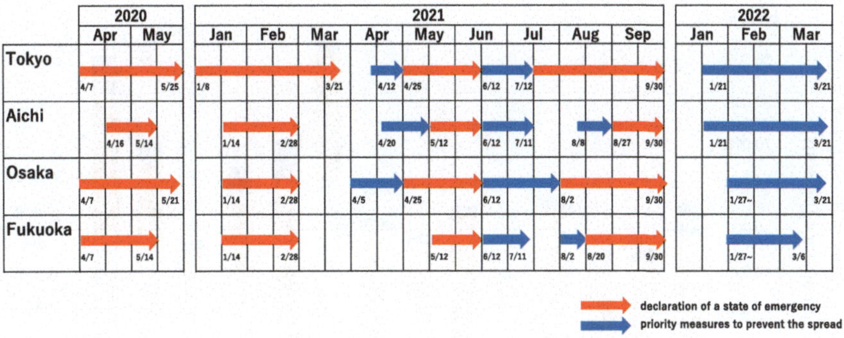

Fig. 6.2 Status of the "declaration of a state of emergency" and "priority measures to prevent the spread" in the four selected prefectures (Tokyo, Aichi, Osaka, and Fukuoka)

2021a). Figure 6.2 shows the status of the declaration and priority measures in the four selected prefectures, i.e., Tokyo, Aichi, Osaka, and Fukuoka.

6.2.2 *Heat Stroke*

In Japan, approximately 40,000–90,000 patients with heat stroke are transported using ambulance each year (FDMA 2023). In 2023, the number of patients with heat stroke transported using an ambulance (PHTA) was 91,467, the second highest after 2018 (95,137) (Fig. 6.3). Regarding age, older people accounted for approximately 55% of the total cases, followed by adults (34%) and juveniles (11%) (Fig. 6.4a). Regarding the location of the cases with ambulance use, residences accounted for approximately 40%, followed by roads (17%) and public (outdoors) areas (13%) (Fig. 6.4b).

The number of mortality cases due to heat stroke recently exceeded 1000 annually (Fig. 6.5). In 2023, the number was 1555, the third highest recorded. The sources of the statistics for the PHTA and mortality cases differ; therefore, mortality cases were not counted among the PHTA. Considering that the number of mortality cases due to natural disasters, such as earthquakes and water disasters, is only a few hundred per year (Ministry of the Environment, Japan [MOEJ] 2022b), the severity of health impacts due to heat is evident. Older individuals account for approximately 80% of total mortality cases (MHLW 2024b). In Japan, with its hyper-aged society, there is concern that the health impacts of heat among older people will increase, and rising temperatures due to climate change will accelerate this increase (Nakamura et al. 2022; Oka et al. 2023, 2024; Ueta et al. 2024).

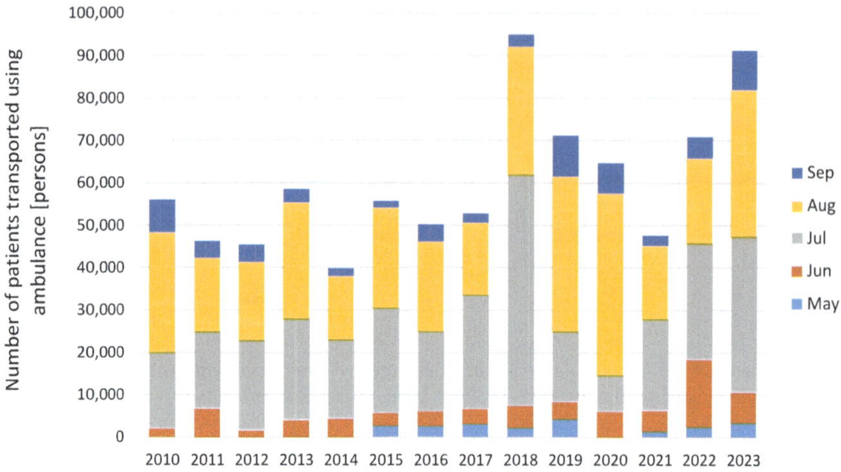

Fig. 6.3 Number of patients with heat stroke transported using ambulance (FDMA 2023)

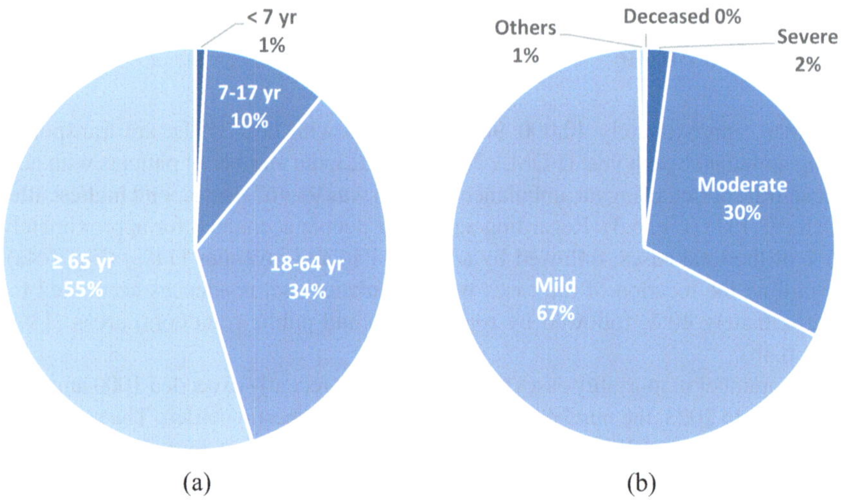

Fig. 6.4 Percentage of patients with heat stroke transported using ambulance by **a** age group and **b** location of occurrence in (FDMA 2023)

6.3 Impact of the COVID-19 Pandemic on Heat Stroke Risk

In Japan, where heat stroke is a significant issue during the summer, measures against the COVID-19 pandemic have drastically changed people's behavior, subsequently altering the risk of heat stroke. Seposo et al. (2021), Hatakeyama et al. (2021), and Hatakeyama and Seposo (2022) demonstrated that the risk of heat stroke, evaluated

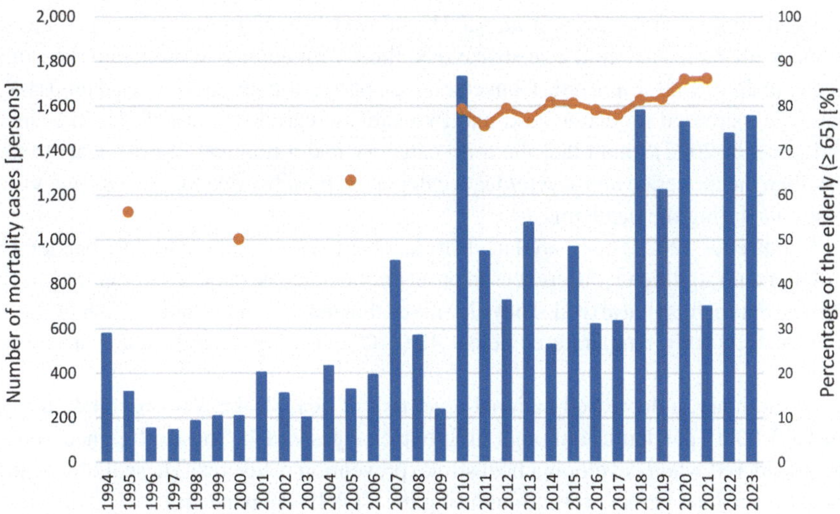

Fig. 6.5 Number of mortality cases for heat stroke (MHLW 2024b)

using the number of PHTA, decreased during the COVID-19 pandemic in 2020 compared with that in the pre-pandemic period at the national level. Similarly, Ke et al. (2024) found a decrease in heat stroke risk, evaluated using the number of PHTA, in the Kansai region from 2020 to 2022.

The decrease in heat stroke risk was attributed to behaviors adopted to prevent COVID-19, such as staying indoors, teleworking, increased use of telemedicine, and home/food delivery (Seposo et al. 2021; Hatakeyama et al. 2021). Ke et al. (2024) also suggested that ambulance congestion due to patients with COVID-19 could have contributed to the decrease in the number of PHTA. In contrast, Uryu et al. (2021) found no statistically significant change in the number of PHTA at the national level; however, they observed decreased risks during some weeks in 14 of the 47 prefectures. They suggested that self-restraint at public events or schools might offset the increased risk of heat stroke associated with mask wearing. Notably, no study reported an increased national-level risk of heatstroke.

By age group, Uryu et al. (2021) found the weeks with decreased heat stroke risk in 15 prefectures for children and adolescents, 16 prefectures for working-age adults, and only three prefectures for older adults. The decrease in heat stroke risk for children and adolescents was partially due to the cancelation or postponement of public events (such as cultural events, sports, and festivals). This was also valid in the working-age group. Only one prefecture reported an increased risk for children and adolescents in one week. Hatakeyama and Seposo (2022) noted a relative decrease in the number of PHTA among adults and adolescents, likely due to restrictions on outdoor activities, whereas older adults showed a relative increase. This increase is reasonable given that heat stroke among older adults mainly occurs indoors (MOEJ 2022a). Hatakeyama and Seposo (2022) noted a relative increase in the risk at home.

Regarding severity, Uryu et al. (2021) showed that 25 prefectures experienced weeks with decreased mild heat stroke risk, three with decreased moderate risk, and five with decreased severe risk. Conversely, one prefecture saw an increased mild risk, two saw increased moderate risk, and two saw increased severe risk. Hatakeyama and Seposo (2022) found that the mild category had a relatively lower heat stroke risk than the moderate and severe/death categories, possibly due to reluctance to visit hospitals during the pandemic.

Kanda et al. (2022) compared heat stroke characteristics pre- and post-pandemic and found no significant differences in onset circumstances, patient demographics, or clinical data at hospital arrival. They also noted that mask wearers had a higher proportion of deep body temperatures below 40.0 °C and a higher in-hospital mortality rate.

In summary, a decreased national-level risk of heat stroke was observed during the COVID-19 pandemic, although subgroup changes varied. These differences must be considered when developing heat stroke prevention measures. These findings are valuable for heat stroke prevention during the COVID-19 pandemic.

6.4 Heat Stroke Prevention Measures During the COVID-19 Pandemic

This section outlines the measures implemented by the Japanese government, local authorities, and recommendations from academic groups to balance COVID-19 and heat stroke prevention.

6.4.1 Government Initiatives

The Japanese government's efforts to balance COVID-19 heat stroke prevention are summarized by the Heat Stroke Prevention Council (2022). Key measures include publicizing compatible prevention strategies and supporting studies on heat stroke factors influenced by changes in environmental factors under "new lifestyles" (Table 6.1).

In June 2020, the MOEJ and MHLW published a leaflet on preventing heat stroke under the "new lifestyle" during the COVID-19 pandemic. This leaflet was partially modified in June 2021 and June 2022 (MOEJ and MHLW 2022). The differences among the leaflets seem to reflect the government's view at that time and the situation surrounding the COVID-19 pandemic and heat stroke. For example, the phrase "wearing a mask increases the risk of heat stroke" was changed to "wearing a mask outdoors increases the risk of heat stroke" and the phrase "be careful on summer days (over 30 °C) and especially on days over 35 °C" was changed to "take preventive action when a heat stroke alert is issued". The expression "be careful on summer

Table 6.1 Government initiatives to balance COVID-19 measures and heat stroke prevention (FY 2022)

Category	Specific measures
Publicizing heat stroke prevention behaviors in "new lifestyles"	• Creation of a leaflet on heat stroke prevention for "new lifestyles" • Alerts and awareness-raising via website and social networking service (SNS) • Intra-administrative communication on measures against heat stroke at vaccination sites
Publicizing the prevention of COVID-19 and heat stroke during school life and sports	• Requests to the board of education to prevent heat stroke accidents • Notification of no necessity to wear masks during physical education classes and in outdoor playgrounds • Notification to sports facility management divisions to take appropriate measures to prevent heat stroke accidents, along with related guidebooks and websites • Alerts and awareness-raising through SNS
Study and survey analysis to evaluate the risk of heat stroke and establish diagnosis, prevention, and treatment methods in line with the "New Lifestyle"	• Working groups to examine new issues in heat stroke countermeasures • Study on prevention methods and initial medical treatment of heat stroke under mask-wearing conditions, funded by a Health, Labour, and Welfare Science Research Grant • Investigation of factors in the onset of heat stroke due to changes in environmental factors in line with "new lifestyles"

days (over 30 °C) and especially on days over 35 °C" has been changed to "take preventive action when a heat stroke alert is announced".

The following is the summary of the efforts and other measures to balance COVID-19 control and heat stroke control described in the latest 2022 leaflet. Some of the listed measures are not limited to COVID-19 countermeasures.

- Remove masks when outdoors, except when talking within 2 m.
- Remove masks during exercise.
- Ventilate frequently while using air conditioning.
- Avoid heat through clothing, shade, and items such as parasols.
- Hydrate frequently when wearing a mask, even if not thirsty.
- Prepare the body for heat and monitor physical condition through exercise, health checks, rest, and relaxation.

Table 6.2 Local authorities' initiatives

Measure	Summary
Awareness-raising on website	• Includes national leaflet, notes on countermeasures, QA by physicians (Ise City 2022)
Awareness-raising through public relations publications	• Awareness-raising through PR magazines, insert national leaflets, or own flyers (Betsukai Town 2021)
Awareness-raising through corporate tie-ups	• Created flyers jointly with companies that have signed comprehensive cooperation agreements, posted at stores of private businesses (Osaka Pref. 2024)
Correction of site management costs for construction work ordered	• Site management cost adjusted according to the guidelines (Ministry of Land, Infrastructure, Transport and Tourism's administrative communication (MLIT) 2020), redefining summer days from "30 °C or higher" to "28 °C or higher" (Shizuoka Pref 2024)

6.4.2 Initiatives by Local Authorities

Local authorities have also taken steps to balance COVID-19 and heat stroke prevention. Table 6.2 summarizes key measures, including the use of leaflets prepared by the government and local authorities to inform the public about heat stroke prevention.

6.4.3 Recommendations from Academic Groups

In June 2020, a working group organized by four academic societies, including the Japanese association for acute medicine, Japanese association for infectious diseases, Japanese society for emergency medicine, and Japanese respiratory society, published "the recommendations for the prevention of heat stroke in light of the outbreak of new Coronavirus infections" (Working group on heat stroke treatment in light of the new Coronavirus epidemic 2020). These recommendations were based on expert consensus due to limited information on heat stroke prevention and treatment related to COVID-19. Key recommendations include:

- Adjust indoor air conditioning temperature frequently and ensure sufficient ventilation.
- Remove masks as needed and take breaks, especially considering physical distancing for infection control. Hydrate frequently, regardless of thirst.
- Gradually acclimatize to heat through moderate exercise, both indoors and outdoors, while avoiding physical distancing.

- Pay special attention to vulnerable populations, such as older people who live alone, and maintain frequent contact to prevent social isolation.
- Monitor daily health conditions and contact the local "Returnee/Contact Center for COVID-19" or the nearest medical institution if feeling unwell.

6.5 Bridging Scientific Findings and Measures Against Heat Stroke

The risk of heat stroke during the COVID-19 pandemic has decreased compared with that during the pre-pandemic period at the national level. One reason for this decrease is the similarity between COVID-19 and heatstroke countermeasures (Seposo et al. 2021; Hatakeyama et al. 2021). However, differences in heat stroke risks were observed between subgroups. Fewer prefectures showed a decrease in heat stroke risk among the older compared to the children, adolescents, and working-age groups (Uryu et al. 2021). Regarding the severity of heat stroke, fewer prefectures showed a decrease in risk among those in the moderate and severe/death categories compared with those in the mild category (Uryu et al. 2021).

The observed differences in each subgroup might not be statistically conclusive due to the short observation period (i.e., the duration of the COVID-19 pandemic) and other factors. Nonetheless, considering appropriate measures for each subgroup based on these differences is crucial. For instance, while refraining from going out is an effective measure against both COVID-19 and heat stroke for children, adolescents, and working-age groups, it may negatively impact older people (MOEJ 2022a). Older people often do not use air conditioners even in hot indoor conditions due to weakened temperature sensitivity, leading to higher indoor temperatures and increased risk of heat stroke (MOEJ 2022a). Thus, a measure effective for one subgroup might have the opposite effect on another, highlighting the importance of tailored countermeasures.

The MHLW's modification of the leaflet descriptions to prevent heatstroke reflects the evolving governmental perspective and the changing circumstances surrounding the COVID-19 pandemic and heat stroke. Scientific findings offer important insights for developing future measures against heat stroke, even though the COVID-19 pandemic has subsided. This chapter focused on the compound impacts of COVID-19 and heat stroke; the experience of integrating scientific findings and practical measures against heat stroke can serve as a reference when considering other compound impacts.

6.6 Concluding Remarks

Compound impacts, such as the interaction between COVID-19 and heatstroke, have already been observed in Japan. This chapter reviewed the impact of COVID-19 on heat stroke and introduced measures implemented to mitigate heatstroke during the pandemic. Additionally, the integration of scientific knowledge and practical measures to prevent heatstroke in the context of COVID-19 was discussed. This case study offers valuable insights for future research on the risk and response to other compound impacts.

Acknowledgements This chapter was partially written using study results from the Climate Change Adaptation Research Program, the Collaborative Study Project of the National Institute for Environmental Studies, the Environment Research and Technology Development Fund (JPMEERF20231007) of the Environmental Restoration and Conservation Agency provided by MOEJ, and relevant information collected during this study.

References

Betsukai Town (2021) Prevention of heat stroke x prevention of Corona infection for a healthy "new lifestyle"! https://betsukai.jp/resources/output/contents/file/release/4189/48270/7.8.pdf (in Japanese)

Cabinet office (2020) 2020 White paper on disaster prevention. https://www.bousai.go.jp/kaigirep/hakusho/r2.html (in Japanese)

Cabinet office (2023) The 6th survey on changes in lifestyle attitudes and behaviors under the influence of new Coronavirus infections. https://www5.cao.go.jp/keizai2/wellbeing/covid/pdf/result6_covid.pdf (in Japanese)

CAICM (2020) New Coronavirus infections-report on the implementation of the declaration of a state of emergency, June 2020. https://www.caicm.go.jp/news/pdf/kinkyujitaisengen_houkoku0604.pdf (in Japanese)

CAICM (2021a) Declaration of a state of emergency and priority measures to prevent the spread. https://www.caicm.go.jp/emergency/pdf/kinkyujitaisochi_20210419.pdf (in Japanese)

CAICM (2021b) New Coronavirus infections-report on the implementation of the declaration of a state of emergency, June 2021. https://www.caicm.go.jp/news/pdf/houkoku_r031008.pdf (in Japanese)

CAICM (2022) Actions based on the basic handling policy. https://www.caicm.go.jp/information/citizen/corona/emergency.html (in Japanese)

FDMA (2019) Damage caused by Boso peninsula typhoon in 2018 and response of fire departments and other organizations (41st report). https://www.fdma.go.jp/disaster/info/items/taihuu15gou41.pdf (in Japanese)

FDMA (2023) Status of Ambulance transport for heat stroke in 2023 (May to September). https://www.fdma.go.jp/pressrelease/houdou/items/d7e69bd2038c29f307d3b517fe2ee80610badb63.pdf (in Japanese)

Hatakeyama K, Seposo X (2022) Heatstroke-related ambulance dispatch risk before and during COVID-19 pandemic: subgroup analysis by age, severity, and incident place. Sci Total Environ 821:153310. https://doi.org/10.1016/j.scitotenv.2022.153310

Hatakeyama K, Ota J, Takahashi Y, Kawamitsu S, Seposo X (2021) Effect of the COVID-19 pandemic on heatstroke-related ambulance dispatch in the 47 prefectures of Japan. Sci Total Environ 768:145176. https://doi.org/10.1016/j.scitotenv.2021.145176

Heat Stroke Prevention Council (2022) Efforts by relevant ministries and agencies based on the action plan for heat stroke prevention (revised in 2022). https://www.wbgt.env.go.jp/pdf/ic_rma/R0402/doc02-1.pdf (in Japanese)

Ise City (2020) Health consultation room. https://www.city.ise.mie.jp/shisei/koho_kocho/koho/kenkousoudansitu/1014437.html (in Japanese). Accessed 1 May 2024

JMA (2023) Climate change monitoring report 2021. https://www.data.jma.go.jp/cpdinfo/monitor/2023/pdf/ccmr2023_all.pdf (in Japanese)

JMA (2024) Climate change monitoring report 2023. https://www.data.jma.go.jp/cpdinfo/monitor/2023/pdf/ccmr2023_all.pdf (in Japanese)

Kanda J, Miyake Y, Umehara T, Yoshiike S et al (2022) Influence of the coronavirus disease 2019 (COVID-19) pandemic on the incidence of heat stroke and heat exhaustion in Japan: a nationwide observational study based on the Heatstroke STUDY 2019 (without COVID-19) and 2020 (with COVID-19). Acute Med Surg 9:e731. https://doi.org/10.1002/ams2.731

Ke D, Ialnazov DS, Takara K (2024) The impact of COVID-19 on heat-related ambulance calls and the challenges for emergency medical transport in the Kansai region, Japan: a mixed methods approach. Environ Res Health 2:035004. https://doi.org/10.1088/2752-5309/ad4f05

MHLW (2020) Heat stroke prevention actions in 2020. https://www.mhlw.go.jp/stf/houdou_kouhou/kouhou_shuppan/magazine/202008_00004.html (in Japanese)

MHLW (2021) History of new coronavirus infections to date (as of March 31, 2021). https://www.mhlw.go.jp/stf/wp/hakusyo/kousei/20/backdata/8-3-1.html (in Japanese)

MHLW (2024a) Information on new coronavirus infection. https://covid19.mhlw.go.jp/ (in Japanese). Accessed 1 May 2024

MHLW (2024b) Deaths due to heat stroke from vital statistics (fixed number). https://www.mhlw.go.jp/toukei/saikin/hw/jinkou/tokusyu/necchusho22/index.html (in Japanese). Accessed 1 May 2024

MLIT (2020) Toward the prevention of heat stroke associated with the new Coronavirus countermeasure (reference). https://www.mlit.go.jp/common/001353620.pdf (in Japanese)

MOEJ (2022a) Heat stroke environmental health manual. https://www.wbgt.env.go.jp/pdf/manual/heatillness_manual_full_high.pdf (in Japanese)

MOEJ (2022b) Heat stroke countermeasures. https://www.wbgt.env.go.jp/pdf/manual/heatillness_manual_full_high.pdf (in Japanese)

MOEJ and MHLW (2022) Prevention of heat stroke x prevention of corona infection for a healthy "New Lifestyle". https://www.wbgt.env.go.jp/pdf/pr/20210625_leaflet_orange.pdf (in Japanese)

Nakamura S, Kusaka H, Sato R, Sato T (2022) Heatstroke risk projection in Japan under current and near future climates. JMSJ 100(4):597–615, 202. https://doi.org/10.2151/jmsj.2022-030

Oka K, Honda Y, Phung VLH, Hijioka Y (2023) Prediction of climate change impacts on heat-stroke cases in Japan's 47 prefectures with the effect of long-term heat adaptation. Environ Res 232:116390. https://doi.org/10.1016/j.envres.2023.116390

Oka K, Honda Y, Hijioka Y (2024) Prediction of ambulance transport system collapse under extremely high temperatures induced by climate change. Environ Health Res 2:035002. https://doi.org/10.1088/2752-5309/ad4581

Osaka Pref (2024) Efforts to prevent heat stroke in FY2022. https://www.pref.osaka.lg.jp/kenkozukuri/nettyusyou/r4_nechusyotorikumi.html (in Japanese). Accessed 1 May 2024

Seposo X, Madaniyazi L, Ng CFS, Honda Y, Hashizume M (2021) COVID-19 pandemic modifies temperature and heat-related illness ambulance transport association in Japan: a nationwide observational study. Environ Health 20:122. https://doi.org/10.1186/s12940-021-00808-w

Shizuoka Pref (2024) Correction of on-site management cost rate to contribute to heat stroke prevention. https://www.pref.shizuoka.jp/kensei/nyusatsukobai/tanka/1053466/1053473.html (in Japanese). Accessed 1 May 2024

The Asahi Shimbun (2019) Chiba has the nation's highest number of heat stroke cases, with 498 cases during a week of prolonged power outages. https://www.asahi.com/articles/ASM9L55C6M9LUTIL03Z.html (in Japanese)

Ueta H, Kodera S, Sugimoto S, Hirata A (2024) Projection of future heat-related morbidity in three metropolitan prefectures of Japan based on large ensemble simulations of climate change under 2 °C global warming scenarios. Environ Res 247:118202. https://doi.org/10.1016/j.envres.2024.118202

Uryu S, Tanoue Y, Nomura S, Matsuura K et al (2021) Trends in emergency transportation due to heat illness under the new normal lifestyle in the COVID-19 era, in Japan and 47 prefectures. Sci Total Environ 768:144723. https://doi.org/10.1016/j.scitotenv.2020.144723

WMO (2024) WMO confirms that 2023 smashes global temperature record. Press release. https://wmo.int/news/media-centre/wmo-confirms-2023-smashes-global-temperature-record. Accessed 1 May 2024

Working group on heat stroke treatment in the light of the new Coronavirus epidemic (2020) The recommendations for the prevention of heat stroke in the light of the outbreak of new Coronavirus infections. https://www.jaam.jp/info/2020/files/info-2020601.pdf (in Japanese)

Chapter 7
Re-emerging Diseases Post-pandemic COVID-19: Indonesia's Responses and Resilience

Budi Haryanto, Hanifa M. Denny, Ira Nurmala, and Rico J. Sitorus

Abstract Indonesia has long faced challenges with communicable diseases such as Dengue Fever, Tuberculosis (TB), and Malaria, alongside the rising threat of air pollution-related diseases. These health issues were exacerbated during the COVID-19 pandemic, disrupting prevention, diagnosis, and treatment programs. Pre-COVID-19, Indonesia recorded one of the highest rates of TB globally, with over 845,000 cases reported in 2019. Dengue fever remained endemic, peaking during the rainy season, with around 138,000 cases annually. Malaria, although largely controlled in certain regions, still impacted provinces in Eastern Indonesia like Papua, with a prevalence of 5.4 per 1000 people. Air pollution, particularly in urban areas like Jakarta, contributed significantly to respiratory diseases, driven by vehicle emissions and industrial activities. The COVID-19 pandemic posed further complications. Health services were redirected to manage the pandemic, leading to a decline in diagnosis and treatment for TB, with an estimated 30% reduction in detected TB cases during 2020. Malaria control programs were disrupted due to movement restrictions, affecting the distribution of bed nets and indoor residual spraying. The incidence of dengue fever saw fluctuations, with initial reports of lower cases, possibly due to underreporting amidst the pandemic. Additionally, air quality in major cities

B. Haryanto (✉)
Department of Environmental Health, Faculty of Public Health, Universitas Indonesia, Jakarta, Indonesia
e-mail: bharyanto@ui.ac.id

Research Center for Climate Change, I-SER Universitas Indonesia, Jakarta, Indonesia

H. M. Denny
Faculty of Public Health, Universitas Diponegoro, Semarang, Indonesia
e-mail: hanifadenny@live.undip.ac.id

I. Nurmala
Faculty of Public Health, Universitas Airlangga, Surabaya, Indonesia
e-mail: iranurmala@fkm.unair.ac.id

R. J. Sitorus
Faculty of Public Health, Universitas Sriwijaya, Palembang, Indonesia
e-mail: rico_januar@fkm.unsri.id

R. Akhtar (ed.), *COVID-19: Impacts on Health and Health Systems*,
https://doi.org/10.1007/978-3-031-90255-0_7

temporarily improved during the early lockdowns, reducing pollution-related respiratory diseases. Post-COVID-19, the healthcare system is gradually recovering, but the pandemic's long-term impact is evident. TB programs are regaining momentum, although cases remain underreported. Dengue and Malaria control initiatives are being reactivated, yet public health efforts are strained by resource allocation. Air pollution remains a significant public health issue, with pollution levels rebounding to pre-pandemic levels as economic activities resume. The post-pandemic recovery requires integrated public health strategies to tackle these endemic diseases alongside emerging global health threats.

Keywords Re-emerging diseases · Pre- and post-COVID-19 pandemic · Indonesia's responses

7.1 Background

The COVID-19 pandemic in Indonesia, first detected in March 2020, had a profound impact on public health, the economy, and social dynamics. Indonesia became one of Southeast Asia's worst-hit nations, with over 6 million confirmed cases and more than 150,000 deaths by 2023. The country's healthcare system struggled with shortages of hospital beds, medical supplies, and healthcare workers during multiple waves of infections, especially with the emergence of the Delta variant in mid-2021. Government responses included nationwide lockdowns, travel restrictions, and mass vaccination campaigns, with vaccines like Sinovac, AstraZeneca, and Pfizer being rolled out to curb the spread. Economic activities slowed, leading to a recession, and millions faced job losses, particularly in sectors like tourism and retail. While vaccinations and public health measures eventually helped control the virus, the pandemic exposed weaknesses in Indonesia's healthcare infrastructure and underscored the need for stronger crisis preparedness.

The COVID-19 pandemic significantly impacted Indonesia's public health landscape, leading to the resurgence of several infectious diseases. The pandemic disrupted routine healthcare services, vaccination programs, and disease surveillance, which contributed to the re-emergence of these diseases. In response, Indonesia has undertaken various measures to strengthen its health system resilience and address the challenges posed by re-emerging diseases. Among others, key re-emerging Diseases in Indonesia Post-COVID-19 include.

7.2 Dengue Fever

Indonesia experienced a resurgence in dengue cases, with some regions reporting higher incidences than pre-pandemic levels. Dengue fever remains a significant public health concern, with the country experiencing periodic outbreaks that strain healthcare resources. The disease, transmitted by the Aedes aegypti mosquito, is endemic in Indonesia, and various factors have contributed to its persistence and periodic intensification. Current issues related to dengue fever in Indonesia are as the followings:

(a) Rising Incidence Rates, includes (1) Seasonal Outbreaks: Indonesia regularly experiences spikes in dengue cases during the rainy season, which typically runs from November to March. The increased rainfall provides ideal breeding conditions for mosquitoes, leading to higher transmission rates; (2) Urbanization and Population Density: Rapid urbanization and high population density in cities like Jakarta, Surabaya, and Bandung have exacerbated the problem. Urban areas with poor sanitation and water storage practices provide ample breeding grounds for mosquitoes.
(b) Healthcare System Strain includes (1) Hospital Overcrowding: During dengue outbreaks, hospitals in affected regions often become overwhelmed with patients, leading to bed shortages and delays in treatment; (2) Limited Resources: Many healthcare facilities, especially in rural areas, struggle with limited resources, including a shortage of trained healthcare workers and diagnostic tools for dengue fever.
(c) Climate Change includes (1) Impact on Mosquito Populations: Climate change has led to shifts in weather patterns, including more frequent and intense rainfall and warmer temperatures, both of which are conducive to mosquito breeding. This has resulted in a more extended transmission season and expanded the geographic range of dengue fever in Indonesia; (2) Newly Affected Areas: Regions previously less affected by dengue, such as higher-altitude areas, are now experiencing outbreaks as warmer temperatures allow mosquitoes to thrive in these new environments.
(d) Public Awareness and Prevention Challenges include (1) Inconsistent Community Engagement: While public health campaigns have been implemented to reduce mosquito breeding sites, community participation has been inconsistent. Many people are not fully aware of the importance of eliminating stagnant water around their homes or are unable to do so due to living conditions; (2) Misconceptions and Late Treatment: There are still widespread misconceptions about dengue fever, and some people delay seeking treatment until symptoms become severe, increasing the risk of complications.
(e) Insecticide Resistance: The Aedes aegypti mosquito population in Indonesia has developed resistance to certain insecticides, reducing the effectiveness of traditional vector control methods such as fogging. This resistance complicates efforts to control mosquito populations and curb dengue transmission.
(f) COVID-19 Impact includes (1) Disruption of Control Programs: The COVID-19 pandemic disrupted routine vector control programs and public health initiatives focused on dengue fever. Resources were diverted to combat COVID-19, leading to less frequent mosquito control efforts and reduced public health outreach; (2) Increased Vulnerability: The pandemic's impact on healthcare access and

the delayed response to other health issues have increased the population's vulnerability to dengue outbreaks.

(g) Vaccine and Treatment Challenges include (1) Limited Vaccine Availability: While the dengue vaccine Dengvaxia is available, its use is limited to individuals with a confirmed previous dengue infection due to the risk of severe dengue in those who have not been previously infected. This restricts its broader application in public health programs; (2) Lack of Specific Antiviral Treatment: There is currently no specific antiviral treatment for dengue fever, with management focused on supportive care to treat symptoms and prevent complications (Fig. 7.1).

The government intensified vector control efforts, including mosquito eradication campaigns, public education on preventing mosquito breeding, and community engagement initiatives. Indonesia is intensifying efforts to combat dengue fever through various strategies. Vector control programs continue with larviciding and fogging campaigns to reduce mosquito populations, despite challenges like insecticide resistance. Public health authorities are involving communities in 'gotong royong' communal activities to eliminate breeding sites. Public awareness campaigns have been amplified using media, schools, and community organizations to educate about dengue prevention. Digital tools such as mobile apps and social media inform citizens about dengue hotspots and encourage preventive actions. Additionally, enhanced surveillance systems allow real-time monitoring and response to outbreaks. Utilizing real-time data collection and analysis, authorities can track the disease's spread more effectively. Early warning systems based on weather patterns and mosquito data are developed to anticipate outbreaks and allocate resources. Research

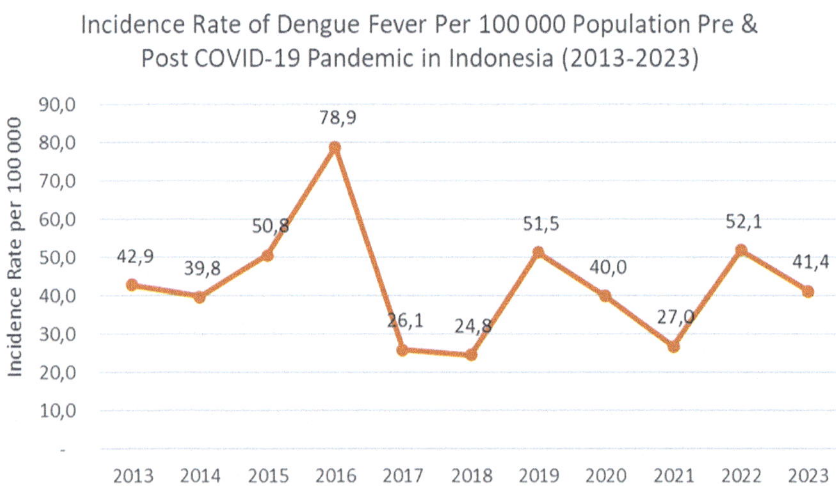

Fig. 7.1 State of dengue fever pre- and post-COVID-19 pandemic in Indonesia. *Source* Indonesia Health Profile 2023

focuses on developing more effective vaccines with minimal severe reactions. Innovative control methods include releasing Wolbachia-infected mosquitoes less capable of transmitting dengue and employing sterile insect techniques to reduce mosquito populations.

7.3 Tuberculosis (TB)

Tuberculosis (TB) remains a critical public health challenge in Indonesia, which is among the countries with the highest TB burdens globally. The disease's persistence and the emergence of drug-resistant strains pose significant challenges to public health efforts. The situation has been exacerbated by the COVID-19 pandemic, which disrupted TB services and led to setbacks in controlling the disease. Indonesia faces significant challenges in its battle against tuberculosis (TB), which remains a major public health concern. The country ranks among the top three globally in TB incidence, with a large number of new cases reported annually. One of the key issues is the high number of undiagnosed cases, often due to social stigma, lack of awareness, and difficulties in accessing health care. This leads to continued transmission of the disease (Fig. 7.2).

Multi-drug-resistant tuberculosis (MDR-TB) presents a growing challenge. MDR-TB is resistant to the two most powerful anti-TB drugs, making it harder and more expensive to treat. Incomplete or improper treatment often leads to drug resistance, and the long, toxic treatment regimens result in higher rates of treatment failure

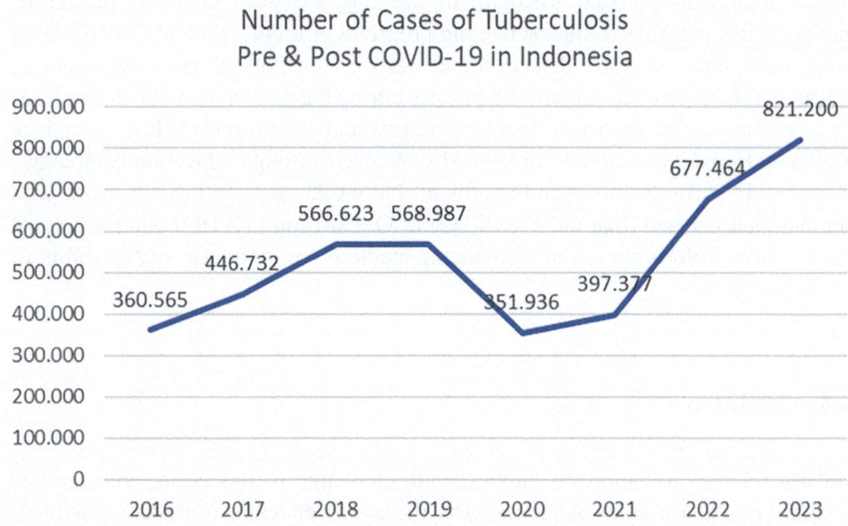

Fig. 7.2 Number of tuberculosis cases pre- and post-COVID-19 pandemic in Indonesia. *Source* Indonesia Health Profile 2023

and mortality. The COVID-19 pandemic severely disrupted TB services in Indonesia. Resources were diverted to the pandemic response, reducing TB testing and treatment follow-ups. Many patients delayed seeking care due to fear of contracting COVID-19, resulting in increased TB transmission. The pandemic also interrupted public health campaigns and outreach efforts. Social stigma surrounding TB remains prevalent, causing isolation and discrimination against those affected, while a general lack of awareness about TB symptoms, transmission, and treatment completion further complicates efforts to control the disease. In addition, Indonesia's healthcare system faces barriers such as limited access to services in rural areas and a shortage of healthcare workers trained in TB care. Lastly, Indonesia's TB control programs suffer from insufficient funding, leading to gaps in diagnosis, treatment, and outreach. The country relies heavily on international support, raising concerns about the sustainability of TB programs in the long term. Addressing these issues is critical to reducing Indonesia's TB burden.

Indonesia has undertaken several government and public health initiatives to combat tuberculosis (TB), with the National TB Program (NTP) at the forefront. The NTP focuses on improving TB prevention, diagnosis, and treatment across the country, particularly in high-burden areas. To expand outreach, community health workers are increasingly involved in TB detection and treatment, especially in rural and hard-to-reach regions, bridging the gap between healthcare services and the population. Addressing the growing issue of multi-drug-resistant TB (MDR-TB), the government is enhancing diagnostics by expanding access to rapid tools like 'GeneXpert machines', which allow for quick detection of TB and drug-resistant strains. Comprehensive care for MDR-TB patients includes support for medication adherence, nutritional help, and management of side effects, ensuring a more holistic treatment approach. Recognizing the impact of the COVID-19 pandemic, Indonesia has introduced 'dual screening programs' for both TB and COVID-19 in some areas, ensuring early detection and treatment of both diseases. Maintaining essential TB services has remained a priority during the pandemic, with an emphasis on uninterrupted access to medications and patient monitoring. Efforts to reduce stigma and raise awareness have been strengthened through education campaigns, encouraging early diagnosis and treatment. Indonesia also continues to work with international partners like the World Health Organization (WHO) and the Global Fund, while exploring innovative financing methods for long-term sustainability of its TB programs.

7.4 Malaria

Malaria remains a significant public health challenge in Indonesia, with marked geographic disparities in incidence. The disease is endemic in eastern provinces such as Papua, West Papua, East Nusa Tenggara, Maluku, and North Maluku, where tropical climates, dense forests, and limited healthcare access contribute to high-transmission rates. In contrast, low-transmission areas like Java and Bali have achieved near-elimination of malaria, though sporadic cases still occur due to migration from endemic regions. One critical issue is drug resistance, particularly in Papua and West Papua, where resistance to chloroquine complicates treatment and necessitates the use of more expensive artemisinin-based combination therapies (ACTs). Insecticide resistance is also emerging, reducing the effectiveness of control measures like insecticide-treated bed nets (ITNs) and indoor residual spraying (IRS) in some regions. Indonesia's vast and varied geography presents additional challenges for vector control. Mosquitoes thrive in different environments, from forests to rice paddies, making it difficult to apply uniform control measures. Environmental changes, such as deforestation and climate change, are also altering mosquito habitats, requiring adaptive, region-specific strategies. Access to health care is limited in many endemic regions, particularly in eastern Indonesia. Poor infrastructure, difficult terrain, and a shortage of healthcare workers hinder timely diagnosis and treatment, leading to higher morbidity and mortality rates. Delays in diagnosis and inconsistent availability of rapid diagnostic tests (RDTs) exacerbate the problem (Fig. 7.3).

The COVID-19 pandemic disrupted malaria control programs, diverting resources and leading to delays in diagnosis and treatment, especially in rural areas. The pandemic also affected the distribution of bed nets and implementation of vector control measures. High levels of migration from endemic areas to malaria-free regions pose further challenges, as migrants can introduce malaria into populations with low immunity. Cross-border movement with neighboring countries like Papua New Guinea adds complexity to control efforts, requiring regional cooperation. Finally, sustaining funding and resources remains crucial. While Indonesia has made

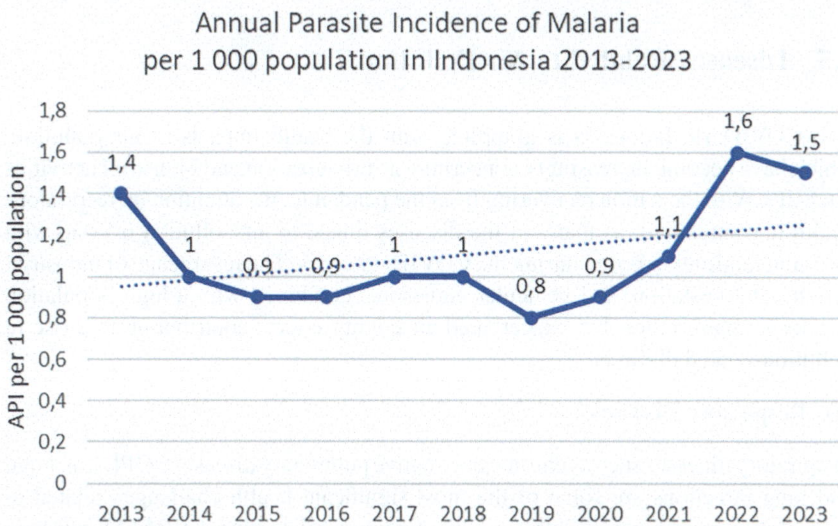

Fig. 7.3 Annual parasite incidence of malaria pre- and post-COVID-19 pandemic in Indonesia. *Source* Indonesia Health Profile 2023

progress with international support, ongoing financial commitment is necessary to ensure continued success in both high-burden and low-transmission regions.

Indonesia's National Malaria Elimination Program aims to eliminate malaria by 2030, focusing on reducing transmission in high-burden areas while preventing its reintroduction in malaria-free regions. Strengthening surveillance systems is key, with real-time data collection helping to detect cases early and monitor progress. Vector control strategies remain central to prevention efforts. The government continues to distribute insecticide-treated bed nets (ITNs) in high-transmission areas, ensuring they are regularly replaced and correctly used. Indoor residual spraying (IRS) is being implemented in areas with high-transmission rates, while larval source management targets mosquito breeding sites through environmental interventions like draining stagnant water. Improving diagnosis and treatment is another priority. The government is expanding access to rapid diagnostic tests (RDTs) and training healthcare workers in their use for timely detection. Ensuring access to artemisinin-based combination therapies (ACTs), especially in drug-resistant areas, is crucial, and efforts to strengthen supply chains are being made to prevent medication shortages.

Community involvement plays a vital role in malaria control. Public health campaigns raise awareness about prevention and treatment, while community-based interventions, involving local leaders, enhance the reach of these efforts in remote areas. Finally, Indonesia is working on cross-border and regional collaborations, coordinating malaria control initiatives with neighboring countries and participating in global elimination efforts, benefiting from international research, funding, and technical support.

7.5 Diseases-Related to Air Pollution

Post-COVID-19, Indonesia is grappling with the health impacts of air pollution, which have become increasingly concerning as urbanization and industrial activities intensify. With the nation recovering from the pandemic, the attention on respiratory health has magnified, as many of the diseases linked to air pollution overlap with the vulnerabilities exposed during the COVID-19 crisis. The resurgence of industrial activity, deforestation, and vehicular emissions, combined with a high population density in major cities, has exacerbated air quality issues, contributing to a rise in pollution-related diseases.

(a) Respiratory Diseases

Respiratory illnesses such as chronic obstructive pulmonary disease (COPD), asthma, and lung infections are some of the most significant health challenges related to air pollution in Indonesia. Pollutants like fine particulate matter (PM2.5), nitrogen dioxide (NO_2), and sulfur dioxide (SO_2) from vehicles, factories, and biomass burning worsen air quality, particularly in densely populated areas such as Jakarta, Surabaya, and Bandung. Prolonged exposure to these pollutants inflames the airways, leading to chronic respiratory problems, and puts people at greater risk for infections like pneumonia. During the COVID-19 pandemic, people with pre-existing respiratory conditions were identified as being particularly vulnerable to severe illness. As the pandemic fades, the continued exposure to polluted air increases the risk of complications for those with already weakened lungs, making respiratory health a critical post-pandemic focus.

(b) Cardiovascular Diseases

There is a strong link between air pollution and cardiovascular diseases (CVD), including hypertension, heart attacks, and strokes. Fine particulate matter (PM2.5) can penetrate deep into the bloodstream, causing inflammation and oxidative stress, which are major contributors to the development of cardiovascular diseases. In Indonesia, where heart disease is already one of the leading causes of death, air pollution exacerbates these conditions. The pandemic underscored the importance of cardiovascular health, as COVID-19 was found to have more severe outcomes in patients with pre-existing heart conditions. This overlap has raised concerns about the long-term health impacts of air pollution in a post-COVID world, where the strain on healthcare systems is heightened due to the increased prevalence of cardiovascular diseases.

(c) Cancers

Another concerning impact of air pollution is its association with lung cancer and other forms of cancer. Prolonged exposure to carcinogenic pollutants, such as benzene, formaldehyde, and polycyclic aromatic hydrocarbons (PAHs), which are emitted from vehicle exhausts, industrial processes, and open burning, can increase the risk of cancer. The high levels of PM2.5 in Indonesian cities have made lung cancer a growing concern, even for non-smokers, as air pollution is a major risk factor.

(d) Post-COVID-19 Vulnerabilities and Long-Term Impacts

The COVID-19 pandemic left many individuals with lingering respiratory and cardiovascular issues, commonly referred to as long COVID-19. The ongoing exposure to poor air quality compounds these post-COVID-19 conditions, worsening symptoms and slowing recovery for many Indonesians. Moreover, the pandemic's economic fallout has slowed efforts to implement stricter air quality regulations, as the government prioritizes economic recovery over environmental reforms, which could have reduced pollution levels.

To combat air pollution-related diseases, Indonesia must prioritize strengthening its public health infrastructure and air quality monitoring systems. Efforts to reduce vehicular emissions and promote cleaner energy represent steps in the right direction. However, more needs to be done to integrate public health concerns with environmental policies, particularly in urban areas where pollution levels are highest. Public education campaigns focusing on the health risks of air pollution, combined with policy changes aimed at reducing emissions, are critical. With the rising awareness of the dangers posed by polluted air in a post-COVID-19 world, Indonesia has an opportunity to reframe its approach to air quality as both an environmental and a public health priority (Figs. 7.4, 7.5, 7.6 and 7.7).

7.6 Responses and Strategies for Resilience

Indonesia's responses and strategies to strengthen its health system, particularly in the face of COVID-19 and re-emerging diseases, have been multifaceted:

1. Strengthening Health Systems: Indonesia has improved integrated disease surveillance, merging data collection for multiple diseases, enabling quicker detection and response to outbreaks. Additionally, healthcare workforce training has been a priority, with the government ensuring healthcare workers are equipped to manage both COVID-19 and other diseases.

The COVID-19 pandemic underscored the importance of strengthening Indonesia's health system, particularly in ensuring access to services during crises. The pandemic disrupted essential healthcare services, increasing the risk of disease burdens. A key lesson was the role of digital health innovations, such as telemedicine and integrated health platforms, which improved healthcare access and effectiveness, especially in low- and middle-income countries. Indonesia has embraced digital health interventions, which played a crucial role in maintaining service delivery during the pandemic. These include medical consultation apps and online hospitals, which have proven effective in improving healthcare accessibility and the capacity of providers. Additionally, these digital tools have enabled faster community responses and social mobilization, essential for disease prevention and control.

Following WHO recommendations, Indonesia implemented National Health Workforce Accounts (NHWA), which uses prioritized health indicators to guide stakeholders in planning. This has improved responses in areas like maternal and child health, HIV/AIDS care, and emergency health care. Post-pandemic, the development of telemedicine services continues to enhance patient follow-ups, while ongoing advocacy and stakeholder collaboration are essential to maintaining improvements in the healthcare system. Preparedness measures and structured responses will ensure Indonesia's health system can meet future public health challenges.

2. Vaccination Campaigns: Restoring routine immunization was critical after pandemic disruptions. The government focused on restoring immunization services, targeting children who missed vaccines. Furthermore, public–private partnerships have played a key role in securing vaccine supplies and expanding coverage, ensuring adequate resources for immunization programs.
3. Community Engagement: Public awareness is essential to disease prevention. The government initiated public awareness campaigns to educate communities on vaccination, vector control, and overall disease prevention. Involving community health workers in door-to-door campaigns helped extend healthcare services, particularly in reaching remote populations.

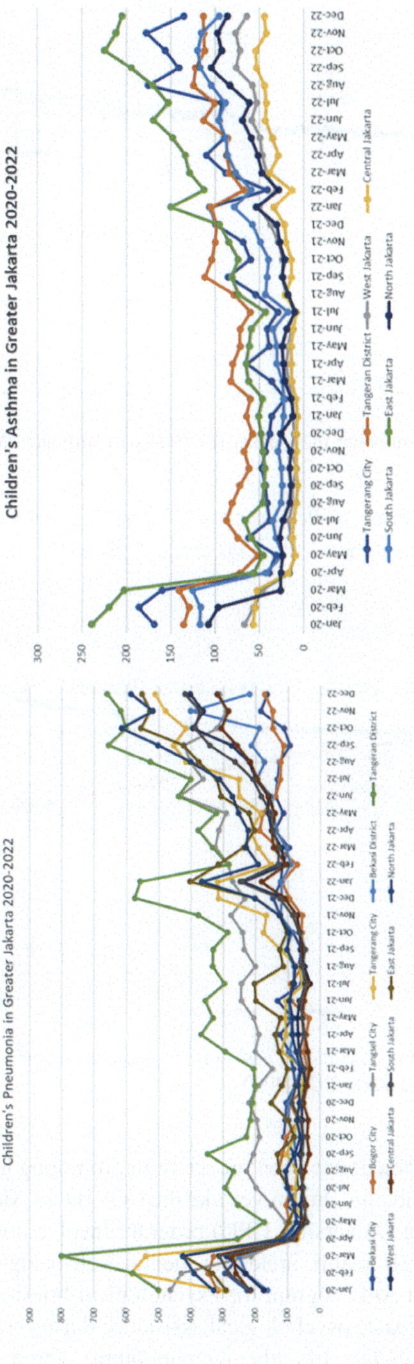

Fig. 7.4 Trend of children's pneumonia and asthma in Greater Jakarta 2020–2022

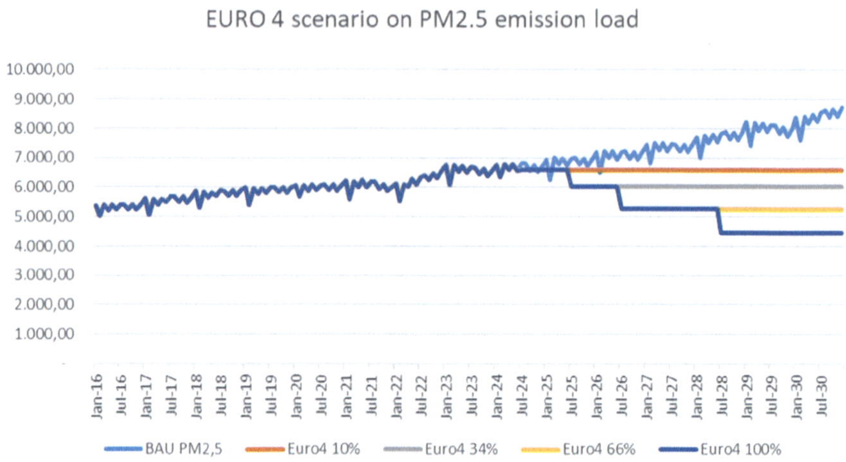

Fig. 7.5 Scenario to implement better fuel quality (EURO4 standard) and PM2.5 impacts

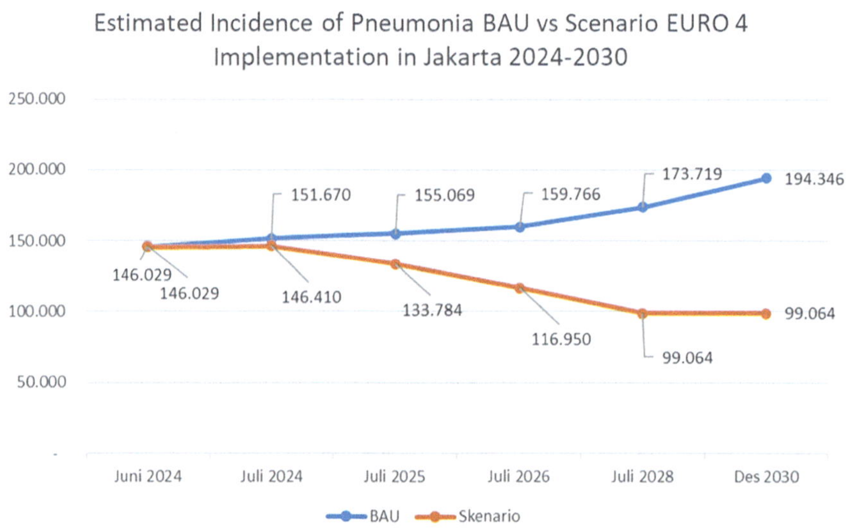

Fig. 7.6 Estimated reduction of Pneumonia cases on the implementation scenario of better fuel quality (EURO 4 standard) in Jakarta 2024–2030

The Indonesian government has been proactive in promoting health, particularly during the COVID-19 pandemic. Initiatives like the COVID-19 Alert Campus Guide and the Health Promoting University (HPU) program involve multiple universities and aim to improve overall health, including mental well-being. Additionally, the Ministry of Health's First Aid Program for Psychological Injuries (P3LP) supports mental health by offering basic psychological assistance during crises. This involves creating healthy public policies, strengthening community action, reorienting health

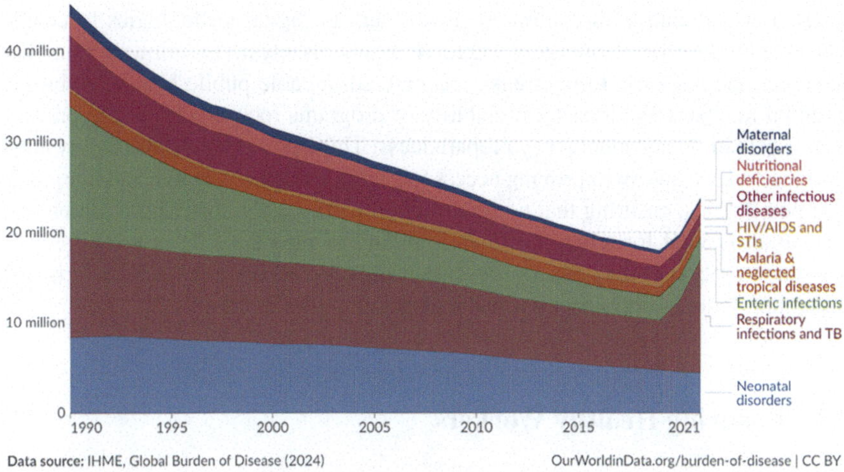

Fig. 7.7 DALY's state of the burden of diseases in Indonesia

services, promoting individual responsibility, and fostering supportive environments to encourage healthy behaviors.

Mental health plays a critical role in these efforts, as it impacts how individuals manage their health and environments. Mental well-being enables individuals to handle stress, realize their potential, and contribute to society. Yet, many mental health issues go unrecognized, especially among teenagers, who often lack access to appropriate care. Studies show that smartphone addiction is increasingly common among teenagers, a trend linked to mental health concerns. In Indonesia, social media use is high, with smartphone addiction rates paralleling those seen in other countries like China and South Korea. Youth empowerment programs, such as peer educator initiatives, are also recommended for promoting mental health. Peer educators are effective due to their relatable experiences and shared understanding with their peers, though they require more support to fully succeed in their roles.

4. Health Infrastructure Investments: Indonesia also invested in healthcare facility upgrades, improving hospitals and clinics to manage both COVID-19 and other infectious diseases. The expansion of telemedicine services further ensured healthcare access during lockdowns and movement restrictions.
5. International Collaboration: Indonesia continues to participate in global health partnerships, collaborating with international organizations to share knowledge and resources. In addition, cross-border disease control efforts have been strengthened with neighboring countries, focusing on managing infectious diseases in border areas. These international collaborations enhance Indonesia's capacity to address global health challenges.

Indonesia's experience with COVID-19 underscored the need for a resilient health system capable of managing diverse public health challenges. To build resilience, the government is prioritizing several key strategies: First, enhanced surveillance systems are crucial, with continuous investment in digital technologies to enable real-time monitoring of emerging and re-emerging diseases. This allows for quicker detection and response to outbreaks; Second, sustainable public health funding is essential to ensure the long-term viability of programs focused on disease prevention, vaccination, and emergency preparedness; Third, a focus on health equity aims to reduce disparities by improving access to healthcare for vulnerable and marginalized populations, ensuring that no one is left behind; Lastly, fostering research and innovation is vital for advancing understanding of infectious diseases and developing new tools and strategies for prevention and control. Together, these efforts will strengthen Indonesia's capacity to respond to future health crises.

7.7 Ensuring Healthy Workers

Ensuring employee well-being and work-life balance in Indonesia's healthcare sector is crucial, especially post-COVID-19. The pandemic placed immense strain on healthcare workers, leading to high levels of stress, burnout, and mental health issues due to long hours, overwhelming workloads, and the constant risk of infection. In Indonesia, 83% of healthcare professionals reported burnout, and 62% showed signs of depression during the pandemic (Anindyajati et al. 2021). To address these challenges, Indonesia's healthcare system is implementing strategies to promote well-being. Mental health support programs like the First Aid for Psychological Injuries (P3LP) help healthcare workers manage stress. Flexible work schedules, including rotating shifts and remote work for administrative tasks, allow workers to recuperate and reduce strain. Strengthening peer support networks provides emotional relief and fosters a sense of community among professionals.

Healthcare organizations must also ensure adequate staffing, PPE supplies, and training to prevent future burnout. These initiatives will create a healthier, more resilient workforce, leading to better patient care, higher job satisfaction, and a sustainable healthcare system prepared for future crises. Research indicates that the emotional strain on healthcare workers has contributed to increased instances of exhaustion and mental health issues. A study found that 83% of healthcare professionals in Indonesia experienced burnout, while 62% showed signs of depression (Anindyajati et al. 2021). The ongoing stress and risks associated with their roles have prompted many healthcare workers to seek support for their well-being. The situation underscores the need for systemic changes to support healthcare professionals and address the mental health challenges they face in the wake of the pandemic. Creating a healthier work-life balance is essential to ensure their resilience and overall well-being as they continue to provide critical care.

Achieving a healthy work-life balance for healthcare professionals in Indonesia, especially post-pandemic, requires a combination of strategies aimed at supporting their well-being. One key approach is providing mental health support through programs like counseling services and First Aid for Psychological Injuries (P3LP). These initiatives help healthcare workers manage stress and emotional strain, improving mental resilience. Flexible work schedules are another essential strategy. Allowing for rotating shifts, shorter working hours, and remote work for administrative tasks can significantly reduce burnout and ensure adequate rest. Giving healthcare workers time to recuperate between shifts prevents exhaustion and helps maintain energy levels for both personal and professional responsibilities. Building peer support networks within healthcare settings also plays a crucial role. These networks provide a platform for workers to share their experiences, reduce feelings of isolation, and foster solidarity, offering emotional and psychological relief in high-stress environments. Additionally, ensuring sufficient staffing levels and access to personal protective equipment (PPE) alleviates the pressure on healthcare workers, allowing them to perform their duties without risking their own well-being. These strategies collectively contribute to a more sustainable work-life balance, enabling healthcare professionals to deliver high-quality care while maintaining their personal health and happiness.

7.8 Conclusion

Following the COVID-19 pandemic, Indonesia strengthened its healthcare system to better combat infectious diseases, including dengue fever, tuberculosis (TB), malaria, and conditions related to air pollution, while also improving worker health. Recognizing its vulnerability, Indonesia invested in health surveillance and disease monitoring systems to provide earlier detection and faster response times for outbreaks, including mosquito-borne diseases like dengue and malaria. Expanded distribution of insecticide-treated nets, environmental management strategies, and public awareness campaigns have helped curb vector populations and reduce transmission.

Indonesia has also scaled up TB prevention and treatment, with a focus on active case finding, the provision of free medication, and the integration of TB care within primary health services. Enhanced testing and treatment programs, supported by partnerships with organizations like the Global Fund, aim to eliminate TB as a major public health threat.

The country tackled air pollution-related diseases by implementing regulations on industrial emissions, expanding green spaces in urban areas, and promoting clean energy use. Efforts to educate communities about reducing exposure to pollutants were also initiated.

To safeguard worker health, Indonesia improved occupational health standards, provided regular screenings, and promoted workplace safety measures. Strengthening the health workforce through training, incentives, and mental health support ensures a resilient frontline in public health crises.

Reference

Anindyajati G, Wiguna T, Murtani BJ, Christian H et al (2021) Anxiety and depression among healthcare workers during the COVID-19 pandemic in Indonesia: a cross-sectional study. BMC Psychiatry 21(1):1–8. https://doi.org/10.1186/s12888-021-03102-6

Chapter 8
Post COVID-19: Transformations in Environment, Health Care, and Disease Patterns in Thailand

Uma Langkulsen and Gamuchirai Munjoma

Abstract The dynamic relationship between diseases and environmental factors, highlighted by the COVID-19 pandemic, underscores the critical role of public health interventions. In Thailand, urbanization, air quality, and climate shaped COVID-19 spread, with densely populated areas bearing the brunt. The government responded with expanded healthcare access and innovative services. Meanwhile, shifts in disease patterns like diabetes and cardiovascular ailments necessitate proactive surveillance. Addressing these complexities demands integrated approaches, considering environmental, social, and medical factors. By prioritizing evidence-based strategies, public health initiatives can effectively navigate this intricate interplay to safeguard communities and enhance overall well-being.

Keywords Disease dynamics · Environmental factors · COVID-19 pandemic · Public health interventions · Healthcare initiatives · Disease patterns

8.1 Introduction

The interplay between diseases and the environment can never be ignored. Different diseases thrive in certain environments hence some diseases are unique to certain countries or places because of the different climatic conditions (Akhtar 2021). The world has been shaken by different catastrophes like diseases, natural disasters, and wars at different points in time. The beginning of 2020 saw the world witnessing the emergence and spread of a devastating disease known as Coronavirus Disease 2019 (COVID-19) caused by a virus termed severe acute respiratory syndrome coronavirus 2 (SARS–CoV–2) (Organization 2020).

U. Langkulsen (✉)
Faculty of Public Health, Thammasat University, Pathum Thani, Thailand
e-mail: u.langkulsen@gmail.com

G. Munjoma
Harare, Zimbabwe

R. Akhtar (ed.), *COVID-19: Impacts on Health and Health Systems*,
https://doi.org/10.1007/978-3-031-90255-0_8

COVID-19 was first discovered in Wuhan city in China late December 2019 and it spread to a significant number of countries by April 2020 as a result of the ease of traveling from one country to another as the world is easily interconnected (Vadiati et al. 2023). COVID-19 is reported to spread through contact with nasal or oral droplets from and infected person (Akhtar 2021). The World Health Organization gave specific guidelines to aid in preventing transmission of the virus from person to person, these include physical distancing, wearing of face masks and use of hand sanitizers (Baker et al. 2022).

Regardless of the universal guidelines in preventing the spread of COVID-19, its spread has not been equal across different borders some countries are shown to have faced more infections and deaths than others (Kong et al. 2021). This is believed to be as a result of the differences in environmental, political, and cultural factors that affect different countries (Kong et al. 2021). For example, Thailand is shown to have far less COVID-19 infections and deaths when compared to countries such as the United States of America and Italy (Azuma et al. 2020). Changes in the severity of deaths and infections have been strongly attributed to environmental factors such as climate, urbanization, and air quality to mention but a few (Luz and Masoodian 2022). Some researchers have discovered that COVID-19 thrives in low temperatures and high pollution levels (Luz and Masoodian 2022). Extreme weather conditions such as floods and heatwaves might also have an impact on the spread or elimination of various diseases such as COVID-19, as reports argue that COVID-19 cannot survive long in very high temperatures (Tidman et al. 2021).

The study aims to explore the relationship between environmental conditions and the spread of COVID-19, as well as the subsequent shifts in disease patterns, within the Kingdom of Thailand. Additionally, it delves into the alterations in healthcare accessibility brought about by the COVID-19 pandemic in Thailand.

8.2 Methods

8.2.1 Study Design

This narrative review utilizes a retrospective observational approach to thoroughly explore the intricate connections between environmental factors, the propagation of COVID-19, and changes in disease patterns across Thailand. Additionally, it delves into the nuanced adjustments in healthcare accessibility that occurred during the COVID-19 pandemic, with the goal of offering a comprehensive understanding of the intricate interplay among environmental conditions, public health dynamics, and healthcare delivery.

8.2.2 Study Area

Figure 8.1 delineates the division of the Kingdom of Thailand into 13 distinct health zones. This strategic arrangement reflects the Thai government's initiative to establish a comprehensive health service network, encompassing primary, secondary, and tertiary healthcare provision.

Guided by the fundamental principle of Seamless Health Service Network, the Thai Ministry of Public Health embarked on a five-year endeavor from 2013 to 2017 to develop 13 health zones across Thailand. The impetus behind establishing these health zones stemmed from the government's commitment to ensuring equitable access to healthcare services throughout the country by leveraging available resources, including finances, equipment, and expertise. To facilitate this initiative, a comprehensive healthcare management system was introduced to facilitate efficient monitoring and management of healthcare provision across regional, provincial, district, and sub-district hospitals, thereby eliminating barriers between different tiers of healthcare facilities.

Health Zone 1 encompasses Chiang Mai, Chiang Rai, Phrae, Nan, Phayao, Lampang, Lamphun, and Mae Hong Son provinces. Health Zone 2 includes Tak, Phetchabun, Phitsanulok, Uttaradit, and Sukhothai provinces. Health Zone 3 comprises Kamphaeng Phet, Phichit, Nakhon Sawan, Chai Nat, and Uthai Thani provinces. Health Zone 4 incorporates Saraburi, Nonthaburi, Lopburi, Ang Thong, Nakhon Nayok, Sing Buri, Phra Nakhon Si Ayutthaya, and Pathum Thani provinces. Health Zone 5 spans Phetchaburi, Samut Sakhon, Samut Songkhram, Prachuap Khiri Khan, Suphan Buri, Nakhon Pathom, Ratchaburi, and Kanchanaburi provinces. Health Zone 6 includes Sa Kaeo, Prachin Buri, Chachoengsao, Samut Prakan, Chonburi, Chanthaburi, Rayong, and Trat provinces. Health Zone 7 comprises Kalasin, Khon Kaen, Maha Sarakham, and Roi Et provinces. Health Zone 8 encompasses Udon Thani, Sakon Nakhon, Nakhon Phanom, Loei, Nong Khai, Nong Bua Lamphu, and Bueng Kan provinces. Health Zone 9 consists of Chaiyaphum, Nakhon Ratchasima, Buriram, and Surin provinces. Health Zone 10 incorporates Ubon Ratchathani, Sisaket, Mukdahan, and Amnat Charoen provinces. Health Zone 11 includes Nakhon Si Thammarat, Surat Thani, Phuket, Krabi, Phang Nga, Ranong, and Chumphon provinces. Health Zone 12 comprises Songkhla, Satun, Trang, Phatthalung, Pattani, Yala, and Narathiwat provinces. Lastly, Health Zone 13 pertains to Bangkok.

8.2.3 Data Collection

Data on COVID-19 cases, environmental factors, and disease patterns were gathered from diverse sources, encompassing official government reports, academic literature, and databases. COVID-19 case information, including new cases and incidence rates, was sourced from the Thai Ministry of Public Health. Environmental metrics, such

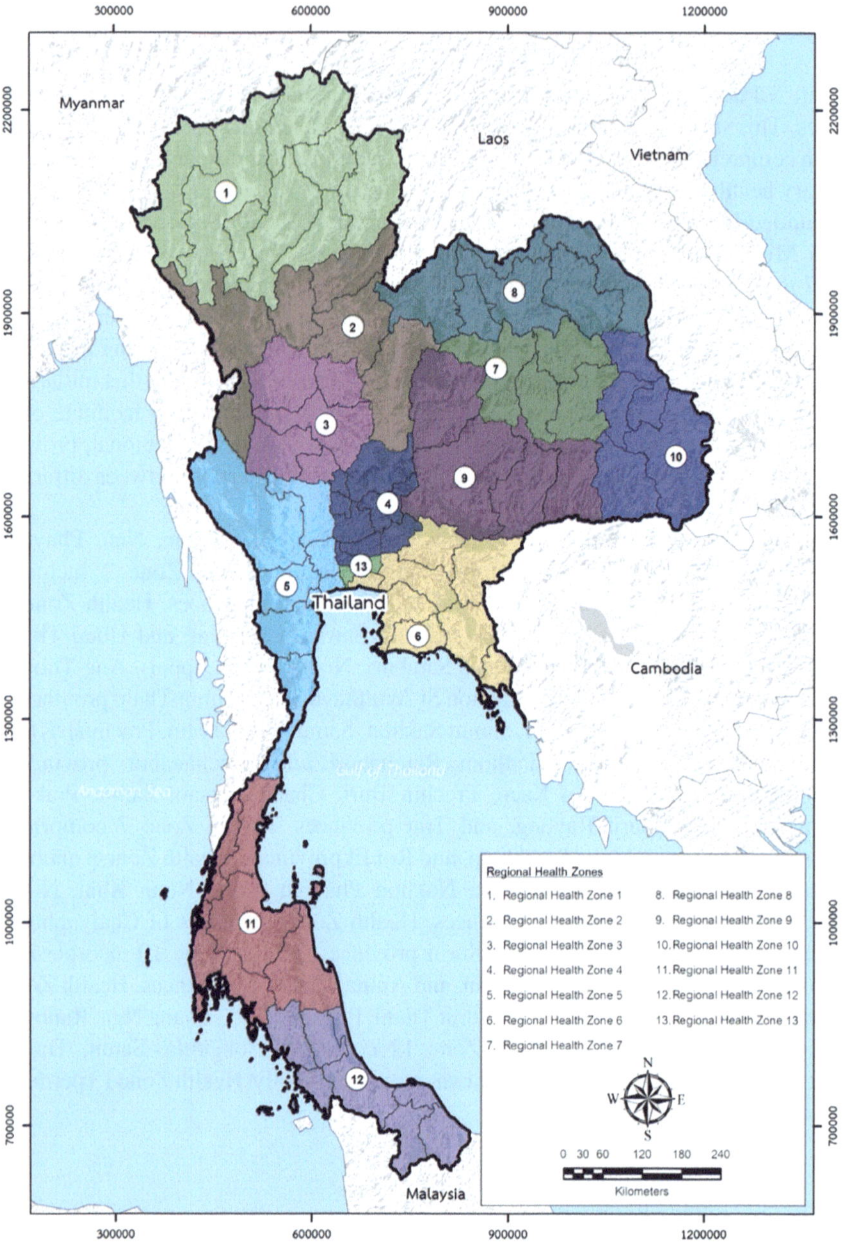

Fig. 8.1 Health zones in Thailand

as air quality indices, temperature, humidity levels, and urbanization rates, were obtained from governmental agencies and research publications. Disease pattern data for conditions like diabetes mellitus (DM), hypertension (HT), cardiovascular diseases (CVDs), and chronic obstructive pulmonary disease (COPD) were extracted from the Health Data Center (HDC) under the Thai Ministry of Public Health.

8.2.4 Data Analysis

The analysis began with the application of descriptive statistics to summarize key aspects of the data related to COVID-19 cases, environmental variables, and disease patterns. For COVID-19 cases, descriptive statistics were used to outline the distribution of new cases over time. Similarly, descriptive statistics were applied to environmental variables such as air quality indices, temperature, humidity, and urbanization rates to provide a comprehensive overview of their distribution and variability across different regions and time periods. Additionally, descriptive statistics were employed to summarize disease patterns for conditions like diabetes mellitus, hypertension, cardiovascular diseases, and chronic obstructive pulmonary disease, focusing on key metrics such as incidence rates and trends over time. These descriptive analyses laid the foundation for further exploration and interpretation of the data to elucidate relationships and trends among the variables of interest.

8.3 Results

8.3.1 COVID-19 in Thailand

Thailand was the first country to record a COVID-19 case after China in January 2020. The first COVID-19 case was that of woman who had traveled from Wuhan China (Tantrakarnapa et al. 2020). As a result of this case, just like most countries around the world, Thailand experienced several COVID-19 waves (Naorungroj et al. 2023). In a bid to control the spread of the virus, the Thai government introduced various COVID-19 containment measures including wearing of face masks when in public, social distancing, frequent hand sanitizing, strict border control measures such as quarantine of all international arrivals and body temperature check points across the country (Wilasang et al. 2022). The timely intervention by the government made COVID-19 infections and deaths in Thailand to be far much less when compared to other countries (Yorsaeng et al. 2022).

8.3.2 *Urbanization*

Regardless of the Thai government's COVID-19 control measures, environmental factors in Thailand also played a significant role in the spread of COVID-19. Studies by some researchers reveal that COVID-19 cases in Thailand and around the world differed from one province to the other because of differences in the environmental conditions (Luz and Masoodian 2022). COVID-19 infections and deaths were reported to be higher in Bangkok metropolitan province, Chiang Mai, Chon Buri, and Samut Prakan because of the effects of urbanization and high population density (Saengchut et al. 2022). The movement of people from other provinces to Bangkok in search for better living conditions and jobs caused urbanization and overcrowding in the capital city (Saengchut et al. 2022). Overcrowding as a result urbanization meant social distancing in public vicinities was difficult hence making it easier for virus transmission from one individual to another (Saengchut et al. 2022). Urbanization has also caused the spread of diseases such as dengue, chikungunya and Zika in urban areas such as Bangkok (Baker et al. 2022). However, it is important to note that the travel restrictions during the pandemic aided in lowering disease spread, not just COVID-19 but also diseases like dengue hence the need to control urbanization in order to curb disease spread (Yorsaeng et al. 2022).

8.3.3 *Air Quality*

Furthermore, climatic conditions such as air quality were also linked to easier spread of COVID-19 and fatality around the world (Azuma et al. 2020). Climatic conditions are reported to have great influence on the emergence and re-emergence of infectious diseases (Thazhathedath Hariharan et al. 2021). Research argues that air quality and windspeed are somewhat intertwined and make it easier for the transmission of COVID-19 (Vadiati et al. 2023). Thailand is reported to have poor air quality that is below the World Health Organization requirements (Nikam et al. 2021). Air pollution is closely linked poor health outcomes when one is infected with respiratory illnesses such as COVID-19 as it weakens the immune system (Weaver et al. 2022). This explains the why most COVID-19 infections and deaths were high in Bangkok as it is one of the provinces with poor air quality in Thailand (Saengchut et al. 2022). However, it is important to note that air pollution or poor air quality does not have detrimental effects on COVID-19 alone but on other health conditions such as diabetes, hypertension and cardiovascular diseases (Patra et al. 2021). Hence, there is need for adequate policy change to lower pollution levels in Thailand.

8.3.4 Temperature

Temperature is also reported to influence the spread of COVID-19. Lower temperatures are shown to increase COVID-19 transmission and death due to COVID-19 as it lowers immunity and the virus survives longer than in higher temperatures (Sajadi et al. 2020). Thailand is a warm country with high temperature ranges (Phosri et al. 2020). High temperatures in Thailand maybe one of the possible explanations to why the country was not hugely burdened by COVID-19 when compared to countries with lower temperatures such as Russia (Pramanik et al. 2022). Exposure to the sun is linked to better immune responses (Susswein et al. 2023). Therefore, high temperatures in Thailand maybe linked to better COVID-19 outcomes.

However, it is important to note that COVID-19 outcomes in Thailand and in other countries are not simply the interplay of environmental and climatic conditions but also of existing comorbidities in a population (Gasmi et al. 2021). Existing comorbidities such as hypertension, cardiovascular diseases, diabetes, smoking and obesity are reported to increase COVID-19 infections and deaths (Gasmi et al. 2021). Following government guidelines such as social distancing, hand sanitization, and wearing of face masks on COVID-19 prevention was shown to reduce transmissions in populations that adhered to them (Meyerowitz et al. 2021).

8.3.5 Shifting Disease Patterns

Table 8.1 illustrates the evolution of disease patterns, including DM, HT, CVDs, and COPD, over a nine-year period.

Diabetes mellitus exhibits a consistent upward trend in both new cases and incidence rates throughout the study period. The peak incidence rate was observed in 2023, with cases rising from 258,669 in 2015 to 384,890 in 2023, indicating a growing incidence of the disease within the population.

Hypertension demonstrates a notable increase in both new cases and incidence rates from 2015 to 2017, with the highest incidence rate recorded in 2017 at 1314.60. However, starting in 2018, there was a slight decline in both new cases and incidence rates. Despite this decline, the incidence rate in 2023 remained relatively high at 1303.65, indicating a persistent burden of hypertension within the population throughout the study period.

Cardiovascular diseases exhibit fluctuations over the years, with an initial rate of 42.94 in 2015 followed by an increase until 2017, succeeded by a subsequent decrease. However, from 2018 onwards, there is a notable decline in both new cases and incidence rates, with the number of new cases dropping to 22,097 in 2018 and the incidence rate decreasing to 36.38 per 100,000 population. This downward trend continues in subsequent years, with further decreases in both new cases and incidence rates, reaching 13,775 new cases and 22.74 incidence rate per 100,000 population in 2023.

Table 8.1 New cases and incidence rate per 100,00 population for non-communicable diseases in Regional Health Zones 1–12, Thailand (2015–2023)

Year	Population	DM[a]		HT[a]		CVDs[a]		COPD[a]	
		New cases	Incidence rate	New cases	Incidence rate	New cases	Incidence rate	New cases	Incidence rate
2015	60,032,689	258,669	430.88	551,409	918.51	25,780	42.94	23,488	39.13
2016	60,244,904	270,820	449.53	639,123	1,060.87	26,278	43.62	25,479	42.29
2017	60,506,088	346,351	572.42	795,414	1,314.60	27,496	45.44	21,709	35.88
2018	60,737,331	326,946	538.29	751,207	1,236.81	22,097	36.38	17,189	28.30
2019	60,892,671	307,964	505.75	698,148	1,146.52	16,273	26.72	13,803	22.67
2020	60,598,505	315,621	520.84	696,498	1,149.36	16,100	26.57	10,918	18.02
2021	60,643,445	323,425	533.32	745,415	1,229.18	13,597	22.42	8,108	13.37
2022	60,595,543	354,772	585.48	720,421	1,188.90	13,586	22.42	9,265	15.29
2023	60,581,027	384,890	635.33	789,765	1,303.65	13,775	22.74	13,367	22.06

[a]DM = diabetes mellitus (ICD10 E10-E14), HT = hypertension (ICD10 I10-I15), CVDs = cardiovascular diseases (ICD10 I00-I09, I20-I28, I30-I52), COPD = chronic obstructive pulmonary disease (ICD10 J44)

Source Thai Ministry of Public Health and Department of Provincial Administration, Ministry of Interior

Chronic obstructive pulmonary disease demonstrates a consistent decrease in incidence rates over the study period. From a rate of 39.13 in 2015, a continual decline was observed each year, reaching 13.37 in 2021. However, there is a slight increase observed in 2022, with the number of new cases rising to 9,265 and the incidence rate increasing to 15.29 per 100,000 population. In 2023, the incidence rate further increased to 22.06 per 100,000 population, indicating a continued upward trend.

Non-communicable diseases (NCDs) in Thailand show varying trends. While some NCDs show an increasing trend in incidence rates (DM and HT), others such as COPD shows a decrease over the years. This mixed pattern suggests that different factors maybe influencing the incidence of the above-mentioned diseases. The NCDs are reported to have detrimental outcomes among affected populations when infected by COVID-19 (Gasmi et al. 2021). Air pollution is quite high in Thailand as the air quality is reported to be below the recommended World Health Organization requirement (Nikam et al. 2021). Poor air quality is a possible explanation to the poor health outcomes and increases in DM and HT incidence (Baker et al. 2022).

The significant decrease in incidence of COPD maybe a result of public health interventions and improvements in access to health care during the COVID-19 pandemic as Universal health Coverage index significantly increased in 2019 (Tangcharoensathien et al. 2023). It is important to note that access to health care and sharing of healthcare resources across the country greatly improved during the COVID-19 era (Langkulsen et al. 2023). Lifestyle changes among the Thai population may also have an impact on the decline in COPD incidence (Liangruenrom et al. 2019). The trends in NCDs provide insights into the overall patterns, however they do not reflect the exact underlying causes hence further in-depth analysis is essential to fully uncover the reasons behind the trends.

8.3.6 COVID-19 and Healthcare Provision in Thailand

Healthcare in Thailand has a long but beneficial history. Decades of hard work in improving the healthcare sector have seen Thailand being ranked sixth in the index of Global health security (Rajatanavin et al. 2021). Before the COVID-19 pandemic struck, great strides in the healthcare sector had been made as a notable increase in healthcare workers was achieved from 0.93 physicians, nurses, and midwives in 1991 to 4.07 in 2019 per 1000 population (Tangcharoensathien et al. 2023). Rural health worker force was also improved through the recruitment and training of persons from rural areas who returned to their home villages upon completing training, thereby strengthening healthcare delivery in rural communities (Nittayasoot et al. 2021). During the COVID-19 pandemic, the number of healthcare workers was improved through the recruitment of migrant health workers and village health volunteers (Langkulsen et al. 2023). However, the pandemic revealed that there was a shortage of intensive care nurses in Thailand and the government aims at recruiting more nurses to feel the gap (Boonmavichit & Hobbs). The government further aims at meeting the sustainable development goal of 44.5 physicians, nurses and midwives per 10,000 population by 2030 (Boonmavichit & Hobbs). Despite some shortages in health workers, Thailand has quite a good number of healthcare workers and volunteers which provide healthcare both in rural and urban areas.

To ensure provide equal access to healthcare for all, Thailand introduced the Universal health coverage. Universal health coverage is a government healthcare insurance cover for all (Rajatanavin et al. 2021). The universal health coverage index increased to 83% in 2019 from a low of 41% in 2000 (Tangcharoensathien et al. 2023). Universal health coverage was also extended to migrant workers and their dependents therefore improving access to healthcare among migrants and Thai citizens alike (Langkulsen et al. 2023). The universal health coverage initiative is shown to aid in achieving better health outcomes among Thai citizens during COVID-19 and after as it reduces the financial burden of getting treatment (Tangcharoensathien et al. 2023).

Thailand's healthcare system is divided into five levels, that is, central, regional, provincial, district, and sub-district all under the Ministry of public health (Langkulsen et al. 2023). During the COVID-19 pandemic, the government established the surveillance and rapid response team (SRRT) (Rajatanavin et al. 2021). The SRRT comprised of nurses, officers and epidemiologists with the support of village health workers (Rajatanavin et al. 2021). SRRT was stationed at all levels of the health care system and they aided not only in preparedness and response to COVID-19 alone but also to diseases such as rabies, foot and mouth, dengue, and meningococcemia (Rajatanavin et al. 2021). The establishment of the SRRT by the government strengthened healthcare provision in Thailand during COVID-19 and beyond.

COVID-19 greatly affected the provision of other healthcare services thereby greatly affecting the Out-Patient-Department (OPD) (Sukmanee et al. 2023). Provision of testing and treatment services of diseases such as human immunodeficiency virus (HIV) and sexually transmitted infections (STIs) was greatly disrupted due to the overwhelming demands placed by COVID-19 on healthcare personnel (Akhtar 2022). To counter the impacts of COVID-19 on the provision of OPD and NCDs, teleconsultation, tele pharmacies, and tele monitoring were introduced by the government making healthcare provision accessible for treatment of other ailments (Pitayarangsarit et al. 2023). The introduction of the tele healthcare services greatly improved healthcare provision in Thailand as one can access healthcare without physically visiting a health care facility.

8.4 Discussion

To optimize public health outcomes in Thailand amidst the dynamic interplay of environmental factors and health, the Ministry of Public Health and relevant agencies should prioritize several key initiatives. Firstly, there is a pressing need to bolster environmental surveillance systems to closely monitor indicators such as air quality, urbanization trends, and climatic conditions. This data can inform targeted interventions aimed at mitigating health risks associated with environmental factors.

Collaborative efforts among government agencies, academic institutions, and NGOs are essential to coordinate comprehensive environmental health assessments and devise evidence-based strategies. Public awareness campaigns should be launched to educate communities about environmental health risks and empower individuals to adopt preventive measures. Concurrently, investments in research and innovation will facilitate a deeper understanding of the complex relationships between environmental factors and health outcomes. Strengthening the resilience of the healthcare system to environmental threats, advocating for policy changes, and promoting interagency collaboration will further bolster Thailand's capacity to safeguard public health in the face of evolving environmental challenges.

Further research in the field of environmental factors and COVID-19 should prioritize longitudinal studies to track the long-term impact of environmental variables on disease transmission and outcomes. Additionally, comparative studies across different countries and regions would offer valuable insights into the generalizability of findings and the influence of contextual factors. Exposure assessment studies are needed to quantify individual-level exposure to environmental factors and their relationship with COVID-19 infection risk and severity, utilizing advanced modeling techniques. Mechanistic studies exploring biological pathways linking environmental exposures to COVID-19 outcomes could provide deeper insights. Intervention studies evaluating the effectiveness of environmental interventions, health equity research examining disparities in COVID-19 outcomes, and behavioral studies on the influence of individual behaviors are crucial. Climate change adaptation research can assess future trends in COVID-19 transmission, while health system resilience studies can identify strategies to strengthen healthcare infrastructure. Finally, community engagement and participatory research can foster collaboration and empowerment, leading to more sustainable public health interventions. These research avenues are essential for informing evidence-based policies and interventions to protect public health and promote resilience in the face of environmental challenges and pandemics.

This study is subjected to several limitations. Firstly, the retrospective nature of the study limits causal inference between environmental factors and disease outcomes. Additionally, reliance on secondary data sources may introduce potential biases or inaccuracies in the study findings. Furthermore, the generalizability of the study findings may be limited to the context of Thailand and may not be applicable to other regions or countries.

8.5 Conclusions

The spread of COVID-19 in Thailand was influenced by a multitude of factors, including climatic and environmental variables such as urbanization, air pollution, temperature, and humidity. This highlights the necessity of integrating environmental considerations into policy formulation processes. Furthermore, the COVID-19 pandemic served as a revealing lens, shedding light on both the strengths and weaknesses of Thailand's healthcare system, as well as those of nations worldwide. In response to identified weaknesses, the Thai government implemented measures to fortify the healthcare system, including the recruitment of village health volunteers, migrant volunteers, and the implementation of tele healthcare services. Additionally, intriguing shifts observed in the disease patterns of NCDs are likely influenced by a combination of factors, including environmental influences, public health interventions, lifestyle modifications, and healthcare accessibility. These findings underscore the importance of holistic approaches to address emerging health challenges and enhance overall population health outcomes.

References

Akhtar R (2021) Coronavirus (COVID-19) outbreaks, environment and human behaviour. Springer

Akhtar R (2022) Coronavirus (COVID-19) outbreaks, vaccination, politics and society: the continuing challenge. Springer, pp 1–14

Azuma K, Yanagi U, Kagi N, Kim H, Ogata M, Hayashi M (2020) Environmental factors involved in SARS-CoV-2 transmission: effect and role of indoor environmental quality in the strategy for COVID-19 infection control. Environ Health Prev Med 25:1–16

Baker RE, Mahmud AS, Miller IF, Rajeev M, Rasambainarivo F, Rice BL, Takahashi S, Tatem AJ, Wagner CE, Wang L-F (2022) Infectious disease in an era of global change. Nat Rev Microbiol 20(4):193–205

Boonmavichit N, Hobbs JD, The future of Thailand's healthcare workforce in light of the covid-19 pandemic: a retroduction analysis

Gasmi A, Peana M, Pivina L, Srinath S, Gasmi Benahmed A, Semenova Y, Menzel A, Dadar M, Bjørklund G (2021) Interrelations between COVID-19 and other disorders. Clin Immunol 224:108651. https://doi.org/10.1016/j.clim.2020.108651

Kong JD, Tekwa EW, Gignoux-Wolfsohn SA (2021) Social, economic, and environmental factors influencing the basic reproduction number of COVID-19 across countries. PLoS ONE 16(6):e0252373

Langkulsen U, Mareke P, Lambonmung A (2023) Migrants and healthcare during COVID-19, the case of Kanchanaburi Province in Thailand. Healthcare

Liangruenrom N, Craike M, Biddle SJ, Suttikasem K, Pedisic Z (2019) Correlates of physical activity and sedentary behaviour in the Thai population: a systematic review. BMC Public Health 19(1):1–26

Luz S, Masoodian M (2022) Exploring environmental and geographical factors influencing the spread of infectious diseases with interactive maps. Sustainability 14(16):9990

Meyerowitz EA, Richterman A, Gandhi RT, Sax PE (2021) Transmission of SARS-CoV-2: a review of viral, host, and environmental factors. Ann Intern Med 174(1):69–79

Naorungroj T, Viarasilpa T, Tongyoo S, Detkaew A, Pinpak T, Wimolwattanaphan R, Ratanarat R, Promsin P, Thamrongpiroj P, Phumpichet A (2023) Characteristics, outcomes, and risk factors for in-hospital mortality of COVID-19 patients: a retrospective study in Thailand. Front Med 9:1061955

Nikam J, Archer D, Nopsert C (2021) Regulating air quality in Thailand–a review of policies. Energy 2015:2036

Nittayasoot N, Suphanchaimat R, Namwat C, Dejburum P, Tangcharoensathien V (2021) Public health policies and health-care workers' response to the COVID-19 pandemic, Thailand. Bull World Health Organ 99(4):312

Organization WH (2020) Mental health and psychosocial considerations during the COVID-19 outbreak, 18 March 2020

Patra SS, Panda S, Das T, Ramasamy B (2021) COVID-19 fatality: statistical evidence to engender the need for focal shift from air pollutants to multi-dimensional intervention. Aerosol Air Qual Res 21(1):200410

Phosri A, Sihabut T, Jaikanlaya C (2020) Short-term effects of diurnal temperature range on hospital admission in Bangkok, Thailand. Sci Total Environ 717:137202

Pitayarangsarit S, Bhagaman N, Yodmai K, Thangsirikul N, Tipayamongkholgul M (2023) The resiliency of noncommunicable diseases services during the public health crisis: a lesson from Bangkok, Thailand. BMC Health Serv Res 23(1):1–10

Pramanik M, Udmale P, Bisht P, Chowdhury K, Szabo S, Pal I (2022) Climatic factors influence the spread of COVID-19 in Russia. Int J Environ Health Res 32(4):723–737

Rajatanavin N, Tuangratananon T, Suphanchaimat R, Tangcharoensathien V (2021) Responding to the COVID-19 second wave in Thailand by diversifying and adapting lessons from the first wave. BMJ Glob Health 6(7)

Saengchut P, Ruksakulpiwat S, Wonglertarak W, Kokaphan C, Phengarree K (2022) Impact of meteorological factors on the COVID-19 pandemic in Thailand. Sci Eng Health Stud 22050024-22050024

Sajadi MM, Habibzadeh P, Vintzileos A, Shokouhi S, Miralles-Wilhelm F, Amoroso A (2020) Temperature, humidity, and latitude analysis to estimate potential spread and seasonality of coronavirus disease 2019 (COVID-19). JAMA Netw Open 3(6):e2011834–e2011834

Sukmanee J, Butchon R, Karunayawong P, Saeraneesopon T, Boonma C, Teerawattananon Y, Isaranuwatchai W (2023) Pattern of OPD utilisation during the COVID-19 pandemic under the universal coverage scheme in Thailand: what can 850 million records tell us? BMC Health Serv Res 23(1):1–7

Susswein Z, Rest EC, Bansal S (2023) Disentangling the rhythms of human activity in the built environment for airborne transmission risk: an analysis of large-scale mobility data. Elife 12

Tangcharoensathien V, Vandelaer J, Brown R, Suphanchaimat R, Boonsuk P, Patcharanarumol W (2023) Learning from pandemic responses: informing a resilient and equitable health system recovery in Thailand. Front Public Health 11:1065883

Tantrakarnapa K, Bhopdhornangkul B, Nakhaapakorn K (2020) Influencing factors of COVID-19 spreading: a case study of Thailand. J Publ Health 1–7

Thazhathedath Hariharan H, Surendran AT, Haridasan RK, Venkitaraman S, Robert D, Narayanan SP, Mammen PC, Siddharth SR, Kuriakose SL (2021) Global COVID-19 transmission and mortality—influence of human development, climate, and climate variability on early phase of the pandemic. Geohealth 5(10):e2020GH000378

Tidman R, Abela-Ridder B, de Castañeda RR (2021) The impact of climate change on neglected tropical diseases: a systematic review. Trans R Soc Trop Med Hyg 115(2):147–168

Vadiati M, Ghasemi L, Samani S, Islam MA, Ahmadi A, Khaleghi S, Movarej M, Gorgij AD, Kisi O, Davoodi S (2023) A sustainable trend in COVID-19 research: an environmental perspective. Front Environ Sci 11:372

Weaver AK, Head JR, Gould CF, Carlton EJ, Remais JV (2022) Environmental factors influencing COVID-19 incidence and severity. Annu Rev Public Health 43:271–291

Wilasang C, Jitsuk NC, Sararat C, Modchang C (2022) Reconstruction of the transmission dynamics of the first COVID-19 epidemic wave in Thailand. Sci Rep 12(1):2002

Yorsaeng R, Suntronwong N, Thongpan I, Chuchaona W, Lestari FB, Pasittungkul S, Puenpa J, Atsawawaranunt K, Sharma C, Sudhinaraset N (2022) The impact of COVID-19 and control measures on public health in Thailand, 2020. PeerJ 10:e12960

Chapter 9
Burden of COVID-19 on Healthcare and Environmental Systems

Nasrin Aghamohammadi, Logaraj Ramakreshnan, Rama Krishna Supramanian, and Yin Cheng Lim

Abstract The COVID-19 pandemic has had a significant impact on healthcare and environmental systems with far-reaching consequences influencing the ecological well-being, economic stability, social cohesion, and institutional trust. The pandemic has worsened prevailing health disparities by disproportionately impacting vulnerable communities with limited access to healthcare resources nationwide. In addition to overwhelming healthcare systems due to escalating infections, hospitalizations, and fatalities, the pandemic has caused significant disruptions to regular healthcare services, led to the postponement of elective surgeries, screenings, and other treatments. The pandemic has led to an increase in mental health issues linked with fear of the virus, social isolation, and economic uncertainty. Lockdowns and social distancing measures have also influenced the physical activity levels, dietary habits, and sleep patterns, potentially affecting the prevalence and management of lifestyle-related diseases. However, the pandemic has highlighted the significance of having a well-prepared public health infrastructure prompting more investments in healthcare systems to effectively address any future health emergencies. The COVID-19 pandemic has had both positive and negative impacts on environmental issues. The implementation of lockdowns and travel restrictions led to a significant decline in energy consumption patterns and air pollution due to reduced use of fossil fuels, passive mobility and industrial emissions. Even though reduced industrial discharges had positive impacts on water bodies, the improper disposal of waste raised concerns about marine pollution and waste management. Changes in human activities benefited some wildlife from reduced human disturbance, whereas the others faced challenges

N. Aghamohammadi (✉)
Curtin University Sustainability Policy Institute, School of Design and the Built Environment, Curtin University, 6031 Bentley, Australia
e-mail: Nasrin@curtin.edu.au

School of Environmental and Conservation Sciences, Murdoch University, Murdoch, Australia

L. Ramakreshnan
Institute for Advanced Studies, University of Malaya, Kuala Lumpur, Malaysia

N. Aghamohammadi · R. K. Supramanian · Y. C. Lim
Department of Social and Preventive Medicine, Faculty of Medicine, University of Malaya, Kuala Lumpur, Malaysia

R. Akhtar (ed.), *COVID-19: Impacts on Health and Health Systems*,
https://doi.org/10.1007/978-3-031-90255-0_9

due to disruptions in food supply chains and conservation efforts. The pandemic also underscored the importance of resilient ecosystems in preventing the emergence and spread of zoonotic diseases, which led to increased emphasis on sustainable land use and wildlife conservation. The full impact of COVID-19 on the healthcare and environmental systems is continuously unfolding, and long-term comprehensive and coordinated efforts are crucial to tackle these dual concerns.

Keywords COVID-19 · Environmental sustainability · Resilient healthcare · Health emergency

9.1 Introduction

The COVID-19 pandemic has had a profound and multifaceted impact on healthcare and environmental systems globally. The most immediate impact has been the surge in COVID-19 cases and the overwhelming healthcare facilities, especially in regions with high infection rates. This strain on resources has led to shortages of hospital beds, medical supplies, and healthcare personnel. Many healthcare systems have had to shift priorities to focus primarily on COVID-19 response, leading to delays or cancellations of non-urgent medical procedures and elective surgeries. A systematic review comparing utilization of healthcare services before and during the COVID-19 pandemic reported that healthcare utilization decreased by about a third during the pandemic, mainly involving patients with less acute illnesses (Moynihan et al. 2021). This has resulted in challenges for patients with chronic conditions or in need of routine care. For individuals with chronic conditions, delays or cancellations of these appointments can lead to worsening health outcomes and elevated risks. Moreover, the postponement of elective surgeries can have implications for patients who require these procedures to alleviate pain, improve mobility, or address other medical issues. Extended wait times can exacerbate discomfort and reduce the quality of life for these patients. From an environmental standpoint, the pandemic has brought about notable environmental effects with both positive and negative outcomes. On the positive side, decreased human activity during lockdowns resulted in decreased energy production and temporary enhancements in air and water quality in numerous areas. However, the heightened usage of clinical waste has placed a strain on waste management systems, and the postponement of environmental regulations and conservation efforts has presented obstacles to achieving long-term sustainability objectives. Hence, striking a balance between these impacts becomes crucial in order to shape post-pandemic environmental policies that prioritize both human health and ecological well-being. The following sections of this chapter will delve into the detailed discussion of the impact of COVID-19 on healthcare and environmental systems.

9.2 Burden of COVID-19 on Healthcare Systems

9.2.1 Non-communicable Diseases

During the containment and mitigation phase of COVID-19, accessibility to healthcare services is affected due to travel limitations, isolation, and quarantine orders resulting in difficulty for patients to receive their routine care for their chronic illness (Luciani et al. 2023). A report by the World Health Organization (WHO) on countries from the sub-Saharan Africa region stated that 22% of countries reported that only emergency and inpatient care services for chronic conditions was made available during the initial phase of the COVID-19 pandemic. This redirection of healthcare resources also affects the provision of specific services related to non-communicable diseases. This report also found that disruption of care and follow-up healthcare services related to hypertension management was found to be disrupted in 59% of the countries and management of diabetic complications has been disrupted in 56% of the countries (Formenti et al. 2022). The disruption on management of these non-communicable diseases may result in a potential rise in complications related to these diseases including microvascular (retinopathy, nephropathy, and neuropathy) and macrovascular complications (coronary artery disease, myocardial infarction, congestive heart failure, stroke, and peripheral vascular disease) (Kendzerska et al. 2021). The initial containment phase of COVID-19 brought about policies like social distancing, travel restrictions, quarantine, and home isolation (Pan et al. 2021). This did serve as a window of opportunity to the growth of telemedicine services as an effective strategy to monitor and manage chronic illnesses which included video or phone consultations and remote monitoring of symptoms and vital signs (e.g., heart rate, blood pressure, and oxygen saturation). It has been reported that utilization of telemedicine services increased by nearly 90% during the pandemic (Anthony Jnr 2020). However, it is important to also highlight that despite these benefits, telemedicine is linked to several implementation barriers such as: (1) limited capacity for conducting physical examinations or ancillary testing; (2) diminished capacity for interpreting verbal and non-verbal cues, potentially leading to more diagnostic errors; and (3) limited access to technology and constraints arising from variations in digital literacy, particularly in older patients with chronic conditions.

The pandemic has also impacted care of patients with chronic kidney disease mainly scheduled appointments, routine dialysis treatments, and kidney transplant evaluation and transplantation have been delayed or canceled, leading to potential exacerbation of their condition. The stringent infection control measures implemented in healthcare facilities to curb the spread of the virus have led to the postponement or cancellation of non-urgent medical procedures and appointments. Data reported by a kidney and transplant center in Iraq showed that weekly hemodialysis sessions and kidney transplants were reduced by 10.3% and 57.2%, respectively, in the transplant center during the early phase of the pandemic (Hussein et al. 2020). For patients relying on dialysis, disruptions in treatment schedules can have dire consequences, potentially exacerbating their condition and leading to complications such as fluid overload, electrolyte imbalances, and increased risk of cardiovascular events. Similarly, delays in transplant evaluations can prolong the wait time for patients awaiting kidney transplantation, worsening their quality of life and increasing the risk of disease progression. Transplant evaluations are a vital part of kidney transplantation as it involves finding a donor that is compatible and with minimal risk of rejection. A delay in the transplant evaluation process will result in delay in the finding a compatible kidney donor and eventually delay transplantation. Meanwhile in the United Kingdom, data from the kidney transplant registry reported a 65% decrease in kidney-alone transplants during the pandemic (Georgiades et al. 2021). This is mainly attributed to a fear of donor-recipient transmission of infectious disease and limited bed availability for organ donation and transplantation due to high bed occupancy rate for COVID-19 patients. The inability to access timely medical care due to COVID-19-related restrictions poses a significant challenge for patients with kidney disease, necessitating innovative solutions such as telemedicine to ensure continuity of care while minimizing the risk of viral transmission. Addressing these disruptions and prioritizing the healthcare needs of patients with kidney disease are essential to mitigate the potential exacerbation of their condition during the ongoing pandemic.

The COVID-19 pandemic not only affected healthcare services at a tertiary level but also at the primary healthcare level. Primary healthcare services encompass preventive care, diagnosis, treatment, and coordination of care for individuals and communities. They focus on promoting health, preventing illness, and managing chronic conditions through screenings, vaccinations, lifestyle counseling, and education (Mak et al. 2022). Screening activities under the primary healthcare facilities mainly covers non-communicable diseases such as diabetes mellitus, mental health disorders, breast and colorectal cancers. Early detection of breast and colorectal cancers is vital for improving treatment outcomes and reducing mortality rates. Screening allows for the identification of cancer at its earliest stages, when it's more likely to be treatable and curable. For breast cancer, regular mammograms can detect tumors before they can be felt, increasing the chances of successful treatment. Similarly, screening tests like colonoscopies can detect precancerous polyps or early-stage colorectal cancer, enabling prompt intervention. Early detection not only improves survival rates but also reduces the need for aggressive treatments and improves overall quality of life for individuals diagnosed with these cancers.

In Malaysia, travel restrictions and temporary suspension of the colorectal (CRC) screening program during the COVID-19 pandemic by the Malaysian government has resulted in an increase in late detection of colorectal cases (Schliemann et al. 2022). In countries where colonoscopy was carried out as part CRC screening also showed similar reduction as this procedure requires utilization of endoscopy rooms. According to modeling of the effects of screening delay for colorectal malignancies (CRC), a wait of more than 4–6 months would result in a 3–6% increase in the likelihood of having advanced CRC, while a 12-month delay might result in a 12% increase in death (Ricciardiello et al. 2021). Breast cancer screening initiatives were also affected during the height of the COVID-19 pandemic. A Dutch study also reported a decrease in mammography screening examinations by 51–96% during the pandemic along with a decrease in the number of newly diagnosed breast cancers

(Filipe et al. 2020). Temporary suspension of breast cancer screening programs and fears of contracting the virus in healthcare settings prompted some individuals to postpone or forgo routine screenings altogether. These disruptions have resulted in a decrease in the number of mammograms performed and consequently, fewer cases of breast cancer being detected at early, more treatable stages. As a result, there is concern that delayed diagnoses may lead to more advanced cancer cases and potentially poorer outcomes for affected individuals.

9.2.2 Elective Surgeries

The COVID-19 pandemic has also had a significant impact on elective surgeries and procedures which were postponed prioritizing diagnosis, treatment, and management of COVID-19 patients. These non-urgent procedures, ranging from joint replacements to cosmetic surgeries, have been postponed or canceled as healthcare systems globally grapple with the challenges posed by the pandemic. The postponement of elective surgeries was a necessary measure to conserve healthcare resources and personnel for the surge in COVID-19 cases. Hospitals needed to redirect intensive care unit (ICU) beds, ventilators, and medical staff to manage the influx of critically ill COVID-19 patients. Consequently, non-urgent procedures were deprioritized to ensure that essential resources were available for pandemic response efforts. A study from a Finnish public hospital reported increase in monthly waiting time for elective surgeries ranging from 7 to 34% for the period between May till November 2020. The increase in waiting time for elective surgeries included surgeries in gastrointestinal (8% increase in waiting time) and genitourinary diseases (19% increase in waiting time). Meanwhile for cancer-related elective surgeries, the waiting time was reported to have increased between 18 to 28% during first year of the COVID-19 pandemic (Uimonen et al. 2021). These postponed surgeries could lead to deterioration of patient's condition diseases and increased mortality due to delayed treatment. In addition to that, the backlog of cases will further strain healthcare systems, escalating resource depletion and exacerbating backlog challenges. Operating room inefficiencies increase, reducing overall surgical capacity and compromising patient access to timely care. This will also increase workload for healthcare workers who are already exhausted with extended services provided during the COVID-19 pandemic and this may have negative impact not only on the physical well-being but also their mental health. The increase in demand for surgical care post-pandemic increases the risk for physician burnout that may affect surgical outcomes as well (Mehta et al. 2022).

9.2.3 Vaccination

Another important area of healthcare services that was deeply affected by the COVID-19 pandemic was disruption of routine vaccination programs to address vaccine-preventable diseases. Globally, primary healthcare facilities were responsible mainly for routine vaccination programs for children in line with driving the goal toward universal health coverage (Jittapunkul 2000). Lockdowns, travel restrictions, and overwhelmed healthcare systems have led to the disruption of routine immunization services. This also has resulted in reduction in vaccination coverage in the population and herd immunity against certain vaccine-preventable diseases to be affected increasing the risk for outbreak of diseases such as measles, pertussis, and rotavirus gastroenteritis. It has been reported that the measles, mumps and rubella (MMR) vaccine uptake in Europe has shown a decline in recent years with reported outbreaks of measles due to poor vaccination coverage (Hungerford and Cunliffe 2020). Disruption of vaccination programs has also resulted in sporadic outbreaks of vaccine-preventable diseases being reported in other countries such as measles in Cambodia, Nepal, Pakistan, Bangladesh, Iraq, Nigeria; cholera in Cameroon, Yemen, Bangladesh; diphtheria in Pakistan, Bangladesh, and Nepal, and the situation may be exaggerated if catch up vaccination efforts are not enhanced (Gavi, the Vaccine Alliance n.d.; UNICEF n.d.). The decline in vaccination coverage of measles vaccination is likely to result in increased mortality and morbidity due to measles. Similar observations were reported in South-East Asia region, where nearly 3 million fewer children received Diphtheria-Tetanus-Pertussis (DTP3) vaccines between January and June 2020 than in the corresponding period in 2019 (Bahl et al. 2021). To mitigate the impact of disrupted vaccination services during the COVID-19 pandemic, it is essential to prioritize and intensify immunization programs, implement strategies to catch up on missed vaccinations, and ensure access to vaccines for all populations. By safeguarding vaccination efforts, we can protect public health and prevent the resurgence of vaccine-preventable diseases.

9.2.4 Mental Health

One of the primary effects of the pandemic on mental health care has been the disruption of treatment continuity. Lockdowns, social distancing measures, and overwhelmed healthcare systems have limited access to mental health services, including therapy sessions, support groups, and psychiatric appointments. For individuals with pre-existing mental health conditions, this interruption in care can lead to worsening symptoms, increased distress, and heightened risk of relapse. Furthermore, the economic repercussions of the pandemic, such as job loss, financial strain, and uncertainty, have contributed to heightened stress, anxiety, and depression in many individuals. Social isolation, another consequence of pandemic-related restrictions, has exacerbated feelings of loneliness and disconnectedness, particularly among vulnerable populations such as the elderly and those living alone (Yeung et al. 2020). In addition to that, a study conducted in United States reported that the fear of increased risk for morbidity and mortality due to COVID-19 especially among people with chronic conditions are subject to higher stress levels (Umucu and Lee 2020).

Moreover, the pandemic has brought to light disparities in access to mental healthcare services, with vulnerable and marginalized communities such as those with poor socioeconomic conditions, homeless, migrant workers, asylum seekers, and refugees to be disproportionately affected. Structural inequalities, including socioeconomic status, race, ethnicity, and access to healthcare, have influenced the differential impact of the pandemic on mental health outcomes. The communities that have already been facing systemic barriers to accessing healthcare services have experienced heightened challenges in accessing mental health resources and support services during the pandemic. This resulted in worsening of existing mental health disorders and higher rates of psychoses among the vulnerable communities (Aragona et al. 2020).

In addition to disrupting care for existing mental health patients, the pandemic has also impacted the screening and early detection of mental health diseases. Routine mental health screenings, particularly in schools, workplaces, and primary care settings, have been disrupted or deprioritized, leading to delays in identifying individuals at risk of mental health disorders (Williams et al. 2020). School closures have interrupted the provision of mental health services for children and adolescents, including counseling, therapy, and behavioral interventions. The transition to remote learning has further compounded mental health challenges for students, contributing to feelings of isolation, anxiety, and academic stress (Gould et al. 2022). Similarly, workplace disruptions and remote work arrangements have limited opportunities for employers to identify and support employees experiencing mental health difficulties. The lack of in-person interactions and informal check-ins may prevent early detection of mental health concerns among workers through workplace support programs such as Employee Assistance Program (EAP), potentially exacerbating issues such as burnout, anxiety, and depression (Brooks and Ling 2020).

9.3 Burden of COVID-19 on Environmental Systems

9.3.1 Air Pollution

Based on the WHO statistics, 4.6 million deaths are attributed to respiratory illnesses like pneumonia, chronic obstructive pulmonary disease, and asthma directly linked to poor air quality (Cohen et al. 2017). The poor air quality is mainly associated with the active anthropogenic, industrial, and transportation activities before the spread of the COVID-19 pandemic. Studies conducted at the onset of the pandemic indicates that air pollution plays a crucial role in exacerbating the effects of COVID-19 (Faruk et al. 2022). A study encompassing nine Asian urban areas revealed a strong association between elevated $PM_{2.5}$ concentrations and the number of COVID-19 fatalities (Gupta et al. 2021). Similar findings were also documented in Malaysia, where $PM_{2.5}$ identified as the primary factor associated with the overall number of COVID-19 cases in Kuala Lumpur and Selangor (Jalaludin et al. 2023).

However, many industrial and transportation sectors were closed during the pandemic due to lockdowns, resulting in lesser emanation of greenhouse gas emissions. The world annual CO_2 emissions from fossil fuels and industries was reported to decrease from 37.04 to 35.01 billion tons during COVID-19 between 2019 and 2020 (Global Carbon Budget 2023). In a multi-country study of various climate using satellite data and a network of over 10,000 air quality stations, Venter et al. (2020) observed a significant decline in population-weighted ground-level NO_2 (60% with 95% CI: 48–72%) and $PM_{2.5}$ (31% with 95% CI: 17–45%) and a marginal increase in O_3 (4% with 95% CI: -2–10%) during the pandemic period. In China, greenhouse gas emissions decreased by approximately 50% due to the shutdown of industrial activities and transportation (Caine 2020). In another populated country, India, the variation in air quality data across metropolitan cities revealed that the concentrations of PM_{10}, $PM_{2.5}$, NO_2, and SO_2 decreased by 55, 49, 60, and 19% in Delhi, and by 44, 37, 78, and 39% in Mumbai during the post-lockdown phase (Kumari and Toshniwal 2022). In the United Kingdom, Jephcote et al. (2021) reported 38.3% reductions in NO_2 and 16.5% in $PM_{2.5}$ concentrations, which were attributed to 69% reduction in traffic activities in the region. Similarly, the air pollution was reduced up to 50% in 2020 when compared with 2019 in New York due to COVID-19-induced lockdown (Singh and Mishra 2021). There were notable decreases in PM_{10}, $PM_{2.5}$, NO_2, and CO levels in Malaysia during the movement control order in the Klang Valley region compared to the corresponding periods in 2019 and 2018 as a result of lesser vehicular emissions (Latif et al. 2021).

9.3.2 Water Pollution

The COVID-19 lockdowns have generally led to improvements in water bodies in various ways. The decrease in air pollutants in the atmosphere during lockdown has led to a reduction in the formation of acid rain, which occurs when NO_2 combines with other substances during precipitation, ultimately enhancing the quality of freshwater worldwide (Zhang et al. 2020a). Ceased industrial operations have resulted in lower levels of pollutants entering water systems (Khan et al. 2021). For instance, there was a noticeable enhancement in the water quality of the Ganga River in India during the lockdown period, characterized by higher levels of dissolved oxygen and lower levels of biological oxygen demand, fecal coliform, and nitrate concentration, especially around industrial hubs and urban regions (Loh et al. 2021). Besides, the lockdown has led to a notable improvement in the condition of rivers and water bodies, leading to the resurgence of various aquatic species in their natural habitats. The decrease in water pollution worldwide due to the halt in marine activities and reduced tourism has created a more favorable environment for both terrestrial and marine creatures to flourish (Mallik et al. 2022).

Nevertheless, improper discarding of sanitary consumables and waste produced from medical activities and personal protection has resulted in substantial harm to the aquatic ecosystem, particularly in the coastal regions of Hong Kong and Canada (Cheval et al. 2020). In addition, aggressive disinfection using various chemicals in commercial spaces, facilities, and streets, coupled with panic conditions, anxiety and unconsciousness, caused the water bodies, soil, and air to be polluted by these disinfectants, forming detrimental and mutagenic secondary products (Bhat et al. 2021). For instance, the application of more than two thousand tons of disinfectants in Wuhan City by China amplified the risk of contaminating aquatic ecosystems and drinking water sources through runoff and secondary drainage as chlorine disinfectants can harm marine life by damaging cell walls, oxidizing proteins, and forming harmful compounds with other materials (Zhang et al. 2020b).

9.3.3 Waste Management

During the COVID-19 pandemic, there was a notable reduction in the generation of municipal solid waste across the globe. For example, Jribi et al. (2020) reported that 85% of respondents of their survey adopted strategies to prevent food waste such as saving, storing, and eating leftovers during the pandemic. In Morocco, North Africa, the cities of Khenifra (2572 tons vs. 2456 tons) and Tighassaline (136 tons vs. 126 tons) saw reductions in municipal solid waste in March 2020 compared to March 2019 (Ouhsine et al. 2020). Although this issue is significant, the pandemic has also placed significant strain on the waste infrastructure of municipalities due to substantial shifts in both quantity and quality of clinical waste that overwhelmed current waste management processes (Jayasinghe et al. 2023). The massive production and improper disposal of facemasks, particularly made from non-biodegradable synthetic polymers derived from petrochemicals, lead to the waste management issues across the world that induce severe ecological and health. The synthesis processes of the polymers from the petrochemical by-products were also not eco-friendly, which releases huge greenhouse gases into the atmosphere. In light of these issues, various research has been conducted to explore the use of biodegradable masks made of biopolymers, which are environment-friendly, zero-hazardous, long-lasting, lightweight, comfortable, and affordable (Pandit et al. 2021). Similarly, Malaysia has documented a 27% rise in the production of clinical waste such as personal protective equipment, primarily linked to COVID-19 (Agamuthu and Barasarathi 2021). Furthermore, protective equipment has become a significant source of microplastic pollution (particle size < 5 mm), which is now ubiquitous in water, soil, air, and even within our bodies (Yang et al. 2022). Despite the prevalence of online food retailing in developed nations, inadequate such facilities coupled with cold storage and supply chain infrastructure at the farmer level in developing countries have led to increased food waste during the pandemic (Zia et al. 2021).

9.3.4 Energy Sector

The slower economic growth associated with COVID-19 has led to a decrease in energy demand and consumption. In 2020, the global energy demand is estimated to decline by 6% compared to 2019, a drop seven times greater than that experienced during the 2009 financial crisis (Aktar et al. 2021). In April 2020, the total mean electricity generation from 16 European countries decreased by 9% (25 GW) compared to the mean value from 2015 to 2019 (Werth et al. 2021). This was attributed to the decrease in fossil energy generation by 28% (24 GW), nuclear energy by 14% (11 GW), while renewable energy increased by 15% (15 GW). In the United States, the pandemic resulted in a 50% reduction in aviation fuel demand and a 30% reduction in gas demand (Gillingham et al. 2020). Nevertheless, the COVID-19 pandemic is reportedly impeding the shift to renewable energy by causing interruptions in clean energy manufacturing facilities, distribution channels, and enterprises. Furthermore, the decrease in energy demand has had adverse effects on renewable energy generation and carbon trading prices (Armeanu et al. 2022). In a comprehensive review of the impact of COVID-19 on renewable energy sector in Malaysia, Hui et al. (2022) revealed that solar energy initiatives faced delays due to the suspension of solar panel imports. In addition, the productivity of biomass suffered as employees were sent home in an effort to control the spread of COVID-19. Similarly, the construction of hydroelectric dams was affected, with a single case potentially leading to a cluster and causing further delays in the project.

9.3.5 Wildlife Conservation

The lockdowns implemented due to COVID-19 resulted in a number of unforeseen benefits for wildlife conservation. There was a notable increase in wildlife sightings in urban environments during the lockdown period. In Chile, where the urban carnivore ecology is largely unknown, Silva-Rodríguez et al. (2021) recorded four native carnivores, including the vulnerable güiña (*Leopardus guigna*) and the endangered southern river otter (*Lontra provocax*) during the pandemic period. According to Rutz et al. (2020), the lockdown helped mitigate the harmful impacts of human activities on animals' mobility, offering valuable insights on enhancing cohabitation between humans and wildlife, particularly in designing a more sustainable and wildlife-friendly environment. On another account, the significant reduction in human activity during the COVID-19 pandemic lockdown led to a months-long decrease of up to 50% in seismic noise (Lecocq et al. 2020). This reduction in noise pollution benefited marine lives that rely on sound for navigation, hunting, and communication such as the North Atlantic Right whales (*Eubalaena glacialis*), which exhibited decreased levels of stress-related fecal hormone metabolites (glucocorticoids), aiding in their population recovery (Loh et al. 2021). In addition, zooplankton and fish species sensitive to noise experienced improved spawning and flourishing due to the quieter underwater environment.

9.3.6 Food Security

The COVID-19 pandemic has directly resulted in heightened concerns regarding food insecurity and the disruption of food supply chains. It exacerbated the existing food shortages and emergencies, expanding food insecurity and malnutrition to billions of people worldwide. Most least developed countries reduced food trade with developed countries due to movement restrictions, lockdowns, and travel bans, which significantly minimized economic activities and disrupted both supply and demand (Aday and Aday 2020). Rising food prices during the pandemic time disproportionately affected people in low- and middle-income countries, which has also made them to cut down on the number of meals and food items per day, posing a significant challenge to agricultural sustainability. Nonetheless, COVID-19 has also yielded several positive effects on food resources. Much like the scenario observed during World War II, the unintended lockdown measures due to COVID-19 facilitated the recovery of fish stocks and their habitats, including coral reefs, mangroves, and riverbeds (Mallik et al. 2022).

9.4 Conclusion

The global healthcare and environmental systems have been significantly impacted by the COVID-19 pandemic, resulting in a mix of positive and negative effects on environmental services, pollution levels, waste management, energy sector, wildlife conservation, food security, non-emergency medical procedures, and the health outcomes of chronic disease patients due to reduced access to regular care. From a health standpoint, elective procedures were postponed, immunization campaigns were suspended, and mental health support services suffered significant harm. Considering these factors, the pandemic underscored the necessity for innovative health screening and delivery methods, yet disparities in technology access hinder widespread telehealth adoption, especially in poorer communities. Addressing these digital disparities and expanding telehealth infrastructure are crucial steps toward ensuring equitable healthcare access moving forward. From an environmental perspective, the pandemic has shown the mankind both the vulnerabilities and opportunities for environmental management and conservation, which has diverse implications on socioecological systems. Moving ahead, it is critical to incorporate resilient and sustainable practices into global recovery efforts to promote a healthier planet and more robust societies.

References

Aday S, Aday MS (2020) Impact of COVID-19 on the food supply chain. Food Qual Saf 4(4):167–180. https://doi.org/10.1093/fqsafe/fyaa024

Agamuthu P, Barasarathi J (2021) Clinical waste management under COVID-19 scenario in Malaysia. Waste Manag Res 39(1_suppl):18–26. https://doi.org/10.1177/0734242X20959701

Aktar MA, Alam MM, Al-Amin AQ (2021) Global economic crisis, energy use, CO_2 emissions, and policy roadmap amid COVID-19. Sustain Prod Consumption 26:770–781. https://doi.org/10.1016/j.spc.2020.12.029

Anthony Jnr B (2020) Use of telemedicine and virtual care for remote treatment in response to COVID-19 pandemic. J Med Syst 44(7). https://doi.org/10.1007/s10916-020-01596-5

Aragona M, Barbato A, Cavani A, Costanzo G, Mirisola C (2020) Negative impacts of COVID-19 lockdown on mental health service access and follow-up adherence for immigrants and individuals in socio-economic difficulties. Public Health 186:52–56. https://doi.org/10.1016/j.puhe.2020.06.055

Armeanu DS, Gherghina SC, Andrei JV, Joldes CC (2022) Modeling the impact of the COVID-19 outbreak on environment, health sector and energy market. Sustain Dev 30(5):1387–1416. https://doi.org/10.1002/sd.2299

Bahl S, Khanal S, Sharifuzzaman M, Liyanage J (2021) Impact of the COVID-19 pandemic on immunization and surveillance of vaccine-preventable diseases in the WHO South-East Asia Region. WHO South-East Asia J Publ Health 10(3):33. https://doi.org/10.4103/2224-3151.309870

Bhat SA, Sher F, Kumar R, Karahmet E, Haq SAU, Zafar A, Lima EC (2021) Environmental and health impacts of spraying COVID-19 disinfectants with associated challenges. Environ Sci Pollut Res. https://doi.org/10.1007/s11356-021-16575-7

Brooks CD, Ling J (2020) "Are we doing enough": an evaluation of the utilization of employee assistance programs to support the mental health needs of employees during the COVID-19 pandemic. J Insur Regul 1–35. https://doi.org/10.52227/23478.2020

Caine P (2020) Environmental impact of COVID-19 lockdowns seen from space. Sci Nat 2 April 2020. https://news.wttw.com/2020/04/02/environmental-impact-covid-19-lockdowns-seen-space

Cheval S, Mihai Adamescu C, Georgiadis T, Herrnegger M, Piticar A, Legates DR (2020) Observed and potential impacts of the COVID-19 pandemic on the environment. Int J Environ Res Public Health 17(11):4140. https://doi.org/10.3390/ijerph17114140

Cohen AJ, Brauer M, Burnett R, Anderson HR, Frostad J, Estep K et al (2017) Estimates and 25-year trends of the global burden of disease attributable to ambient air pollution: an analysis of data from the Global Burden of diseases study 2015. Lancet 389(10082):1907–1918. https://doi.org/10.1016/S0140-6736(17)30505-6

Faruk MO, Rahman MS, Jannat SN, Arafat Y, Islam K, Akhter S (2022) A review of the impact of environmental factors and pollutants on covid-19 transmission. Aerobiologia 38(3):277–286. https://link.springer.com/article/10.1007/s10453-022-09748-5

Filipe MD, van Deukeren D, Kip M, Doeksen A, Pronk A, Verheijen PM, Heikens JT, Witkamp AJ, Richir MC (2020) Effect of the COVID-19 pandemic on surgical breast cancer care in the Netherlands: a multicenter retrospective cohort study. Clin Breast Cancer 20(6):454–461. https://doi.org/10.1016/j.clbc.2020.08.002

Formenti B, Gregori N, Crosato V, Marchese V, Tomasoni LR, Castelli F (2022) The impact of COVID-19 on communicable and non-communicable diseases in Africa: a narrative review. Infezioni in Medicina 30(1):30–40. https://doi.org/10.53854/liim-3001-4

Gavi The Vaccine Alliance. (n.d.). COVID-19: massive impact on lower-income countries threatens more disease outbreaks. Retrieved 20 March 2024, from https://www.gavi.org/news/media-room/covid-19-massive-impact-lower-income-countries-threatens-more-disease-outbreaks

Georgiades F, Summers DM, Butler AJ, Russell NKI, Clatworthy MR, Torpey N (2021) Renal transplantation during the SARS-CoV-2 pandemic in the UK: experience from a large-volume center. Clin Transplant 35(1):1–10. https://doi.org/10.1111/ctr.14150

Gillingham KT, Knittel CR, Li J, Ovaere M, Reguant M (2020) The short-run and long-run effects of Covid-19 on energy and the environment. Joule 4(7):1337–1341. https://doi.org/10.1016/j.joule.2020.06.010

Global Carbon Budget (2023) Annual CO_2 emissions. Restrieved from https://ourworldindata.org/grapher/annual-co2-emissions-per-country?country=~OWID_WRL

Gould JB, Walter HJ, Bromberg J, Correa ET, Hatoun J, Vernacchio L (2022) Impact of the coronavirus disease 2019 pandemic on mental health visits in pediatric primary care. Pediatrics 150(6). https://doi.org/10.1542/peds.2022-057176

Gupta A, Bherwani H, Gautam S, Anjum S, Musugu K, Kumar N et al (2021) Air pollution aggravating COVID-19 lethality? Exploration in Asian cities using statistical models. Environ Develop Sustain 23:6408–6417. https://link.springer.com/article/10.1007/s10668-020-00878-9

Hui LC, Meng GTK, Jue HL, Qian EVW, Hassan A, Afrouzi HN, Mehranzamir K (2022) Impact of Covid-19 on renewable energy sector and lessons learned: a case study on Malaysia. Future Energy 1(3):44–50. https://doi.org/10.55670/fpll.fuen.1.3.5

Hungerford D, Cunliffe NA (2020) Coronavirus disease (COVID-19)—impact on vaccine preventable diseases. In: Eurosurveillance. European Centre for Disease Prevention and Control (ECDC), vol 25, Issue 18. https://doi.org/10.2807/1560-7917.ES.2020.25.18.2000756

Hussein NR, Zana ZS, Ibrahim N, Musa DH, Naqid IA (2020) The impact of COVID-19 pandemic on the care of patients with kidney diseases in Duhok City, Kurdistan Region of Iraq. Diab Metab Syndr 14(6):1551–1553. https://doi.org/10.1016/j.dsx.2020.08.013

Jalaludin J, Wan Mansor WN, Abidin NA, Suhaimi NF, Chao HR (2023) The impact of air quality and meteorology on COVID-19 cases at Kuala Lumpur and Selangor, Malaysia and prediction using machine learning. Atmosphere 14(6):973. https://doi.org/10.3390/atmos14060973

Jayasinghe PA, Jalilzadeh H, Hettiaratchi P (2023) The impact of COVID-19 on waste infrastructure: lessons learned and opportunities for a sustainable future. Int J Environ Res Public Health 20(5):4310. https://doi.org/10.3390/ijerph20054310

Jephcote C, Hansell AL, Adams K, Gulliver J (2021) Changes in air quality during COVID-19 'lockdown' in the United Kingdom. Environ Pollut 272:116011. https://doi.org/10.1016/j.envpol.2020.116011

Jittapunkul S (2000) Primary health care and health care reform. Chulalongkorn Med J 44(2):69–73. https://doi.org/10.58837/chula.cmj.44.2.1

Jribi S, Ben Ismail H, Doggui D, Debbabi H (2020) COVID-19 virus outbreak lockdown: What impacts on household food wastage?. Environ Develop Sustain 22(5):3939–3955. https://link.springer.com/article/10.1007/s10668-020-00740-y

Kendzerska T, Zhu DT, Gershon AS, Edwards JD, Peixoto C, Robillard R, Kendall CE (2021) The effects of the health system response to the covid-19 pandemic on chronic disease management: a narrative review. Risk Manag Healthc Policy 14:575–584. https://doi.org/10.2147/RMHP.S293471

Khan I, Shah D, Shah SS (2021) COVID-19 pandemic and its positive impacts on environment: an updated review. Int J Environ Sci Technol 18:521–530. https://link.springer.com/article/10.1007/s13762-020-03021-3

Kumari P, Toshniwal D (2022) Impact of lockdown measures during COVID-19 on air quality—a case study of India. Int J Environ Health Res 32(3):503–510. https://doi.org/10.1080/09603123.2020.1778646

Latif MT, Dominick D, Hawari NSSL, Mohtar AAA, Othman M (2021) The concentration of major air pollutants during the movement control order due to the COVID-19 pandemic in the Klang Valley, Malaysia. Sustain Cities Soc 66:102660. https://doi.org/10.1016/j.scs.2020.102660

Lecocq T, Hicks SP, Van Noten K, Van Wijk K, Koelemeijer P, De Plaen RS et al (2020) Global quieting of high-frequency seismic noise due to COVID-19 pandemic lockdown measures. Science 369(6509):1338–1343. https://doi.org/10.1126/science.abd2438

Loh HC, Looi I, Ch'ng ASH, Goh KW, Ming LC, Ang KH (2021) Positive global environmental impacts of the COVID-19 pandemic lockdown: a review. GeoJournal 1–13.https://doi.org/10.1007/s10708-021-10475-6

Luciani S, Caixeta R, Chavez C, Ondarsuhu D, Hennis A (2023) What is the NCD service capacity and disruptions due to COVID-19? Results from the WHO non-communicable disease country capacity survey in the Americas region. BMJ Open 13(3):1–8. https://doi.org/10.1136/bmjopen-2022-070085

Mak IL, Wan EYF, Wong TKT, Lee WWJ, Chan EWY, Choi EPH, Chui CSL, Ip MSM, Lau WCS, Lau KK, Lee SF, Wong ICK, Yu EYT, Lam CLK (2022) The spill-over impact of the novel coronavirus-19 pandemic on medical care and disease outcomes in non-communicable diseases: a narrative review. Public Health Rev 43(April):1–13. https://doi.org/10.3389/phrs.2022.1604121

Mallik A, Chakraborty P, Bhushan S, Nayak BB (2022) Impact of COVID-19 lockdown on aquatic environment and fishing community: boon or bane? Mar Policy 141:105088. https://doi.org/10.1016/j.marpol.2022.105088

Mehta A, Awuah WA, Ng JC, Kundu M, Yarlagadda R, Sen M, Nansubuga EP, Abdul-Rahman T, Hasan MM (2022) Elective surgeries during and after the COVID-19 pandemic: case burden and physician shortage concerns. Ann Med Surg 81(August):104395. https://doi.org/10.1016/j.amsu.2022.104395

Moynihan R, Sanders S, Michaleff ZA, Scott AM, Clark J, To EJ, Jones M, Kitchener E, Fox M, Johansson M, Lang E, Duggan A, Scott I, Albarqouni L (2021) Impact of COVID-19 pandemic on utilisation of healthcare services: a systematic review. BMJ Open 11(3). https://doi.org/10.1136/bmjopen-2020-045343

Ouhsine O, Ouigmane A, Layati E, Aba B, Isaifan RJ, Berkani M (2020) Impact of COVID-19 on the qualitative and quantitative aspect of household solid waste. Glob J Environ Sci Manag 6(Special issue), 41–52. https://doi.org/10.22034/GJESM.2019.06.SI.05

Pan XF, Yang J, Wen Y, Li N, Chen S, Pan A (2021) Non-communicable diseases during the COVID-19 pandemic and beyond. Engineering 7(7):899–902. https://doi.org/10.1016/j.eng.2021.02.013

Pandit P, Maity S, Singha K, Uzun M, Shekh M, Ahmed S (2021) Potential biodegradable face mask to counter environmental impact of Covid-19. Cleaner Eng Technol 4:100218. https://doi.org/10.1016/j.clet.2021.100218

Ricciardiello L, Ferrari C, Cameletti M, Gaianill F, Buttitta F, Bazzoli F, Luigi de'Angelis G, Malesci A, Laghi L (2021) Impact of SARS-CoV-2 pandemic on colorectal cancer screening delay: effect on stage shift and increased mortality. Clin Gastroenterol Hepatol 19(7):1410–1417.e9.https://doi.org/10.1016/j.cgh.2020.09.008

Rutz C, Loretto MC, Bates AE, Davidson SC, Duarte CM, Jetz W et al (2020) COVID-19 lockdown allows researchers to quantify the effects of human activity on wildlife. Nat Ecol Evol 4(9):1156–1159. https://doi.org/10.1038/s41559-020-1237-z

Schliemann D, Ramanathan K, Ibrahim Tamin NSB, O'Neill C, Cardwell CR, Ismail R, Nanu P, Mohd Ghazali AN, Bin Kee F, Su TT, Donnelly M (2022) Implementation of a colorectal cancer screening intervention in Malaysia (CRC-SIM) in the context of a pandemic: study protocol. BMJ Open 12(9):1–11.https://doi.org/10.1136/bmjopen-2021-058420

Silva-Rodríguez EA, Gálvez N, Swan GJ, Cusack JJ, Moreira-Arce D (2021) Urban wildlife in times of COVID-19: what can we infer from novel carnivore records in urban areas? Sci Total Environ 765:142713. https://doi.org/10.1016/j.scitotenv.2020.142713

Singh V, Mishra V (2021) Environmental impacts of coronavirus disease 2019 (COVID-19). Bioresour Technol Rep 15:100744. https://doi.org/10.1016/j.biteb.2021.100744

Uimonen M, Kuitunen I, Paloneva J, Launonen AP, Ponkilainen V, Mattila VM (2021) The impact of the COVID-19 pandemic on waiting times for elective surgery patients: a multicenter study. PLoS ONE 16(7 July):1–8. https://doi.org/10.1371/journal.pone.0253875
Umucu E, Lee B (2020) Examining the impact of COVID-19 on stress and coping strategies in individuals with disabilities and chronic conditions. Rehabil Psychol 65(3):193–198. https://doi.org/10.1037/rep0000328
Venter ZS, Aunan K, Chowdhury S, Lelieveld J (2020) COVID-19 lockdowns cause global air pollution declines. Proc Natl Acad Sci 117(32):18984–18990. https://doi.org/10.1073/pnas.2006853117
Werth A, Gravino P, Prevedello G (2021) Impact analysis of COVID-19 responses on energy grid dynamics in Europe. Appl Energy 281:116045. https://doi.org/10.1016/j.apenergy.2020.116045
Williams R, Jenkins DA, Ashcroft DM, Brown B, Campbell S, Carr MJ, Cheraghi-sohi S, Kapur N, Thomas O, Webb RT, Peek N (2020) Diagnosis of physical and mental health conditions in primary care during the COVID-19 pandemic: a retrospective cohort study. Lancet Publ Health 5(10):e543–e550. https://doi.org/10.1016/S2468-2667(20)30201-2
Yang S, Cheng Y, Liu T, Huang S, Yin L, Pu Y, Liang G (2022) Impact of waste of COVID-19 protective equipment on the environment, animals and human health: a review. Environ Chem Lett 20(5):2951–2970. https://doi.org/10.1007/s10311-022-01462-5
Yeung S, Wong S, Zhang D, Wing R, Sit S, Hon B, Yip K, Chung RY, Ka C, Wong M, Cheong D, Chan C, Sun W, Kwok KO (2020) Impact of COVID-19 on loneliness, mental health, and health service utilisation: a prospective cohort study of older adults with multimorbidity in primary care. Br J General Pract 817–824
Zhang R, Zhang Y, Lin H, Feng X, Fu TM, Wang Y (2020a) NOx emission reduction and recovery during COVID-19 in East China. Atmosphere 11(4):433
Zhang H, Tang W, Chen Y, Yin W (2020b) Disinfection threatens aquatic ecosystems. Science 368(6487):146–147. https://www.science.org/doi/full/10.1126/science.abb8905
Zia M, Ahmed S, Kumar A (2021) Environmental impact of food, fruit and vegetable waste during COVID-19 pandemic: a review. J Clin Diagn Res 15(6):LC01–LC04. https://doi.org/10.7860/JCDR/2021/45721.14961

Chapter 10
Recent Waves, Variants and Trends for COVID-19 in India and Canada: The Way Forward

Hem H. Dholakia and Malav D. Jhala

Abstract In Canada and India, indicators show that COVID-19 activity has been stable or decreasing. Vaccination efforts, awareness creation, quarantine measures and travel advisories in both countries have contributed to this. Over time, new variants of COVID-19 have emerged in both countries. Currently, the BA.2.86 variant is dominant in Canada while BA.2.75 (Omicron) has remained the dominant variant in India. It is expected that new variants may emerge in future and surveillance efforts need to continue. Elderly or people with pre-existing diseases constitute the majority of hospital admissions suggesting that containment efforts should be directed toward vulnerable populations.

Keywords India · Canada · COVID-19 variants · Vaccination · Medical surveillance

10.1 Introduction

December 2019 marked an inflection point in the history of global health. It was the time when the first cases of COVID-19 were detected in Wuhan, China (CDC 2023). Over subsequent months, the disease spread to all parts of the world (Spiteri et al. 2020). Globally, the pandemic has claimed over 7 million lives and infected more than 775 million people (as on 16 June 2024), and remains the deadliest outbreak the world has seen in recent decades (WHO 2024).

And while it may appear that COVID-19 has somewhat loosened its grip (at least the acute phase seems to have passed), there is enough evidence to show that the virus is still evolving. New infections are on the rise. Scientists are contemplating whether COVID-19 represents a single rare event in history or whether it embodies

H. H. Dholakia (✉)
Independent Environment and Health Expert, Mumbai, India
e-mail: drdholakia@gmail.com

M. D. Jhala
Consulting Cardiologist, Mumbai, India

R. Akhtar (ed.), *COVID-19: Impacts on Health and Health Systems*,
https://doi.org/10.1007/978-3-031-90255-0_10

the new changing reality of public health (i.e. more frequent and deadlier pandemics). Whereas it may be too early to answer this question, it remains instructive to understand the current evolution of the virus. This chapter, therefore, looks at some of the trends and variants that have had a dominant presence in Canada and India.

In doing so, it aims to compare some of the differences and understand the commonalities across two different countries and health systems. The rest of the chapter is organized as follows: first it provides a brief overview of trends across both countries. Second, it discusses the emergence of new variants. Third it covers medical surveillance and finally provides insights for the way forward.

10.2 COVID-19 Trends

There has been great heterogeneity in the observed trends of COVID-19 across different countries. Yet, a common pattern has been the exponential increase in cases within a very short of time. This sudden spike has challenged and sometimes overwhelmed health systems globally. Figure 10.1 shows the total number of cases recorded in Canada from January 2020 to February 2023. The figure shows a sharp increase in the number of cases in October 2020 and November 2021 following which the rate of growth has been slower. Similarly, Fig. 10.2 shows the total number of cases recorded in India. There were exponential increases around July 2020 and March 2021.

In addition to the number of cases, another important metric was the number of deaths. Figure 10.3 compares the number of deaths per 100,000 people across

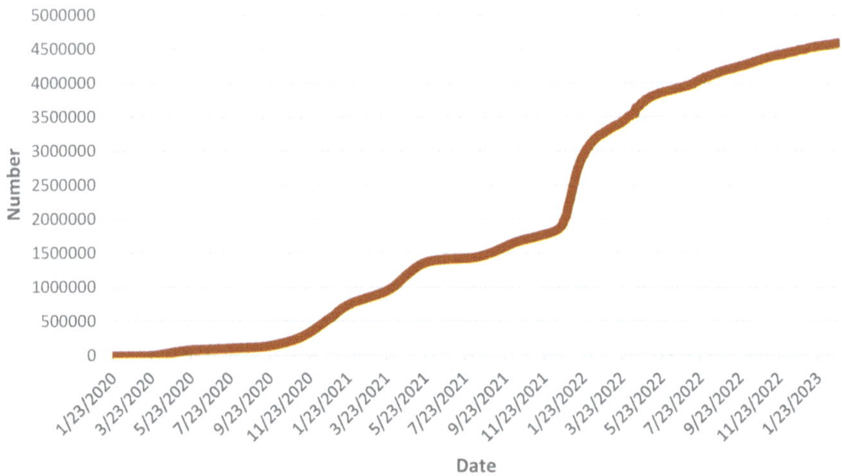

Fig. 10.1 Total cases of COVID-19 in Canada from Jan 2020 to Feb 2023. *Source* Data obtained from the Public Health Agency of Canada

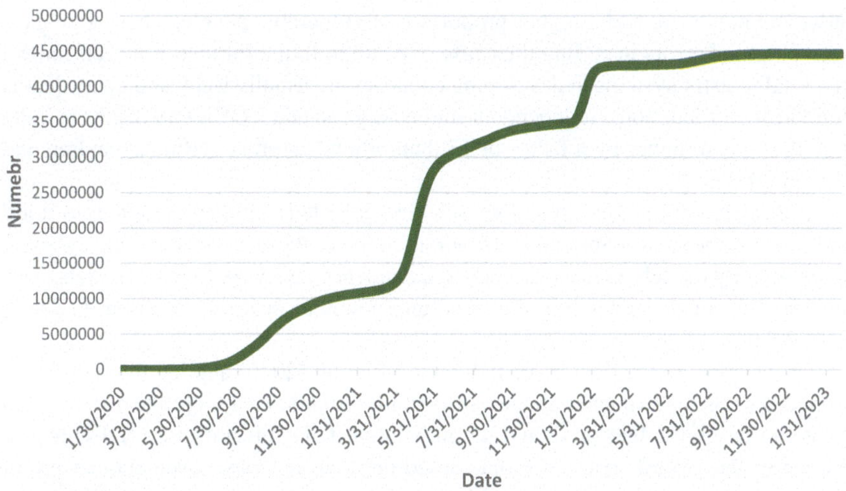

Fig. 10.2 Total cases of COVID-19 in India from Jan 2020 to Feb 2023. *Source* Data obtained from the Public Health Agency of Canada

Canada and India. The data are normalized by population to be comparable across the two countries.

It is interesting to note that India recorded fewer number of deaths per 100,000 people when compared with Canada. However, this does not necessarily imply that one country fared better at managing the disease than the other. There may be several reasons for these differences. Canada has a population with a higher median age than

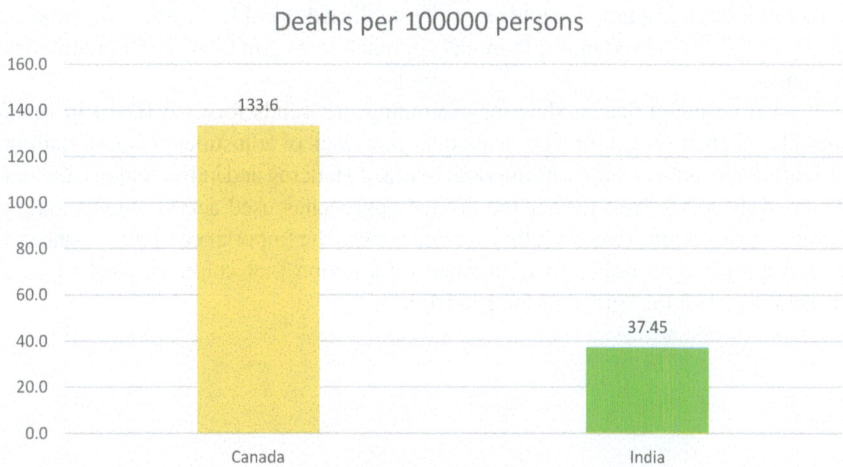

Fig. 10.3 COVID-19 deaths per 100,000 people in Canada and India (2020–2023). *Source* Data obtained from the Public Health Agency of Canada

India and may have had a higher proportion of vulnerable people. There may have been potential errors in coding the cause of death in India. Further, it is well-known that vital registry systems in developing countries are usually weak and many deaths don't get recorded. Some independent studies suggest that COVID-19 deaths in India may have been higher by a factor of 13 than official estimates (Ramachandran and Malani 2021).

As the pandemic started spreading, scientists adopted a number of models to study and forecast the spatiotemporal distribution. Some of these models include dynamic epidemiology models, Gaussian functions, incidence decays or logistic models, with the intention of understanding the trend and potential peak of the disease (Zhang et al. 2020).

For instance, in the context of India, Ghosh et al. (2020) applied three different models—logistic, exponential and susceptible-infectious-susceptible models—to create an ensemble that predicted the maximum daily infection rate (DIR) across sixteen of the largest states of India. Based on their analysis, seven states were in the severe category, four in the moderate and the remaining in the controlled category. Others used deep learning models to predict the curve of new COVID-19 cases versus the total number of tests conducted in India (Shrivastava et al. 2021). Studies such as the one by Song et al. (2021) used the published mortality data to calculate the effective reproduction number and to model wave patterns in India. Their model simulation matched the reported deaths accurately and this was then used to predict future mortality. Murugesan et al. (2020) analyzed the spatial distribution of COVID-19 using Geographical Information System (GIS) software. Using inverse weighted distance and interpolation techniques, their study found that the states of Maharashtra and Tamil Nadu were at highest risk.

For Canada, we find that similar models and approaches were adopted to identify trends as those seen for India. This is expected as model selection comes from several years of research and practice that are well established in public health. For instance, Zhang et al. (2020) used multiple-model approach to inform COVID-19 predictions in Canada.

It must be noted that models for examining the trends for COVID-19 in India have also been critiqued for significant data gaps, lack of adjustment of confounding variables such as the lockdowns imposed, social distancing and individual risk factors (Shah et al. 2020). And further the diverse approaches used across studies makes comparison difficult. However, they all underscore the importance of good data and using multiple approaches to triangulate information that could be used to plan measure to arrest the spread of the pandemic.

10.3 Medical Surveillance

Surveillance has been a well-recognized public health strategy for the systematic collection, analysis and interpretation of data for timely decision-making (Nsubuga et al. 2006). For COVID-19 surveillance has been a key tool especially for ascertaining the true incidence as well as identify the beginning of an outbreak.

The Public Health Agency of Canada set up the novel Canadian COVID-19 Outbreak Surveillance System (CCOSS) to monitor the frequency and severity of outbreaks. The CCOSS partnered with the different provinces and territories and agreed to share data around outbreak details, number of conformed cases, hospitalizations, deaths (McGill et al. 2023). This effort had several implications such as detecting outbreaks, building highly skilled and qualified surveillance teams and planning responses for the timely implementation of interventions (McGill et al. 2023).

Interestingly, both in Canada and India, syndromic surveillance using phone and internet methods have been used for collecting information from voluntary participants (Bassi et al. 2020; Lapointe-Shaw et al. 2020). For instance, Lapointe-Shaw et al. (2020), found that 44% of Canada-wide respondents had at least one symptom in the previous month during the period of March–April 2020 based on telephone web or phone-based polling. Of these, women, children, Indigenous and First Nations persons were more likely to report one symptom.

Similarly, there were several mobile health applications that were developed. For example, the study by Bassi et al. (2020) found that there were at least 50 applications designed for COVID-19 surveillance. Of these, > 60% were targeted toward the general public, 38% toward quarantined individuals and others for care providers, police and researchers (Bassi et al. 2020). The Government of India and States and Union territories had also invested in the development of some of these applications.

The use of lockdowns or mobility restrictions were also imposed across the two countries. For instance, India adopted a series of complete and partial lockdowns as both a surveillance and disease control mechanism in the early and middle stages of the pandemic. Similarly Canada too imposed restrictions such as school and business closures, regional lockdowns (that differed across provinces), social distancing and the compulsory use of masks (Chum et al. 2021).

Some studies from India found that the lockdown had the potential of being effective in areas with a higher percentage of symptomatic infections (Sardar et al. 2020). The lockdowns were also associated with a reduction in the positive trend of new case reports suggesting that the measure helped reigning in the spread (Thayer et al. 2021). However the lockdown also had unintended consequences such as reduced mobility of people, reduced access to food and education (Alvi and Gupta 2020) and increased symptoms of anxiety and depression among people (Grover et al. 2020).

Another measure was that Canada set up the Acute Care Enhanced Surveillance (ACES) system to collect data related to COVID-19 across 131 hospitals across Ontario (Papadomanolakis-Pakis et al. 2021). This approach helped correlate laboratory-confirmed hospital admissions for COVID-19 with suspected COVID-19 admissions (Papadomanolakis-Pakis et al. 2021).

As our understanding of COVID-19 has evolved, so has our ability to use novel methods to monitor potential outbreaks. Whereas initial efforts around surveillance, was focused on patients (Jahan et al. 2021), new evidence has helped uncover transmission dynamics from animal vectors. Some research indicates that animal vectors such as ferrets, cats and dogs may be permissive to transmission of COVID-19 (Opriessnig and Huang 2020). However, no significant evidence has been found in other animals such as pigs, chickens and ducks which are often found in close human proximity, especially on farms (Opriessnig and Huang 2020). Experts have advocated that although animal vectors may have a minor role to play in transmission, tracking of animal vectors is important for early identification of potential new strains of the virus that may emerge in animal reservoirs (Valencak et al. 2021).

10.4 Emergence of New Variants

Virus mutations are a well-known phenomenon. Primarily they relate to genetic information that is passed on to subsequent generations. Many researchers have looked at the underlying mechanisms of virus mutation (Sanjuán and Domingo-Calap 2016) that are often driven by multiple virus and host-dependent processes. Others have studied the mutation rates that are an estimate of genetic diversity within a population of offspring (Peck and Lauring 2018). These mutations are responsible for the emergence of new variants and are more common across RNA viruses (such as COVID-19) as compared to DNA viruses. Globally, several variants have been labeled as 'variants of concern'.

In India, the B.1.617.2 (VOC delta) and B.1.617.1 (kappa) were initially predominant variants (Cherian et al. 2021). In fact, the delta variant has been the major reason behind the morbidity and mortality seen during the second wave of COVID-19 (Alexandar et al. 2021). For instance, the city of Chennai, which was one of the most severely affected cities during the second wave reported nearly 6000 daily cases during the first three weeks of May 2021 (Hsu et al. 2022). Interestingly one study found that both vaccinated and unvaccinated individuals had the same prevalence of the delta variant (Hsu et al. 2022) which is suggestive of its highly infectious nature. After the delta variant, it was the Omicron variant that dominated the third wave of COVID-19 infections in India (Ranjan 2022). Whereas the Omicron variant resulted in many more infections, the silver lining was that the it was milder than previous variants. Studies found that the probability of hospitalizations was 15–20% less with Omicron than the Delta variant (Christie 2021). Presently, Omicron has moved to the 'de-escalated' category.

In Canada too, the Alpha, Gamma, Delta and Omicron varieties have been dominant. Trends with respect to the variants and their infectiousness and severity have been similar to those observed in India. For instance, Harrigan et al. (2024) retrospectively studied 91,000 individuals infected with Alpha (22.28%), Gamma (16.55%),

Delta (53.46%) and Omicron (7.71%) variants across Canada. They found that Delta was associated with the most severe disease in terms of hospitalizations, ICU admissions and deaths relative to other variants. Currently, many variants are being tracked across Canada. For the ease of classification, these have been grouped into lineages, i.e., variants which share a common ancestry. Presently in the last two months (July to September 2024), the KP3 variant accounts for nearly 50% of cases (Government of Canada 2024).

10.5 The Way Forward

Nearly five years have passed since the beginning of the COVID-19 pandemic. Over this time period, the virus has successfully mutated, several times, and often managed to evade host's immune defenses. It has also challenged vaccine makers to find an adequate response and often overwhelmed health systems around the world (both in developed and developing countries).

While the gravity of the suffering caused cannot be overstated, there has been a silver lining. COVID-19 has resulted in the rapid adoption of technology for surveillance as well as vaccine development. One such technology development is the Internet of Things (IoT) in healthcare and physical distance monitoring for pandemic situations (Vedaei et al. 2020). This combines a smart phone application, IoT node and a machine learning model for data analysis. The setup aims to transmit real-time risk information while reducing exposure. However, in several places such as Europe, the use of technology have brought a focus on sovereignty challenges, data protection and cyber security (Francisco de Abreu Duarte and Francesca Palmiotto Ettorre 2022). All of these issues are likely to evolve in the coming years as technology plays a more prominent role in healthcare. In addition to IoT, wastewater surveillance is being seen as instrumental in understanding virus behavior and transmission across different settings (Kumar et al. 2024).

On the diagnosis and treatment dimension, technology is assisting the identification of new biomarkers and drug targets for COVID-19. At the heart of these are advances seen across genomics, proteomics, transcriptomics and metabolomics (Guest 2023). Advances in these fields are expected to generate new cures for viral and bacterial infections in future. Some areas that are being explored also include approaches for multi-drug-resistant pathogenic bacteria (Corredor and Munoz-Gomez 2020). Other technologies such as machine learning are also playing a prominent role for the rapid classification of novel pathogens. Combined with genomics, these approaches may open a new paradigm of diagnosis and treatment in the coming years.

In addition to providing a boost to technologies, the pandemic has also pushed policymakers and health professionals to improve and upgrade the quality of health infrastructure in their respective countries. It has also emphasized the need to think about global health beyond national boundaries. An unanticipated change has been a behavioral shift to a remote working culture. This change has played a role in reducing global carbon-dioxide emissions, although some rebound was seen after restrictions were removed (Tollefson 2021).

While it appears that the worst for COVID-19 may be over, it remains important for health professionals around the world to remain vigilant. Medical surveillance will remain the cornerstone of anticipating, early detection and interventions for any potential new waves or strains that may emerge.

References

Alexandar DS, Ravisankar M, Kumar DRS, Jakkan DK (2021) A comprehensive review on covid-19 delta variant. Int J Pharmacol Clin Res 5. https://www.researchgate.net/profile/Ravisankar-Mathesan/publication/353179027_A_Comprehensive_Review_on_Covid-19_Delta_variant/links/60ec151630e8e50c01fbf87c/A-Comprehensive-Review-on-Covid-19-Delta-variant.pdf

Alvi M, Gupta M (2020) Learning in times of lockdown: how Covid-19 is affecting education and food security in India. Food Secur 12(4):793–796. https://doi.org/10.1007/s12571-020-01065-4

Bassi A, Arfin S, John O, Jha V (2020) An overview of mobile applications (apps) to support the coronavirus disease-2019 response in India. Indian J Med Res. https://doi.org/10.4103/ijmr.IJMR_1200_20

CDC (2023) CDC Museum COVID-19 Timeline [US Government]. Center for Disease Control. https://www.cdc.gov/museum/timeline/covid19.html

Cherian S, Potdar V, Jadhav S, Yadav P, Gupta N, Das M, Rakshit P, Singh S, Abraham P, Panda S, Team N (2021) SARS-CoV-2 Spike Mutations, L452R, T478K, E484Q and P681R, in the second wave of COVID-19 in Maharashtra, India. Microorganisms 9(7):1542.https://doi.org/10.3390/microorganisms9071542

Christie B (2021) Covid-19: early studies give hope omicron is milder than other variants. BMJ n3144. https://doi.org/10.1136/bmj.n3144

Chum A, Nielsen A, Bellows Z, Farrell E, Durette P-N, Banda JM, Cupchik G (2021) Changes in public response associated with various COVID-19 restrictions in Ontario, Canada: observational Infoveillance study using social media time series data. J Med Internet Res 23(8):e28716. https://doi.org/10.2196/28716

Corredor M, Munoz-Gomez A (2020) Pan-genomics of multidrug-resistant human pathogenic bacteria and their resistome. In: Pan-genomics: applications, challenges, and future prospects. Elsevier

Francisco de Abreu Duarte, Francesca Palmiotto Ettorre (2022) Sovereignty, technology and governance after Covid-19: legal challenges in a post-pandemic Europe (First edition.). Hart Publishing

Ghosh P, Ghosh R, Chakraborty B (2020) COVID-19 in India: Statewise analysis and prediction. JMIR Public Health Surveill 6:e20341. https://doi.org/10.2196/20341

Government of Canada (2024) COVID-19 epidemiology update: testing and variants. Health Infobase Canada. https://health-infobase.canada.ca/covid-19/testing-variants.html

Grover S, Sahoo S, Mehra A, Avasthi A, Tripathi A, Subramanyan A, Pattojoshi A, Rao G, Saha G, Mishra K, Chakraborty K, Rao N, Vaishnav M, Singh O, Dalal P, Chadda R, Gupta R, Gautam S, Sarkar S et al (2020) Psychological impact of COVID-19 lockdown: an online survey from India. Indian J Psychiatry 62(4):354. https://doi.org/10.4103/psychiatry.IndianJPsychiatry_427_20

Guest PC (ed) (2023) Application of Omic techniques to identify new biomarkers and drug targets for COVID-19 (1st ed.). Springer, Cham. https://link-springer-com.proxy.bib.uottawa.ca/book/10.1007/978-3-031-28012-2#bibliographic-information

Harrigan SP, Velásquez HA, Abdia Y, Wilton J, Prystajecky N, Tyson J, Fjell C, Hoang L, Kwong JC, Mishra S, Wang L, Sander B, Janjua NZ, Sbihi H (2024) The clinical severity of COVID-19 variants of concern: retrospective population-based analysis. JMIR Publ Health Surveill 10:e45513. https://doi.org/10.2196/45513

Hsu L, Wisplinghoff H, Kossow A, Hurraß J, Wiesmüller GA, Grüne B, Hoffmann D, Lüsebrink J, Demuth S, Schildgen O, Schildgen V (2022) Limited protection against SARS-CoV-2 infection and virus transmission after mRNA vaccination. J Infect 84(1):94–118. https://doi.org/10.1016/j.jinf.2021.06.023

Jahan N, Rubeshkumar P, Karuppiah M, Sambath I, Sendhilkumar M, Ilangovan K, Ongesh R, Sakthivel M, Mohankumar R, Santhosh Kumar M, Ganeshkumar P, Ponnaiah M, Kaur P (2021) Entry and initial spread of COVID-19 in India: epidemiological analysis of media surveillance data, India, 2020. Clin Epidemiol Glob Health 9:347–354. https://doi.org/10.1016/j.cegh.2020.10.008

Kumar M, Kuroda K, Mukherjee S, Ngiehm LD, Vithanage M, Tyagi VK (2024) Wastewater surveillance for Covid-19 management. Springer

Lapointe-Shaw L, Rader B, Astley CM, Hawkins JB, Bhatia D, Schatten WJ, Lee TC, Liu JJ, Ivers NM, Stall NM, Gournis E, Tuite AR, Fisman DN, Bogoch II, Brownstein JS (2020) Web and phone-based COVID-19 syndromic surveillance in Canada: a cross-sectional study. PLoS ONE 15(10):e0239886. https://doi.org/10.1371/journal.pone.0239886

McGill E, Coulby C, Dam D, Bellos A, McCormick R, Patterson K (2023) Canadian COVID-19 outbreak surveillance system: implementation of national surveillance during a global pandemic. Can J Publ Health 114(3):358–367. https://doi.org/10.17269/s41997-023-00766-5

Murugesan B, Karuppannan S, Mengistie AT, Ranganathan M, Gopalakrishnan G (2020) Distribution and trend analysis of COVID-19 in India: geospatial approach. J Geogr Stud 4(1):1–9. https://doi.org/10.21523/gcj5.20040101

Nsubuga P, White ME, Thacker SB, Anderson M, Blount SB, Broome C, Chiller TM, Espitia V, Imtiaz R, Sosin D, Stroup D, Tauxe R, Vijayaraghavan M, Trostle M (2006) Public health surveillance: a tool for targeting and monitoring interventions. In: Disease control priorities in developing countries, 2nd edn. Oxford University Press. https://www.ncbi.nlm.nih.gov/books/NBK11770/

Opriessnig T, Huang Y (2020) Update on possible animal sources for COVID-19 in humans. Xenotransplantation 27(3):e12621. https://doi.org/10.1111/xen.12621

Papadomanolakis-Pakis N, Maier A, Van Dijk A, VanStone N, Moore KM (2021) Development and assessment of a hospital admissions-based syndromic surveillance system for COVID-19 in Ontario, Canada: ACES pandemic tracker. BMC Publ Health 21(1):1230. https://doi.org/10.1186/s12889-021-11303-9

Peck KM, Lauring AS (2018) Complexities of viral mutation rates. J Virol 92(14):e01031-e1117. https://doi.org/10.1128/JVI.01031-17

Ramachandran S, Malani A (2021) All-cause mortality during SARS-CoV-2 Pandemic in India: Nationally-representative estimates independent of official death registry. Medrxiv. https://doi.org/10.1101/2021.07.20.21260577

Ranjan R (2022) Omicron impact in India: analysis of the ongoing COVID-19 third wave based on global data. https://doi.org/10.1101/2022.01.09.22268969

Sanjuán R, Domingo-Calap P (2016) Mechanisms of viral mutation. Cell Mol Life Sci 73(23):4433–4448. https://doi.org/10.1007/s00018-016-2299-6

Sardar T, Nadim SS, Rana S, Chattopadhyay J (2020) Assessment of lockdown effect in some states and overall India: a predictive mathematical study on COVID-19 outbreak. Chaos, Solitons Fractals 139:110078. https://doi.org/10.1016/j.chaos.2020.110078

Shah K, Awasthi A, Modi B, Kundapur R, Saxena D (2020) Unfolding trends of COVID-19 transmission in India: critical review of available mathematical models. Indian J Commun Health 32(2):206–214

Shrivastava VK, Kumar A, Shrivastava A, Tiwari A, Thiru K, Batra R (2021) Study and trend prediction of covid-19 cases in India using deep learning techniques. J Phys: Conf Ser 1950(1):012084. https://doi.org/10.1088/1742-6596/1950/1/012084

Song H, Fan G, Zhao S, Li H, Huang Q, He D (2021) Forecast of the COVID-19 trend in India: a simple modelling approach. Res Square. https://doi.org/10.21203/rs.3.rs-502990/v1

Spiteri G, Fielding J, Diercke M, Campese C, Enouf V, Gaymard A, Sognamiglio P, Moros MJS, Riutort AN, Demina YV, Mahieu R, Bengnér M, Buda S, Schilling J, Filleul L, Lepoutre A, Saura C, Levy-Bruhl D, Coignard B et al (2020) First cases of coronavirus disease 2019 (COVID-19) in the WHO European Region, 24 January to 21 February 2020. Eurosurveillance. https://www.ncbi.nlm.nih.gov/pmc/articles/PMC7068164/pdf/eurosurv-25-9-1.pdf

Thayer WM, Hasan MZ, Sankhla P, Gupta S (2021) An interrupted time series analysis of the lockdown policies in India: a national-level analysis of COVID-19 incidence. Health Policy Plan 36(5):620–629. https://doi.org/10.1093/heapol/czab027

Tollefson J (2021) COVID curbed carbon emissions in 2020—but not by much. Nature 589(7842):343

Valencak TG, Csiszar A, Szalai G, Podlutsky A, Tarantini S, Fazekas-Pongor V, Papp M, Ungvari Z (2021) Animal reservoirs of SARS-CoV-2: calculable COVID-19 risk for older adults from animal to human transmission. GeroScience 43(5):2305–2320. https://doi.org/10.1007/s11357-021-00444-9

Vedaei SS, Fotovvat A, Mohebbian MR, Rahman GME, Wahid KA, Babyn P, Marateb HR, Mansourian M, Sami R (2020) COVID-SAFE: an IoT-based system for automated health monitoring and surveillance in post-pandemic life. IEEE Access 8:188538–188551. https://doi.org/10.1109/ACCESS.2020.3030194

WHO (2024) WHO COVID-19 dashboard. World Health Organization. https://data.who.int/dashboards/covid19/deaths?n=o

Zhang W, Zhao W, Wu D, Yang Y (2020) Predicting COVID-19 trends in Canada: a tale of four models. Cogn Comput Syst 2(3):112–118

Chapter 11
Epidemiology of COVID-19 and Health Policy Response in Nepal

Yogendra Shah, Bipin Kumar Acharya, Megha Raj Banjara, Janak Koirala, and Meghnath Dhimal

Abstract The first case of COVID-19 in Nepal was reported on January 23, 2020. Nation-wide lockdowns were imposed in March 2020 to curb the spread and prevent the morbidity and mortality due to pandemic. Among total deaths due to COVID-19 in Nepal, three times more deaths occurred in the second wave as compared to the first wave. Males, aged 60 years or older, patients from hills and patients with at least one comorbid condition were associated with a higher mortality. The Government of Nepal took several measures for the prevention and control of COVID-19 and provided support to vulnerable groups. The High-Level Coordination Committee for COVID-19 Prevention and Control coordinated the response and provide necessary policy directions. The COVID-19 Crisis Management Centre monitored the essential supplies, and conducted other necessary activities for prevention and control of COVID-19. Nepal started COVID-19 vaccination in early 2021. The coordinated response activities resulted in a relatively a smaller number of mortalities due to COVID-19 in Nepal. Taking lessons from COVID-19, there are several policy responses and activities to address pandemics in the future. This chapter aims to summarize epidemiology of COVID-19 and Policy Response including, post-COVID-19 health situation in Nepal.

Keywords COVID-19 · Nepal · Post-COVID-19 · Epidemiology · Policy response · Pandemic preparedness

Y. Shah · B. K. Acharya
Planetary Health Research Centre, Kathmandu, Nepal

M. R. Banjara
Central Department of Microbiology, Tribhuvan University, Kathmandu, Nepal

J. Koirala
Patan Academy of Health Sciences, Lalitpur, Nepal

M. Dhimal (✉)
Nepal Health Research Council, Kathmandu, Nepal
e-mail: meghdhimal@gmail.com

R. Akhtar (ed.), *COVID-19: Impacts on Health and Health Systems*,
https://doi.org/10.1007/978-3-031-90255-0_11

11.1 Introduction

Human coronaviruses were first identified in 1960, approximately 23 years after their initial description in birds. Coronaviruses belong to the order Nidovirales, family Coronaviridae and the subfamily Coronavirinae with an enveloped, single-stranded RNA viruses with a genome size ranging between 27 and 30 kilobases (Lu et al. 2020; Fehr and Perlman 2015; Li 2016). The subfamily is further categorized into four genera: Alpha, Beta, Gamma, and Delta coronaviruses. The membrane (M) and envelope (E) proteins facilitate virus assembly, while the spike (S) protein enables host cell entry. Coronaviruses cause diseases in humans and animals, with over 80% of infections being zoonotic, indicating their ability to transmit from animals to human origin. Seven human CoVs have been identified: HCoV-OC43, HCoV-HKU1, HCoV-229E, HCoV-NL63, SARS-CoV, MERS-CoV, and SARS-CoV-2 (Lu et al. 2020; Fehr and Perlman 2015; Li 2016).

The most recent and impactful, Severe Acute Respiratory Syndrome Coronavirus-2 (SARS-CoV-2), triggered the global COVID-19 pandemic. SARS-CoV-2, the causative agent of COVID-19, was first reported on the basis of laboratory confirmation in Wuhan, China, in December 2019. The virus spread rapidly causing outbreaks in multiple countries prompting World Health Organization (WHO) to declare a global pandemic in March 2020. SARS-CoV-2 has infected more than 776 million people with 7 million deaths globally as of September 15, 2024. Between August 19, and September 15, 2024, global SARS-CoV-2 PCR positivity declined from 8.2% to 5.8%, based on the data from 89 reporting countries. WHO is tracking two variants of interest (BA.2.86, JN.1) and seven variants under monitoring, with JN.1 remaining the most prevalent variant (16% in week 37), while KP.3.1.1 and XEC are increasing. During this period, global COVID-19 cases rose by 10% and deaths by 11%, with 270,000 new cases and 5,700 deaths reported. Hospitalizations and ICU admissions increased by 14% and 52%, respectively, in consistently reporting countries. As of December 2023, WHO reported 13.64 billion doses of COVID-19 vaccine was administered covering 67% of world's population with primary series and 32% with at least one booster dose (WHO 2024).

Despite extensive efforts to manage COVID-19, the pandemic morbidity and mortality significantly disrupted daily life. SARS-CoV-2 evolved into various variants, including Omicron (B.1.1.529), identified amid a case surge and classified by WHO as a Variant of Concern on 26 November 2021, joining earlier variants like Alpha, Beta, Gamma, and Delta. The continuous evolution of the virus, coupled with limited vaccine access and recurrent outbreaks, posed severe challenges for resource-constrained countries like Nepal (WHO 2020, 2024).

In Nepal, the total number of confirmed COVID-19 cases from January 2020 to December 9, 2024 reached 1,001,143, according to the Epidemiology and Disease Control Division (EDCD). Among them, 989,117 people have recovered, with a recovery rate of 98.8%. However, 12,020 individuals lost their lives secondary to COVID-19 infection resulting in a fatality rate of 1.2%. The gender distribution of all 1,001,143 cases shows 58.7% of the infected individuals were male and 41.3%

were female. In terms of provincial distribution, Bagmati Province was the most affected province reporting the highest number of cases 542,832 (54.22%) in Nepal. Additional details, including a breakdown of the cases by province and other factors, are presented in the accompanying figure reflecting the pandemic's impact across the country (Fig. 11.1) (EDCD 2024).

Although the first case of COVID-19 in Nepal was reported on January 23, 2020 in a traveler from China, the local transmission did not occur until second half of March 2020 (MOHP 2021a; b; Koirala and Tamrakar 2020). Therefore, to curb the pandemic's impact on morbidity and mortality, nationwide lockdowns were enforced beginning on March 24, 2020 (Basnet et al. 2020). Transmission of COVID-19 and deaths during the first wave were at a low level during the first wave of COVID-19. However, the mortality during the second wave caused by the Delta variant was three times higher than those recorded in the first wave (Bastola et al. 2020; Panthee et al. 2020a, b; Gupta et al. 2021; Gyanwali et al. 2021; Sah et al. 2020; Adhikari et al. 2021a, b; Pandey et al.2022a, b). Mortality was notably higher among males,

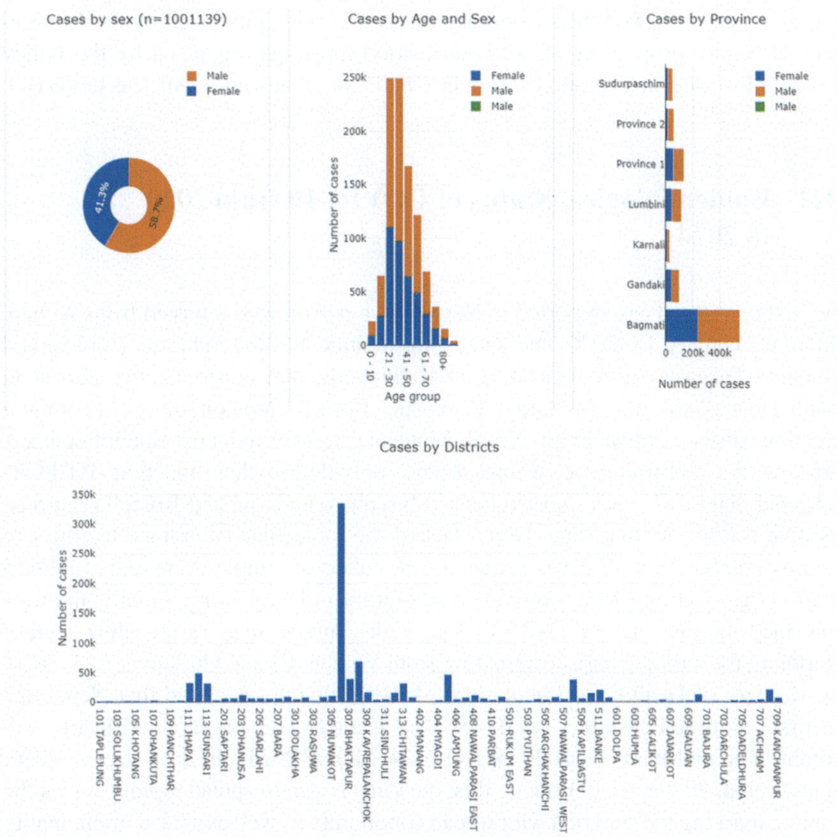

Fig. 11.1 Epidemiological status of COVID-19 from 2020 to 2024]

individuals aged 60 or older, residents from hilly regions, and patients with at least one comorbid condition. The Government of Nepal (GON) implemented numerous measures to prevent and control the spread of COVID-19 while extending support to vulnerable populations. Nepal government's high-Level Coordination Committee for COVID-19 Prevention and Control provided strategic policy directions, while the COVID-19 Crisis Management Centre (CMC) oversaw the supply of essential resources and coordinated critical activities for pandemic prevention and management. Nepal launched its COVID-19 vaccination program in early 2021, and the collective response efforts contributed to a relatively lower mortality rate compared to global averages. Drawing on lessons from the pandemic, Nepal has devised several policy responses and strategies to effectively address future pandemics and other health emergencies. This chapter aims to provide an overview of the epidemiology of COVID-19 and Policy Response including Post-COVID-19 health situation in Nepal.

Methods: A comprehensive literature search on the epidemiology of COVID-19 and policy responses in Nepal from 2020 to 2024 was gathered from official sources, including peer-reviewed index journals, research reports published by the Government of Nepal, proceedings from coordination meetings organized by the Policy Research Institute, and resources from international organizations such as the WHO.

11.2 Epidemiological Status of COVID-19 from 2020 to 2024

The first confirmed case reported in Nepal was a person who returned from Wuhan, China, on January 13, 2020, and sought for treatment at the Sukraraj Tropical and Infectious Disease Hospital (STIDH) in Kathmandu after experiencing a persistent cough. He was admitted, and initial diagnostic efforts focused on ruling out common infectious diseases prevalent in Nepal. Molecular test for diagnosis the influenza A and B with a real-time reverse transcriptase polymerase chain reaction (RT-PCR) and rapid diagnostic tests for dengue, scrub typhus, malaria, and Brucella returned negative results. At that time, Nepal lacked the molecular laboratory facilities to diagnosis for SARS-CoV-2. Consequently, the collected samples were sent to a World Health Organization (WHO) reference laboratory in Hong Kong. Finally, the tests confirmed the presence of SARS-CoV-2, with genome analysis revealing a close relation to the strain of virus originating from Wuhan, China. On January 23, 2020, the Ministry of Health and Population (MoHP) officially declared that Nepal had the first confirmed case of COVID-19. The second case was reported nearly two months later, on March 17, 2020 (WHO 2024; MOHP 2021a, b; Bastola et al. 2020; Panthee et al. 2020a, b). Following this, the virus began to spread rapidly across the country, marking the start of a widespread community as well as a local municipality level transmission. However, Nepal did not have a policy regarding the control and prevention of pandemic diseases, such as COVID-19, MERS and Ebola Virus.

11.3 Early Detection of COVID-19 Cases in Nepal

The study on clinical and epidemiological features of COVID-19 Deaths in Nepal, based on official media releases from the Government of Nepal (GON), analyzes the characteristics of individuals who died from COVID-19 between 23 January and 10 August 2020. The study found that nearly half of the deaths occurred in individuals under 50 years of age, highlighting the impact of the disease on younger populations. Additionally, being female was associated with an increased risk of death. Co-morbidities such as cardiovascular diseases, diabetes, and respiratory conditions were commonly linked to these deaths, underscoring the vulnerability of individuals with underlying health issues. As the festive season approaches and the lockdown measures ease, the findings of the study emphasize the need for continued vigilance. Although the initial response was successful, the country ability to manage the pandemic weakened due to limited testing capacity, poor quarantine management, and insufficient adherence to physical distancing. With the easing of restrictions and the upcoming festive season, there is a heightened risk of increased transmission. Therefore, both the government and citizens must prioritize preventive measures and strengthen their preparedness to avoid overwhelming the healthcare system and mitigate the impact of the disease (Panthee et al. 2020a, b).

11.4 Government Initial Response

The Ministry of Health and Population (MoHP) implemented measures like establishment of health desks at airports and border points, travel restrictions, and repatriation of 175 Nepalese citizens from Wuhan, all quarantined for 14 days. The government activated the Incident Command System, Health Emergency Operation Center, and a High-Level Coordination Committee on 2 March 2020 to manage and contain the virus's spread (MOHP 2021a, b; Bastola et al. 2020).

11.5 Detection of a Second Case and Lockdown Announcement

Nepal second COVID-19 case was confirmed on March 17, 2020 in a person returning from France. Public concerns arose over limited diagnostic capabilities due to no reported cases since January. In response, schools closed on 18 March, and restrictions on gatherings, international flights, and border crossings followed. A nationwide lockdown began on March 24, 2020, lasting until July 21, 2020. The reproduction number (R0) typically reflects the average number of secondary cases produced by an infected individual in a susceptible population, and factors like population density and contact frequency can influence it. During that period, the reproduction number was calculated in various research studies conducted in Nepal to estimate the average number of secondary COVID-19 transmission cases per individual within a susceptible population (Adhikari et al. 2021a, b; Bastola et al. 2020; MOHP 2021a, b).

11.6 System Strengthening

Initially, COVID-19 testing with RT-PCR assay for detection of SARS-CoV-2 was limited to the National Public Health Laboratory (NPHL) in Kathmandu. Testing facilities expanded nationwide, reaching over 100 labs within a year. Initial Rapid diagnostic tests (RDTs) used for screening were later discontinued due to inaccuracies. The COVID-19 Crisis Management Centre (CCMC) was established on March 29, 2020 to coordinate response efforts (MOHP 2021a; b; Bastola et al. 2020).

11.7 Information Management and Communication

The government emphasized risk communication through daily press briefings, hotlines, and social media platforms. The Health Cluster for COVID-19, co-led by WHO, facilitated regular coordination meetings to enhance response strategies at federal and provincial levels (Bastola et al. 2020; MOHP 2021a, b).

11.8 Case and Data Management

STIDH was the primary treatment center, with additional hospitals designated as cases surged. Quarantine centers were established, though some faced challenges in management. Over 150 hospitals, including private facilities, were involved in COVID-19 care, with services offered free of charge. Gaps in ICU capacity, infection prevention, and logistics were addressed through assessments and training programs (MOHP 2021a; b; Bastola et al. 2020; Panthee et al. 2020a, b; Gupta et al. 2021; Gyanwali et al. 2021; Sah et al. 2020; Adhikari et al. 2021a, b; Pandey et al.2022a, b). Initially, COVID-19 data in Nepal were managed manually in Excel by the NPHL and sent to the EDCD for reporting. However, as cases surged and testing expanded, handling the data became challenging, leading to the adoption of tools like Go Data from WHO's GOARN. Conflicting data from multiple centers, including the CCMC and Home Ministry, caused discrepancies. In April 2020, a uniform reporting template was introduced, but compliance issues and varied data formats complicated the process. Provincial agencies often hesitated to share data, citing jurisdictional concerns. To streamline efforts, an Information Management Unit was established, but discrepancies in reporting COVID-19 deaths persisted between health facilities and the Nepali Army, due to differing definitions of COVID-19-related deaths (Pandey et al. 2022a, b).

11.9 Case Investigation and Contact Tracing

Nepal confirmed its first local COVID-19 transmission on April 4, 2020, in Kailali District, with further clusters linked to travelers and religious gatherings. Initial contact tracing teams, coordinated by the EDCD, faced challenges as cases rose. Localized contact tracing plans were developed with municipal health coordinators. By May 2020, a Preparedness and Response Plan was implemented, creating 1075 local-level CICT teams. Quarantine and isolation centers were established, primarily near the Indo-Nepal border, but logistical issues persisted. The surge from March to July 2020 saw daily cases rise to over 600, but deaths remained minimal (MOHP 2021a; b; Gyanwali et al.2021; Sah et al. 2020; Adhikari et al. 2021a, b; Pandey et al.2022a, b; Bajracharya et al. 2024).

11.10 First and Second Wave

The first COVID-19 wave in Nepal peaked in October 2020, with 5743 daily cases and a case fatality rate of less than 1%. Challenges included delays in managing returnees from India, contributing to case surges, especially in border provinces. Despite a lockdown lifted in July 2020, cases spiked, highlighting difficulties in systematic quarantine management (MOHP 2021a; b; Bastola et al. 2020; Gyanwali et al.2021; Sah et al. 2020; Adhikari et al. 2021a, b; Pandey et al. 2022a, b).

During the COVID-19 pandemic, the spread of dengue in Nepal was suppressed. However, in 2022, there was a surge in cases of dengue mainly in Bagmati Province, leading to a dual burden of both COVID-19 and dengue (Bajracharya et al. 2024; Pandey et al. 2022a, b; Pandey et al. 2021; Lamsal 2021; Koirala and Basnet 2020). The overlap of these diseases complicated diagnosis and treatment, heightening the risk of severe cases, and deaths due to misdiagnosis or delayed treatment (Pandey et al. 2022a, b). Furthermore, other diseases, such as scrub typhus, were overshadowed during the COVID-19 pandemic (Lamsal 2021). The pandemic highlights the need for the Government of Nepal's Ministry of Health and Population (MOHP) to enhance case detection for these diseases, strengthen disease control mechanisms, and improve preparedness for future pandemics. Therefore, the GON needed to strengthen diagnostic tools, train healthcare workers, and improve disease surveillance to effectively manage potential outbreaks and address simultaneous disease threats through the formation and strict implementation of health policies and systems in Nepal (Pandey et al. 2022a, b; Pandey et al. 2021; Lamsal 2021; Koirala and Basnet 2020).

The second wave began in April 2021, driven by the Delta variant. Large gatherings and festivals exacerbated the crisis, peaking at over 9000 daily cases and 246 deaths per day in May. Hospitals faced critical shortages of oxygen and medical supplies. A nationwide seroprevalence survey later showed that two-thirds of the population had antibodies by August 2021 (Sah et al. 2020; Relief web 2022, NHRC 2022; Kandikattu et al. 2021; Dhakal and Karki 2020; MOHP 2021a, b).

11.11 Omicron Variant

Nepal's third wave, fueled by the Omicron variant, peaked in January 2022 with over 10,000 daily cases. However, fatalities were significantly lower than in the second wave. Improved vaccination coverage and lessons from earlier waves aided response efforts. Public health protocols along the Indo-Nepal border remained weak (Pandey et al. 2022a, b).

As of December 2024, several variants of SARS-CoV-2 have been categorized as Variants Under Monitoring (VUMs) or Variants of Interest (VOIs) based on their genetic features and epidemiological significance. Among the VUMs, KP.2 (24B), documented since January 2024, carries mutations such as S:R346T, S:F456L, and S:V1104L, while KP.3 (24C), first reported in February 2024, includes additional changes like S:Q493E. Variants like KP.3.1.1 and LB.1 exhibit further mutations, indicating evolutionary diversification. Notably, the variant XEC, identified in June 2024 and designated in September 2024, combines multiple mutations (S:T22N, S:F59S, S:F456L, S:Q493E, S:V1104L), warranting ongoing risk assessments. Among the VOIs, JN.1#, initially evaluated in December 2023 and updated in early 2024, includes the S:L455S mutation derived from BA.2.86, reflecting its potential significance in disease transmission and vaccine response. These classifications highlight the dynamic nature of variant monitoring to ensure timely public health interventions (WHO 2020, 2024).

11.12 Vaccination

Nepal launched its vaccination campaign on January 27, 2021, with a Recombinant ChAdOx1 nCoV-19 Corona Virus Vaccine (COVISHIELD, Serum Institute, India). By April 2022, over 75% of the population had received at least one dose, and 66% were fully vaccinated. Vaccines used included inactivated vero cell vaccine (Sinopham, China), mRNA vaccines (Moderna and Pfizer, USA), and Janssen vaccine (Janssen, USA). Vaccinations for children under 12 were not initiated until later due to the unavailability of vaccine for children. Booster doses were also distributed, enhancing immunity against subsequent waves (Relief web 2022, NHRC 2022; Kandikattu et al. 2021; Dhakal and Karki 2020; MOHP 2021a, b; Lama et al. 2023; SR 2019; Bhattarai and Dhungana 2021). The details timeline reflects Nepal ongoing efforts to manage the pandemic, adapt to emerging variants, and strengthen public health systems given in Table 11.1.

Table 11.1 Timeline of major COVID-19 Events in Nepal from 2020 to 2024

Event	Description
First case detection	First case on January 23, 2020, confirmed through RT-PCR after a man returned from Wuhan. Symptoms appeared on 3 January, sought medical care on 13 January
Government response (Initial)	MoHP implemented health desks at airports, travel restrictions, repatriation of 175 citizens from Wuhan. Activated Incident Command System and Health Emergency Operation Center (March 2, 2020)
Second case and lockdown	Second case on March 17, 2020. Schools closed on 18 March, international flights restricted, and nationwide lockdown started on March 24, 2020, lasting until July 21, 2020
Testing and system strengthening	Initial testing at NPHL, expanded to over 100 labs by 2021. Rapid diagnostic tests (RDTs) discontinued. CCMC formed on March 29, 2020
Information management	Risk communication via daily press briefings, hotlines, and social media. WHO led Health Cluster for COVID-19
Case management	STIDH was primary treatment center, over 150 hospitals involved. Free care provided but gaps in ICU and infection prevention capacity addressed through assessments and training
Data management	Initially handled manually by NPHL. Adopted WHO's Go Data, but discrepancies arose from multiple reporting agencies. Information Management Unit established to streamline processes
Contact tracing and investigation	First local transmission in Kailali on April 4, 2020. Contact tracing and isolation plans developed with municipal health coordinators. By May 2020, 1075 local-level CICT teams established
First wave	First wave peaked in October 2020 with 5743 daily cases, a case fatality rate of $< 1\%$. Border management and quarantine system challenges contributed to case surges
Second wave	Second wave in April 2021, driven by Delta variant. Peak of over 9000 daily cases and 246 deaths per day in May. Shortages of oxygen and medical supplies, but widespread seroprevalence by August 2021
Omicron variant	Third wave in January 2022, with Omicron variant. Over 10,000 daily cases but lower fatalities. Enhanced vaccination coverage and lessons from earlier waves aided response
Vaccination campaign	Vaccination began on January 27, 2021 with COVISHIELD. By April 2022, over 75% had one dose, 66% fully vaccinated. Vaccines included Vero Cell, Moderna, Pfizer, and Janssen. Booster doses were distributed. No vaccination for children under 12

Bastola et al. (2020), Panthee et al. (2020a, b), Gupta et al. (2021), Gyanwali et al. (2021), Sah et al. (2020), Adhikari et al. (2021a, b), Pandey et al. (2022a, b)

11.13 Policy and Guidelines Formation and Implementation of Tackle Against the Pandemic COVID-19 Nepal

In Nepal, National Disaster Response Framework (2013) established the humanitarian cluster approach, with the Ministry of Health and Population (MoHP) as the lead agency for the Health Pillar, supported by the WHO as co-lead. During the COVID-19 pandemic, this model was implemented for health partner coordination, with the MoHP's Health Coordination Division (HCD) leading and WHO Nepal as co-lead. Coordination at national and subnational levels was managed through the Health Emergency Operation Center (HEOC), with information management and secretariat support. The HCD chaired national coordination meetings, the first of which was held on April 9, 2020. These virtual meetings, which involved both national and subnational participation, included information sharing, gap analysis, and discussions on next steps. Action points were transparently shared and followed up by focal points. As the response demand decreased, the HCD, supported by WHO, reviewed the effectiveness of the coordination platform to identify areas for improvement. Their study evaluates the performance of the Health Partner Coordination during Nepal's COVID-19 response (GONMHA 2013).

A study by Mishra et al. (2024) on effectiveness of health partners coordination for COVID-19 pandemic response in Nepal revealed that following findings: Over three-fifths of the meetings demonstrated positive outcomes in terms of conduction, process, participation, and documentation of action points, though follow-up (22.2%) showed areas for improvement. When evaluating health partner coordination subfunctions, results were generally rated as 'good' (> 75.0%) or 'satisfactory' (50.1–75.0%), with the exception of prioritization based on analyses, which was rated as 'unsatisfactory' (< 50%). Partners acknowledged the successful coordination practices, raised some concerns, and offered recommendations for further enhancement (Mishra et al. 2024).

Mishra et al. (2024) findings concluded that COVID-19 underscored the critical importance of effective coordination in pandemic response. Nepal demonstrated success in managing a coordinated response through early health coordination meetings and the smooth operation of health coordination mechanisms. The study recommended measures to enhance coordination, including continuing regular meetings, adopting a hybrid format (virtual and in-person), creating clear Terms of Reference (ToR), regularly monitoring and sharing progress, and developing an information-sharing website. The most frequently emphasized recommendation was the need to clearly define roles and responsibilities among humanitarian health partners through the formulation of detailed ToRs.

11.14 The Government of Nepal Response to the COVID-19 Pandemic

The Government of Nepal (GoN) implemented several measures such as stay-at-home orders and mass quarantines to slow the spread of SARS-CoV-2. Movement restrictions have proven to be one of the most effective interventions, especially in resource-limited countries like Nepal. Despite ongoing efforts, the country experienced COVID-19 cases due to challenges in fully implementing necessary strategies. However, the GoN remains committed to controlling the virus's transmission at the community level through testing, contact tracing, public awareness campaigns, lockdowns, and economic support packages (Rayamajhee et al. 2021).

11.15 Priorities for the Government of Nepal

The GoN's priorities include leadership and management, securing essential supplies, improving media and quarantine management, revitalizing healthcare delivery, and advancing the COVID-19 vaccination strategy.

Before the onset of the COVID-19 pandemic, the Government of Nepal (GoN) lacked well-defined policies and comprehensive guidelines specifically tailored to control and prevent public health emergencies within the framework of the six-health system building blocks. These building blocks-leadership and governance, health workforce, service delivery, health information systems, financing, and medical products are critical components for a resilient and effective health system. The absence of structured policies and coordinated strategies left Nepal vulnerable to handling large-scale outbreaks and public health emergencies effectively (Rayamajhee et al. 2021; NRO 2022; Srivastav et al. 2021; PRI 2022).

However, the challenges posed by the COVID-19 pandemic acted as a catalyst for significant policy development. Recognizing the need for a cohesive and well-structured approach, the Ministry of Health and Population (MoHP), under the leadership of the GoN, initiated the formulation of new policies and guidelines to mitigate the impacts of the pandemic. One notable development was the National Guidance Document for Coordination among Humanitarian Health Partners, designed to prepare for and respond to disasters and public health emergencies using a cluster-based approach (Rayamajhee et al. 2021; NRO 2022; Srivastav et al. 2021; PRI 2022).

The post-pandemic policy highlights robust leadership and coordination across government levels, clear roles in emergency management, and a cluster approach for effective partner collaboration. It integrates disaster preparedness with public health responses, streamlines resource allocation, and strengthens health systems through training, infrastructure upgrades, and surveillance. Thus, community engagement and awareness are also prioritized to enhance education, preparedness, and trust in public health measures (Rayamajhee et al. 2021; NRO 2022; Srivastav et al. 2021; PRI 2022).

11.16 Coordination Strategies to Improve COVID-19 PCR Laboratory Testing Scale up in Nepal

Bakker et al. (2024) study on coordination strategies to scale up COVID-19 PCR laboratory testing in Nepal, highlighting that centralized coordination and strategic sample transfer partnerships significantly improve diagnostic network performance under high demand. However, collaboration between public and private laboratories had limited impact due to geographic and demand constraints. Furthermore, the study also provides valuable recommendations for policymakers to adopt effective strategies to strengthen diagnostic networks in developing countries like Nepal for tackle against the future pandemics.

The study findings of Adhikari et al. (2021b) on transmission dynamics of COVID-19 in Nepal: Insights into Mathematical Model showed the effectiveness of lockdown strategies implemented in three times (the first lockdown, second lock down (border screening-quarantine), and third lock down (isolation measures) in Nepal during the pandemic. During the controlled phase, most cases were imported, with limited local transmission, aligning with observed data. These measures maintained the reproduction number at approximately 0.21, preventing an estimated 442,640 cases and saving over 1200 lives. However, during the outgrown phase, the relaxation of these interventions led to a reproduction number increase to 1.8, driving exponential case growth. Model projections indicate that, without interventions, Nepal could face 18.76 million cases (10.7 million detected and 8.06 million undetected) and 89,000 deaths by the end of 2021. Therefore, it is strongly recommended that these findings offer essential policy guidance for designing effective control strategies to help Nepal recover from the devastating COVID-19 pandemic (Adhikari et al. 2021b).

Pantha et al. (2021) study conducted on the inter-provincial disparity of COVID-19 transmission and control in Nepal revealed that significant variation across provinces in epidemic trends, growth rates, and reproduction numbers. Additionally, the effectiveness of control strategies differed by province, highlighting the need for both province-specific and national-level approaches to effectively manage and control local outbreaks.

The study revealed a multi-phasic epidemic with three surge periods linked to policy changes. Daily exponential growth rates varied temporally and geographically, with a 4% national median (Pantha et al. 2021). The basic reproduction number (R0) was 1.083 nationally, ranging from 0.707 to 1.205 across provinces. Among the seven provinces, the highest reproduction number (R0) was observed in Province 3 (1.191–1.219, 95% CI), followed by Province 1 (1.107–1.166, 95% CI), and Province 4 (1.009–1.085, 95% CI). The higher population densities and contact frequencies in these provinces likely contributed to their elevated R0 values indicating that the effective reproduction number (Rt) fluctuated generally remaining below 1 during policy implementation but rising above 1 after the relaxation of restrictions. Thus, all four policies (lockdown, testing, isolation, and quarantine) had a significant impact on case numbers, with lockdown generally being the most effective. However, the effectiveness of these policies varied across provinces, influenced by geographical and demographic factors (Pantha et al. 2021). Similarly, province-level COVID-19 data from Nepal shows that national policies impact epidemic indicators differently at the provincial level due to geographic and demographic factors. Therefore, local-level analyses are crucial for accurately estimating epidemic indicators and designing effective non-pharmaceutical interventions to control COVID-19 outbreaks (Pantha et al. 2021).

11.17 Post-COVID Health Facilities Development in Nepal

COVID-19 pandemic exposed weaknesses in Nepal healthcare system, especially in rural and remote hilly areas. Since then, efforts have been made to improve healthcare infrastructure and prepare for future health crises. On the other hand, a significant number of health professionals are at risk of facing various post-COVID health complications following COVID-19 infection (Sapkota et al. 2022). In response, substantial efforts have been made to strengthen healthcare infrastructure, including upgrades to tertiary care hospitals, provincial hospitals, and both urban and remote health posts with laboratory services and enhanced the preparedness for future health emergency in Nepal. As district hospitals have also expanded their critical care facilities, incorporating oxygen plants, vaccine cold chain infrastructure, intensive care units and high dependency unit with ventilators, and essential emergency equipment in key locations across Nepal (WHO 2021; GON 2022; World Bank 2020).

Furthermore, molecular laboratories with Biosafety Level-2 have been established alongside other services, including general laboratories, mobile health clinics, and telemedicine, significantly improving healthcare access in Nepal hilly and remote regions. In addition, essential instruments and equipment such as real-time polymerase chain reaction thermocyclers (RT-PCR), automated nucleic acid extraction systems, protein analyzers, immunology assay machines, consumables, and necessary reagents for RT-PCR have been provided with support from the Ministry of Health and Population (MOHP), provincial governments (Ministry of Social Development), as well as national and international organizations. These efforts have been further strengthened by the deployment of trained technical staff, training for healthcare workers, and the provision of emergency medical equipment, enhancing pandemic preparedness. As a result, collaboration among public, private, and international organizations, along with community-based programs that promote health awareness and accessibility, has driven these advancements forward. Moreover, establish emergency response plans with clear protocols for crises and essential services, enhance healthcare waste management, implement telemedicine and innovative delivery methods to improve resilience, and strengthen local health service systems.

Additionally, the pandemic fund received a $19 million grant, supplemented by $6 million in co-financing, to bolster Nepal pandemic preparedness. This initiative was made possible through a partnership with FAO, UNICEF, and WHO, particularly through the Strengthening Pandemic Preparedness for Early Detection (SPEED) Project. In summary, the main key components of the project included surveillance and early warning systems, improved laboratory capabilities, workforce development, and a sustainable health security framework (The Pandemic Fund. n.d.).

In conclusion, COVID-19 developments of remarkable progress, sustained investment and equitable healthcare delivery, particularly in hard-to-reach areas, are essential to building resilience against future pandemics. Nepal post-COVID-19 reforms have introduced emergency response plans, strengthened molecular and laboratory services, expanded telemedicine services, upgraded vaccine cold chain infrastructure, improved waste management, and enhanced healthcare access in remote regions. Therefore, molecular and general laboratories, emergency public health responses, and the formulation and implementation of health system policies based on the six building blocks of the health system have been strengthened, establishing a resilient and equitable healthcare system to enhance preparedness for future pandemics in Nepal (WHO 2021; GON 2022; World Bank 2020).

11.18 Spatiotemporal Distribution

A total of 100,450 COVID-19 infections and 12,013 deaths have been reported in Nepal since the first case was detected on January 23, 2020. The immediate implementation of restrictive measures, including lockdowns, played a critical role in slowing the spread of the virus. As a result, the second case was reported nearly two months later, on March 17, 2020. However, following the detection of the second case, the virus began spreading rapidly across the country. The figure below illustrates the temporal dynamics of the COVID-19 spread in Nepal during the pandemic. As shown in Fig. 11.2a, Nepal experienced three major waves of infections. The peaks of the first, second, and third waves occurred on October 21, 2020; May 7, 2021; and January 20, 2022, respectively. The first wave lasted relatively longer compared to the second and third waves, while the third wave was the shortest. The second wave, however, was more fatal than both the first and third waves.

The spatial distribution patterns of COVID-19 incidence in Nepal have been uneven and heterogeneous since the very beginning of the pandemic. From the onset, certain areas experienced higher rates of infection compared to others, reflecting distinct regional variations in how the virus spread. Kathmandu, the capital and largest city of Nepal, consistently remained a hotspot throughout all three waves of the pandemic (Fig. 11.3). As the central hub for political, economic, and social activities, Kathmandu was particularly vulnerable to rapid transmission due to its dense

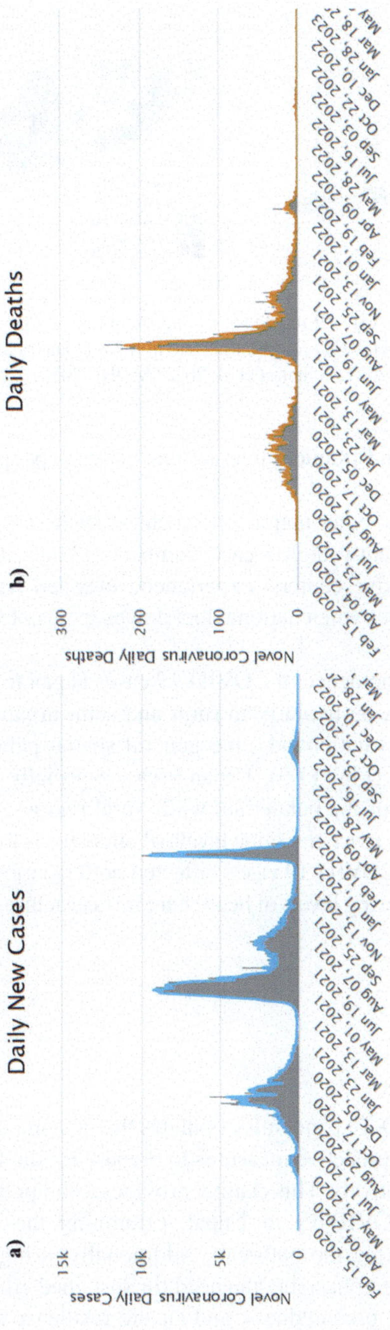

Fig. 11.2 Epidemic curve of **a** COVID-19 reported cases and **b** Death during the three major outbreak in Nepal . *Source* https://www.worldometers.info/coronavirus/country/nepal/)

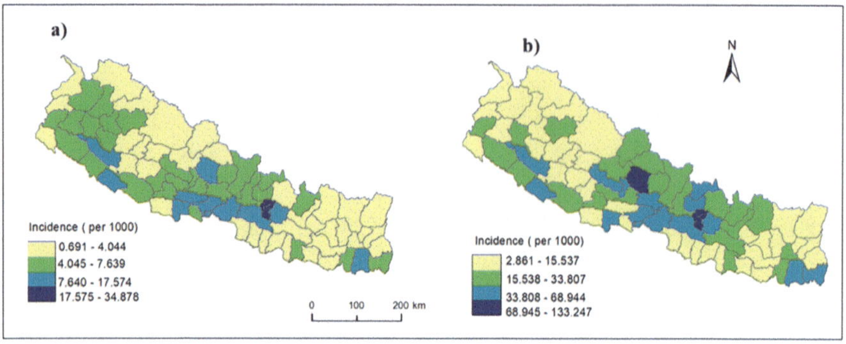

Fig. 11.3 Spatial distribution COVID-19 incidence during the **a** end of first wave (March 2021) at the end of third wave (December 2022) representing the initial and late phase of the COVID-19 Pandemic in Nepal (Basnet et al. 2021; Acharya et al. 2022; MOHP 2022)

population, high mobility, and frequent interactions between people from different regions of the country.

In addition to Kathmandu, other major urban centers such as Pokhara and Biratnagar also reported significantly higher case numbers. These cities, being major commercial and administrative centers, experienced elevated transmission rates, particularly in the initial waves when national lockdowns were not yet fully effective in limiting movement.

As time progressed, the incidence of COVID-19 cases began to increase in other parts of the country as well, particularly in rural and semi-urban areas. However, despite the increase in cases nationwide, the general spatial patterns of infection remained largely consistent (Fig. 11.3). Urban areas, especially the major cities, continued to bear the brunt of the pandemic, while rural regions, although experiencing a rise in cases, did not reach the same levels of intensity as the urban hotspots. This uneven distribution of COVID-19 cases reflected both the movement of people between regions and the differing levels of healthcare infrastructure and public health interventions across Nepal.

11.19 Conclusions

The lessons from the COVID-19 pandemic stipulates that it is imperative to develop robust policy responses and proactive measures to prepare for the future pandemics and health emergencies effectively. This chapter provides a comprehensive overview of the epidemiology of COVID-19 in Nepal, examining the country's policy responses and strategies during the outbreak. Additionally, it highlights the post-COVID-19 health scenario, emphasizing the need for sustained efforts to strengthen healthcare systems, enhance preparedness, and ensure resilience against emerging health threats. By understanding past challenges and successes, this analysis aims to guide the formulation of a more adaptive and coordinated approach to managing pandemics in the future. Nepal's Ministry of Health and Population has developed the post-COVID-19 guidelines.

The Government of Nepal (GON) should develop specific guidelines and policies for disease outbreak response and post-pandemic strategies, similar to those implemented during the COVID-19 pandemic. These should be applied at the central, provincial, and local levels of the government and include public health response, surveillance, mass surveillance for the detection of pandemic diseases, and molecular surveillance. This would involve the development of testing and reporting algorithms, coordination, collaboration, and the sharing of data through a standardized portal to understand the true situation of diseases. The lessons learned from this processes should inform future strategies to mitigate and interrupt the transmission of pandemic diseases across Nepal.

Additionally, the GON, Ministry of Health and Population (MOHP), National Public Health Laboratory (NPHL), and Nepal Health Research Council (NHRC) should work together to establish national testing algorithms that include serum, stool, urine, and wastewater sampling from nearby rivers to determine the environmental status of COVID-19. The existing molecular laboratories with PCR facilities across Nepal should be utilized to strengthen preparedness for future pandemics. From a public health perspective, antigen, antibody, and PCR testing have played vital roles in tracing, testing, and detecting COVID-19 in both pre and post-pandemic periods. Results from RT-PCR cycle threshold (CT) values, along with radiology, hematology, biochemistry, and immunology, should continue to be used as markers for detection (Joshi et al. 2024). Furthermore, the GON should establish research centers dedicated to the development of antiviral drugs, vaccines, and research on pandemic viruses. These centers should incorporate advanced molecular techniques such as whole genome sequencing and virus culture laboratories to enhance testing and detection capabilities, preparing Nepal for any future unknown viral outbreaks.

In conclusion, Government of Nepal (GON) should place a high priority on developing and implementing robust pandemic policy responses while simultaneously addressing the long-term health challenges that have emerged in the post-COVID-19 era. To achieve this, adopting the One Health approach is essential, as it fosters a holistic strategy to mitigate the risks associated with infectious diseases by integrating human, animal, and environmental health sectors (Shah and Joshi 2022). Nepal can draw valuable lessons from initiatives such as the collaboration between the Indian Council of Medical Research (ICMR) and the Integrated Disease Surveillance Program (IDSP), which have partnered with veterinary agencies like the National Centre for Disease Control (NCDC) to enhance the management of zoonotic diseases. By leveraging the One Health model, Nepal can build a comprehensive framework for integrated disease control and prevention, ensuring more effective responses to emerging public health threats and improving overall health resilience (NCDC 2023; ICMR 2017; Joshi et al. 2023; Shah and Joshi 2022).

References

Acharya BK, Adhikari SK, Pandit S, Neupane BK, Paudel BK, Khanal L (2022) An initial spatiotemporal assessment of COVID-19 clusters in Nepal. J Inst Sci Technol 27(1):37–49

Adhikari B, Devkota A, Budhathoki SS, Pinder R, Basnet LB (2021a) COVID-19 crisis in Nepal: a case of systems and governance failure in a low-income country. J Glob Health Econ Policy 1:262996. https://doi.org/10.22022/JGHEP/01.2021.262996

Adhikari K, Gautam R, Pokharel A, Uprety KN, Vaidya NK (2021b) Transmission dynamics of COVID-19 in Nepal: mathematical model uncovering effective controls. J Theor Biol 521:110680

Bahadori M, Khankeh HR, Zaboli R, Raiszadeh M, Malmir I (2016) Health sector coordination in disasters: barriers & facilitators. Health Emergencies Disasters Quarterly 1(4):177–186. Available from: https://hdq.uswr.ac.ir/article-1-119-fa.pdf

Basnet S, Koirala S, Pandey B et al (2020) COVID-19 containment efforts of a low-resource nation: the first four months in Nepal. Cureus 12(7):e8946. https://doi.org/10.7759/cureus.8946

Basnet BB, Bishwakarma K, Pant RR, Dhakal S, Pandey N, Gautam D et al (2021) Combating the COVID-19 pandemic: experiences of the first wave from Nepal. Front Public Health 9:613402. https://doi.org/10.3389/fpubh.2021.613402

Bajracharya B, Shrestha P, Shrestha D, Karna AK, Mahato RK, Shrestha S et al (2024) Co-infection of COVID-19 with malaria and dengue: a hospital-based study in Kathmandu, Nepal. Kathmandu Univ Med J 22(3):132–137

Bakker H, Govindakarnavar A, Krishnan Krishnakumari P, Gromicho J, Côté FL, Lahrichi N et al (2024) Coordination strategies to improve COVID-19 PCR laboratory testing scale-up in Nepal: an analysis. PLoS ONE 19(12):e0314746. https://doi.org/10.1371/journal.pone.0314746

Bastola A, Sah R, Rodriguez-Morales AJ, Lal BK, Jha R, Ojha HC, Shrestha B, Chu DKW, Poon LLM, Costello A et al (2020) The first 2019 novel coronavirus case in Nepal. Lancet Infect Dis 20(3):279–280. https://doi.org/10.1016/S1473-3099(20)30067-0

Bhattarai S, Dhungana J (2021) A million-dose success for Nepal: insights from Nepal's national vaccine deployment plan against COVID-19. J Travel Med 28(taab027). https://doi.org/10.1093/jtm/taab027

Castañer X, Oliveira N (2020) Collaboration, coordination, and cooperation among organizations: establishing the distinctive meanings of these terms through a systematic literature review. J Manag 46(6):965–1001. Available from: https://journals.sagepub.com/doi/10.1177/0149206320901565

Dhakal S, Karki S (2020) Early epidemiological features of COVID-19 in Nepal and public health response. Front Med 7:524. https://doi.org/10.3389/fmed.2020.00524

Epidemiology and Disease Control Division (EDCD) (2024) COVID-19 statistics: Nepal. Retrieved from https://portal.edcd.gov.np/covid19/index.html

Fehr AR, Perlman S (2015) Coronaviruses: an overview of their replication and pathogenesis. Methods Mol Biol 1282:1–23. https://doi.org/10.1007/978-1-4939-2438-7_1

Gupta GP, Shah Y, Pant DK, Lekhak S, Shahi R, Mandal A, Poudel P, Dumre SP, Pandey K, Pandey BD (2021) Preparatory phase for clinical trials of COVID-19 vaccine in Nepal. Hum Vaccin Immunother 17(2):418–419. https://doi.org/10.1080/21645515.2020.1805510

Gyanwali P, Bista NR, Khadka M, Vaidya A, Mahato NK, Karn MK, Pant S, Ghimire N, Pokhrel A, Dhimal M (2021) Assessment of preparedness of Government of Nepal in COVID designated hospitals and clinics for pandemic response. J Nepal Health Res Counc 19(1):48–54. https://doi.org/10.33314/jnhrc.v19i1.3367

Government of Nepal, Ministry of Health and Population (2022) Post-pandemic recovery plan. Nepal, Kathmandu

Indian Council of Medical Research (2018) Annual report 2017–2018. Retrieved 28 June 2023, from https://niirncd.icmr.org.in/annualreport/en/AR_English_2017_18_final.pdf

Joshi J, Shah Y, Pandey K, Ojha RP, Joshi CR, Bhatt LR, Dumre SP, Acharya PR, Joshi HR, Rimal S, Shahi R (2023) Possible high risk of transmission of the Nipah virus in South and Southeast Asia: a review. Trop Med Health 51(1):44

Joshi J, Joshi OP, Kamar SB et al (2024) Association of RT-qPCR Ct values and disease severity among COVID-19 patients visiting a tertiary care hospital in Nepal. J Pure Appl Microbiol 18(1):662–671. https://doi.org/10.22207/JPAM.18.1.51

Kandikattu HK, Yadavalli CS, Venkateshaiah SU, Mishra A (2021) Vaccine efficacy in mutant SARS-CoV-2 variants. Int J Cell Biol Physiol 4:1–12

Koirala P, Tamrakar D (2020) Threat of dengue outbreak in Nepal in the context of COVID-19 pandemic. J Lumbini Med Coll 8(1):169–170. https://doi.org/10.22502/jlmc.v8i1.392

Koirala J, Basnet S (2020) COVID-19: a global pandemic causing local threat in Nepal. J Adv Intern Med 9(1):1–3. https://doi.org/10.3126/jaim.v9i1.29154

Lama G, Shrestha L, Karmacharya N, Manandhar R, Jha R (2022) Determination of sero-prevalence of SARS-CoV-2 antibody among immunized individuals with COVID-19 vaccine in a tertiary center, Nepal. Int J Multidiscip Res Growth Eval 3:62–65

Lamsal M (2021) The risk of dengue and scrub typhus overshadowed by COVID-19 in Nepal. Asia Pac J Publ Health 33(5):679–680. https://doi.org/10.1177/1010539521990605

Li F (2016) Structure, function, and evolution of coronavirus spike proteins. Ann Rev Virol 3(1):237–261. https://doi.org/10.1146/annurev-virology-110615-042301

Lu R, Zhao X, Li J, Nie P, Yang B, Wang H, Wang W, Song H, Huang B, Zhu N et al (2020) Genomic characterisation and epidemiology of 2019 novel coronavirus: implications for virus origins and receptor binding. Lancet 395(10224):565–574. https://doi.org/10.1016/S0140-6736(20)30251-8

Ministry of Health and Population (MOHP) (n.d.) COVID-19. Retrieved 15 Nov 2021, from https://covid19.mohp.gov.np/

Ministry of Health and Population (MoHP) (2021) Enhanced surveillance on sero prevalence of SARS-CoV-2 in the general population of Nepal. MoHP, Kathmandu, Nepal

Ministry of Health and Population (2021) Responding to COVID-19: health sector preparedness, response and lessons learnt. Available from: https://old.mohp.gov.np/attachments/article/703/Responding%20to%20COVID19,%20Health%20sector%20preparedness,%20response%20and%20lessons%20learnt.pdf

Ministry of Health and Population, Nepal (2022)

Mishra SK, Adhikari SK, Sah PK, Gocotano AE, Neupane S, Thapa B et al (2024) Effectiveness of health partners coordination for COVID-19 pandemic response in Nepal. PLoS ONE 19(10):e0308941

National Center for Disease Control (2023) Nipah virus guidelines. Retrieved 7 July 2023, from https://ncdc.gov.in/index1.php?lang=1&level=1&sublinkid=238&lid=242

National Human Rights Commission (NHRC) (2022) Assessment of compliance with SMS measures against COVID-19 in Kathmandu Valley 2020, health sector response to COVID-19 pandemic in Nepal. NHRC, New Delhi, India

Nepal Risk Outlook (2022) COVID-19 pandemic and Nepal's road to post-pandemic recovery: June 2022 policy compendium

Pandey K, Dumre SP, Dhimal M, Pun SB, Shah Y, Fernandez S, Morita K, Pandey BD (2021) The double burden of COVID-19 and dengue in Nepal: the challenges ahead. Kathmandu Univ Med J 19(1):140–142

Pandey BD, Ngwe Tun MM, Pandey K, Dumre SP, Nwe KM, Shah Y, Culleton R, Takamatsu Y, Costello A, Morita K (2022a) How an outbreak of COVID-19 circulated widely in Nepal: a chronological analysis of the national response to an unprecedented pandemic. Life 12:1087. https://doi.org/10.3390/life12071087

Pandey BD, Tun MM, Pandey K, Dumre SP, Bhandari P, Pyakurel UR, Pokhrel N, Dhimal M, Gyanwali P, Culleton R, Takamatsu Y (2022b) Has COVID-19 suppressed dengue transmission in Nepal? Epidemiol Infect 150:e196. https://doi.org/10.1017/S0950268822001553

Pantha B, Acharya S, Joshi HR, Vaidya NK (2021) Inter-provincial disparity of COVID-19 transmission and control in Nepal. Sci Rep 11(1):13363

Panthee B, Dhungana S, Panthee N, Gyawali S, Paudel A, Panthee S (2020a) Clinical and epidemiological features of COVID-19 deaths in Nepal. New Microb New Infect 38:100797. https://doi.org/10.1016/j.nmni.2020.100797

Panthee B, Dhungana S, Panthee N, Paudel A, Gyawali S, Panthee S (2020b) COVID-19: the current situation in Nepal. New Microb New Infect 37:100737. https://doi.org/10.1016/j.nmni.2020.100737

Policy Research Institute (2022) A study of Nepal's public policy processes: policymakers' perception and use of information in decision-making during the COVID-19 pandemic. Policy recommendations for Nepal's public policy processes

Rayamajhee B, Pokhrel A, Syangtan G, Khadka S, Lama B, Rawal LB, Mehata S, Mishra SK, Pokhrel R, Yadav UN (2021) How well the government of Nepal is responding to COVID-19? An experience from a resource-limited country to confront unprecedented pandemic. Front Public Health 9:597808. https://doi.org/10.3389/fpubh.2021.597808

ReliefWeb (2022) Rapid assessment on impacts of COVID-19 on returnee migrants and responses of local governments of Nepal. Available online: https://reliefweb.int/report/nepal/rapid-assessment-impacts-covid-19-returnee-migrantsand-responses-local-governments. Accessed on 10 June 2022

Sah R, Rodriguez-Morales AJ, Jha R, Chu DKW, Gu H, Peiris M, Bastola A, Lal BK, Ojha HC, Rabaan AA et al (2020) Complete genome sequence of a 2019 novel coronavirus (SARS-CoV-2) strain isolated in Nepal. Microbiol Resour Announcements 9(e00169–20). https://doi.org/10.1128/MRA.00169-20

Sapkota K, Dangal G, Koirala M, Sapkota K, Poudel A, Dhital SR (2020) Strategies for prevention and control of COVID-19 in Nepal. J Patan Acad Health Sci 7(1):85–88. Available from: https://doi.org/10.3126/jpahs.v7i1.28879

Sapkota B, Dhakal G, Chalise G, Shrestha S (2022) Post COVID health problems faced by COVID-19 infected healthcare professionals of a tertiary level hospital. MJSBH 21(2):1–7. https://doi.org/10.3126/mjsbh.v21i2.47361

Shah Y, Joshi J (2022) Introductory chapter: plan and preparedness to prevent and combat against the pandemic RNA virus infection. In: RNA viruses infection. IntechOpen. https://doi.org/10.5772/intechopen.106743

Shah Y, Paudel S, Pandey K, Gupta GP, Solo ES, Joshi J et al (2022) Insights into transmission dynamics of Mycobacterium tuberculosis complex in Nepal. Trop Med Health 50(1):8. https://doi.org/10.1186/s41182-022-00400-z

Shoaf K (2014) Organizing the health sector for response to disasters. Ciencia Saúde Coletiva 19(9):3705–3715. Available from: https://www.scielo.br/j/csc/a/4QVJq7GgxmkCdYBbFJZrnYB/?format=pdf&lang=en. https://doi.org/10.1590/1413-81232014199.03722014

Srivastava N, Nepal B, Gauchan B, Kennedy A (2021) Policy responses to the COVID-19 pandemic in Nepal. International Food Policy Research Institute.

The Global Cluster Coordination Group (2014) Cluster coordination performance monitoring: guidance note. Available from: https://reliefweb.int/report/world/cluster-coordination-performance-monitoring-guidance-note

The Government of Nepal, Ministry of Home Affairs (2013) National disaster response framework. Available from: http://drrportal.gov.np/uploads/document/113.pdf

The Pandemic Fund (n.d.) Strengthening pandemic preparedness for early detection [Project overview]. https://www.thepandemicfund.org/projects/NEPAL-strengthening-pandemic-preparedness-early-detection

World Bank (2020) Nepal development update: post-pandemic Nepal—charting a resilient recovery and future growth directions. World Bank. https://www.worldbank.org/en/news/feature/2020/07/23/nepal-development-update-post-pandemic-nepal-charting-a-resilient-recovery-and-future-growth-directions

World Health Organization (WHO) Country Office for Nepal (2021) Situation update-84: Coronavirus Disease 2019 (COVID-19) reporting. Available online: https://covid19.mohp.gov.np/situation-report. Accessed on 10 Nov 2021

World Health Organization (WHO) Country Office for Nepal (2022) Situation report-800: Coronavirus Disease 2019 (COVID-19) reporting. Available online: https://covid19.mohp.gov.np/situation-report. Accessed on 20 June 2022

World Health Organization (2016) The International Health Regulations (2005)—Third edition. Available from: https://iris.who.int/bitstream/handle/10665/246107/9789241580496-eng.pdf?sequence=1

World Health Organization (2019) Health cluster coordination guidance for heads of WHO country offices as cluster lead agency. Available from: http://apps.who.int/bookorders

World Health Organization (n.d.) COVID-19 IHR emergency committee statements. Available online: https://www.who.int/groups/covid-19-ihr-emergency-committee

World Health Organization (2020) COVID-19 strategic preparedness and response plan: operational planning guidelines to support country preparedness and response. Available from: https://www.who.int/docs/default-source/coronaviruse/covid-19-sprp-operational-planning-guidelines-to-support-country-preparedness-and-response-(22may20).pdf

World Health Organization (2020) Statement on the second meeting of the International Health Regulations (2005) Emergency Committee regarding the outbreak of novel coronavirus (2019-nCoV). Retrieved 15 Nov 2021, from https://www.who.int/news/item/30-01-2020-statement-on-the-second-meeting-of-the-international-health-regulations-(2005)-emergency-committee-regarding-the-outbreak-of-novel-coronavirus-(2019-ncov)

World Health Organization (2021) Health systems resilience during COVID-19. Retrieved from https://www.who.int

World Health Organization (2024) COVID-19 epidemiological update: edition 173. Retrieved from https://www.who.int/emergencies/diseases/novel-coronavirus-2019/situation-reports

Worldometer (n.d.) Coronavirus cases: Nepal. Worldometer. Retrieved 16 Dec 2024, from https://www.worldometers.info/coronavirus/country/nepal

Chapter 12
Impact of COVID-19 on Health Care and Human Adaptation: Snapshots from Kuwait

Mona Al-Ahmad and Hamad Bastaki

Abstract The COVID-19 pandemic profoundly affected healthcare systems, highlighting vulnerabilities with far-reaching implications beyond health disciplines. In response, Kuwait invested significantly in pandemic containment and mitigation strategies, bolstering its health system's resilience. This chapter explores the adaptational changes within Kuwait's healthcare landscape post-COVID-19. It highlights the nation's strategic initiatives, including community-based surveys for SARS-CoV-2, the implementation of an electronic vaccine registration system, and the establishment of a vaccine adverse event reporting system. These measures represent a paradigm shift in Kuwait's healthcare approach, signifying a broader impact of the pandemic on healthcare policy and practice.

Keywords COVID-19 · SARS-CoV-2 · Healthcare · Human behavior · Vaccine · Adverse events · Kuwait

12.1 Introduction

Originating in China, COVID-19 rapidly spread around the world in early 2020. As such, governments, business leaders, and citizens were required to adapt to constantly evolving socio-economic conditions and respond to unforeseen challenges. While no country could escape the negative effects of the COVID-19 pandemic, Kuwait's sound macroeconomic fundamentals allowed it to withstand some of the most severe challenges presented by the crisis. Kuwait had invested in a resilient healthcare system that withstand the crisis in an adaptable way. In this chapter, we discuss some of Kuwait strategic adaptations during the COVID-19 pandemic.

M. Al-Ahmad (✉)
Microbiology Department, College of Medicine, Kuwait University, Kuwait City, Kuwait
e-mail: mona.alahmad@ku.edu.kw

H. Bastaki
Head of Communicable Disease Control Division, Ministry of Health, Kuwait City, Kuwait

R. Akhtar (ed.), *COVID-19: Impacts on Health and Health Systems*,
https://doi.org/10.1007/978-3-031-90255-0_12

12.2 Community Survey of SARS-CoV-2 in Kuwait

12.2.1 Introduction

The SARS-CoV-2 pandemic highlighted the limitations of conventional symptom-based testing and contact tracing, particularly with the virus's ability to spread asymptomatically (Gandhi et al. 2020). In Kuwait, a community survey utilizing both molecular (RT-PCR) and serological testing provided valuable insights into SARS-CoV-2 transmission, particularly in identifying asymptomatic cases that typical surveillance might miss. By employing random community screening and time-series data collection, this survey aimed to capture a holistic view of disease prevalence, offering strategic benefits over other surveillance methods.

12.2.2 Objectives of the Community Survey

The primary objectives of the survey were to:

1. Estimate the SARS-CoV-2 seroprevalence, stratified by demographics such as age and sex, to assess community-level immunity.
2. Quantify current infections using RT-PCR to gauge disease activity, including asymptomatic and pre-symptomatic cases not captured through traditional testing.
3. Provide real-time, regional infection rates to guide targeted interventions.

12.2.3 Methodology

The survey covered five health regions in Kuwait, incorporating a multi-stage probability sampling strategy to ensure representative coverage across demographic and geographic segments. Participants were selected randomly and screened weekly, with samples taken via nasopharyngeal swabs for RT-PCR and blood samples for serological testing (Padula 2020; WHO 2020). Both RT-PCR and serology tests were conducted to distinguish between active infections and past exposures, using IgM and IgG antibodies as indicators (WHO 2020; ECDC 2020). The operational structure of the survey included mobile testing units staffed with public health doctors, nursing staff, health inspectors, and IT support to ensure effective data collection and processing. Data analysis employed statistical techniques to compare infection rates across regions, demographics, and time.

12.2.4 Advantages of Community Surveys in Infectious Disease Monitoring

1. **Capturing Asymptomatic Spread**: Unlike facility-based or symptom-based testing, random community screening detects asymptomatic and subclinical cases, providing a more accurate picture of SARS-CoV-2 transmission. As asymptomatic individuals can silently fuel the spread, identifying these cases is crucial for informed public health decisions (Lohse et al. 2020).
2. **Evaluating Immunity Levels**: Serological testing within the survey framework helped estimate community-level immunity, crucial for understanding the pandemic's progression and planning vaccine rollouts. Estimating immunity among different demographic groups also aids in identifying potential immunity gaps.
3. **Real-Time, Regional Data**: By performing regular, region-specific testing, the survey tracked fluctuations in infection rates across Kuwait. This data enabled the Ministry of Health to identify and prioritize high-burden areas, guiding resource allocation and tailored interventions.
4. **Adaptable to Changing Public Health Needs**: Community surveys can quickly respond to emerging variants or shifts in infection rates (Padula 2020). The survey's design allowed it to scale up or down as needed, maintaining a balance between comprehensive surveillance and operational feasibility.
5. **Supporting Policy Development and Intervention**: The real-time data and demographic insights gathered from the survey informed policy decisions, such as implementing or lifting movement restrictions in high-prevalence areas. The survey also identified specific socio-demographic factors associated with increased infection risk, contributing to targeted public health campaigns.

12.2.5 Comparative Analysis: Community Surveys Versus Other Surveillance Methods

Community surveys stand out for their ability to detect asymptomatic carriers, estimate population-level immunity (Lohse et al. 2020; Gudbjartsson et al. 2020) and monitor disease trends within distinct communities, Table 12.1.

12.2.6 Findings and Impact

Data collected during the survey showed varying infection rates across Kuwait's health regions, with particular hotspots identified in high-density areas. Furthermore, in terms of temporal trends, peaks in test positivity occurred at different times of year and did not follow a seasonal pattern. When combined with data from genomic

Table 12.1 Comparative analysis: community surveys vs. other surveillance methods

Surveillance method	Key characteristics	Limitations
Community surveys	Random, representative sampling, captures asymptomatic cases	Resource-intensive, requires mobile and logistical support
Facility-based testing	Quick access to symptomatic patients	Misses asymptomatic cases, limited to those seeking care
Contact tracing	Identifies clusters, links cases to sources	Time-consuming, dependent on accurate reporting. Positivity of result affected by timing of test post-exposure
Sentinel surveillance	Monitors specific high-risk sites (facility or community)	Limited scope, may not represent broader community dynamics

sequencing, peaks in test positivity were always preceded by the detection of new variants indicating an emerging variant driven surge in cases. Additionally, serology results revealed a significant portion of the population had either active or past infections, indicating ongoing community transmission despite symptom-based testing and contact tracing efforts.

12.3 The Electronic Vaccine Registration System (EVRS): A Digital Leap Forward in Kuwait's Public Health Infrastructure

12.3.1 Introduction: Adapting to the Challenges of COVID-19

The COVID-19 pandemic catalyzed the transformation of healthcare systems globally. In Kuwait, one of the key responses to managing the pandemic was the development of an electronic vaccine registration system. Faced with a limited supply of vaccines during the initial rollout, the country needed a method to identify and prioritize high-risk groups, schedule vaccinations efficiently, and reduce on-site registration times. The *Electronic Vaccine Registration System (EVRS)* was introduced in December 2020 specifically to address these challenges during the COVID-19 vaccine rollout. This initial success paved the way for the system to be further developed and expanded to include all routine vaccinations in November 2021.

12.3.2 Initial Rollout: Managing the COVID-19 Vaccination Campaign

When the COVID-19 vaccines became available, there was a pressing need for an efficient, scalable system to manage vaccine distribution. The EVRS was quickly deployed by the Ministry of Health in collaboration with the Communicable Diseases Control Division. The system was initially designed to streamline several key functions in the COVID-19 vaccine distribution process:

- **Identification and Prioritization of Risk Groups**: Given the limited supply of vaccines during the early phases of the pandemic, it was critical to identify and prioritize those most at risk, such as the elderly, frontline workers, and individuals with chronic illnesses. The EVRS allowed healthcare authorities to systematically register individuals from these groups, ensuring they received priority access to the vaccines.
- **Appointment Scheduling**: To manage the large demand for vaccines while ensuring safety measures such as social distancing, the EVRS provided an integrated appointment scheduling feature. This ensured that vaccinations could be administered in a controlled and orderly fashion.
- **Streamlining Registration and Reducing Wait Times**: The EVRS enabled individuals to register online before their appointment, allowing key information such as personal details, medical history, and civil ID numbers to be pre-filled. This significantly reduced the time spent on administrative tasks during the vaccine administration process, helping to increase the efficiency of vaccination centers.

12.3.3 Expansion of the System: Incorporating All Vaccinations

Following the success of the EVRS in managing the COVID-19 vaccination campaign, the system was further developed to include all vaccinations by November 2021. This expansion reflected a long-term strategy to digitize and modernize Kuwait's public health infrastructure. The extended system now manages vaccinations for all age groups and includes seasonal immunizations such as the winter flu vaccine as well as childhood vaccines.

12.3.4 Key Features and Functions of the Expanded EVRS

The enhanced EVRS introduced new functionalities that made it a critical tool for the broader public health landscape:

- **Comprehensive Vaccine Registration Interface**: The expanded system now allows healthcare professionals to register individuals for all types of vaccines and serums. The same user-friendly portal, initially designed for COVID-19 vaccines, was adapted to support routine immunization schedules and seasonal vaccination programs.
- **Vaccination History and Default Notifications**: The system's ability to track and maintain vaccination histories is crucial for ensuring individuals remain up to date on required vaccines. If a patient fails to receive a scheduled vaccination within two weeks, their details are flagged, and a defaulter list is generated for follow-up. This function has proven instrumental in maintaining high vaccination coverage.
- **Daily Vaccination Records and Reporting**: Medical staff can easily track daily vaccination records and generate detailed reports by health center, vaccine type, or individual identifiers such as civil ID numbers. These records are critical for public health officials in monitoring vaccination rates and ensuring resource allocation.
- **Adaptability to Various Population Groups**: The EVRS was designed to handle a variety of population groups, including those without valid civil IDs, such as newborns and visitors. Temporary IDs can be generated to ensure that everyone, regardless of their status, is included in the vaccination program.

12.3.5 Impact of EVRS on Kuwait's Public Health Response

The EVRS has had a profound impact on Kuwait's ability to manage the COVID-19 pandemic and beyond. Its initial success during the COVID-19 vaccination campaign demonstrated the benefits of a centralized, electronic system in improving public health efficiency. By automating appointment scheduling, reducing registration times, and enabling real-time data tracking, the EVRS allowed Kuwait to manage its vaccine rollout effectively under challenging conditions.

As the system expanded to include all vaccinations, it has further reinforced the country's public health infrastructure, ensuring a seamless transition from crisis management during the pandemic to routine health services. The integration of winter vaccinations, as well as vaccines for diseases such as HPV, highlights the versatility and scalability of the EVRS.

12.3.6 Challenges and Ongoing Improvements

Despite its successes, the EVRS has encountered challenges. System downtimes have, at times, required healthcare workers to revert to paper-based documentation. However, the Ministry of Health has implemented protocols for ensuring that any data recorded manually is promptly uploaded into the EVRS once the system is back online.

Ongoing improvements to the system, including enhanced cybersecurity measures and further integration with other health databases, will help solidify its role in Kuwait's public health strategy for the future.

12.3.7 Vaccine Adverse Event Reporting System

Up to December 31, 2023, different types of coronavirus disease 2019 (COVID-19) vaccines have been authorized for use in the state of Kuwait. The pharmaceutical and Herbal Medicines Registration and Control Administration issued Emergency Use Authorization (EUA) for Pfizer-BioNTech Pediatric COVID-19 vaccine for individuals 5 to 11 years, Pfizer-BioNTech COVID-19 vaccine for individuals 12 years and above, Pfizer-BioNTech COVID-19 Bivalent vaccine for individuals 12 years and above, Spikevax Monovalent (Moderna) COVID-19 Vaccine for individuals 6 years and above, Spikevax Bivalent original/Omicron BA.1 and Spikevax Bivalent Original/Omicron BA.4/BA.5 for individuals 12 years and above and Spikevax XBB 1.5 for individuals 6 months and above, AstraZeneca / Oxford COVID-19 Vaccine for individuals 18 years and above. (Emergency Use Authorization is valid for one year, in case of non-renewal, the authorization is considered invalid).

During the period covered by this report, the vaccines that were used in COVID-19 immunization program were for individuals 5 years and above, booster dose was administered after 3 months from primary series and as recommended for high-risk or immunocompromised individuals.

The clinical trials of COVID-19 vaccines have shown them to be safe and effective; however, safety monitoring is a continuous and ongoing process to ensure that vaccines benefits continue to outweigh any risk.

This safety monitoring strategy is achieved by continuous, proactive surveillance which is based on a wide range of information sources, with a dedicated team of scientists reviewing information daily to look for safety issues or unexpected, rare events.

Since the start of Kuwait National COVID-19 Vaccination Campaign on December 27, 2020, Surveillance for adverse events following immunization (AEFI) was an integral part of the National Immunization Program and reinforces the safe use of all vaccines in the state of Kuwait while also helping to maintain public confidence in its immunization program.

12.4 Safety Surveillance Tools

12.4.1 Vaccine Adverse Events Reporting System

Ministry of Health has created different electronic platforms for reporting of any suspected adverse events following immunization, where the vaccinated individual can report through an electronic link to the adverse events reporting form received by text message after vaccination.

The healthcare professionals in Ministry of Health hospitals and clinics can report via the COVID-19 vaccine electronic system, moreover, the reporting form has been added to Ministry of Health website to enable healthcare professionals in private sector to easily report any suspected adverse events following immunization.

12.4.2 AEFI Hospitals Focal Points

A team of healthcare professionals have been assigned representing all ministry of health main hospitals as focal points to report and collect data on adverse events following COVID-19 vaccines. For serious cases, the focal points provide the adverse events following immunization committee with the patient medical reports, laboratory tests results, imaging reports and all the needed information to investigate and assess the cases.

12.4.3 Safety Communication Team

In collaboration with Public Relations Department, a safety communication team has been created consisting of physicians and pharmacists to communicate with the reporters of adverse events following COVID-19 vaccination to obtain more information on the reported adverse events and also to conduct active safety surveillance for special sub-population or to monitor adverse events of special interest.

12.4.4 Adverse Events Following Immunization

An adverse event following immunization is any untoward medical occurrence (unfavorable or unintended sign, abnormal laboratory finding, symptom or disease) which follows immunization, and which does not necessarily have a causal relationship with the usage of the vaccine. Reported adverse events can either be true adverse events—i.e., resulting from the vaccine or immunization process—or coincidental events that are not due to the vaccine or immunization process but are temporally associated with immunization.

Reported adverse events are being continually assessed to identify any new safety signals. Serious adverse events or unexpected adverse events undergo an investigation process by specialized medical team followed by causality assessment (as per WHO AEFI causality assessment criteria) by AEFI committee to determine if the adverse event is related to the vaccine or not related.

Statistical techniques are applied to provide information on the numbers and rates of events to guide us to detect clusters or if reports of adverse events received are more than expected, which could indicate a new safety concern.

Reporting of AEFI and subsequent investigation may trigger regulatory action including withdrawing the marketing authorization of a vaccine, instructing vaccine manufacturers to change their product labels, restricting the use of vaccines to specific patient groups, or recalling defective vaccine batches from the market.

12.4.5 Vaccine Reactions

Based specifically on cause, seriousness and frequency, vaccine reactions may be grouped into two broad categories:

1. Cause-specific vaccine reactions:
   - vaccine product-related reaction and
   - vaccine quality defect-related reaction
2. Vaccine reactions by seriousness and frequency:
   - common or minor reactions.
   - rare or serious reactions.
3. Immunization error-related reaction.
4. Immunization anxiety-related reaction.
5. Coincidental events.

12.4.6 Total Number of Adverse Events Reports Received

From December 27, 2020 to December 31, 2023, the total number of adverse events reports received for all vaccines were 450,054 reports. After filtration of the repeated reports (reports sent by the reporter more than one time for the same dose), the total number of reports were 143,799 with a reporting rate of 1.9%. Most of which were local and mild.

12.5 Conclusion

Kuwait acted quickly to limit the spread of COVID-19 within its borders. Kuwait's community survey of SARS-CoV-2 infections exemplifies the advantages of combining molecular and serological testing in a representative, random sampling approach. It provided a comprehensive view of infection dynamics, supports informed decision-making, and aids in targeted interventions. Furthermore, Kuwait's experience with the EVRS underscores the importance of digital infrastructure in managing public health emergencies and routine healthcare services. The system's ability to adapt from the specific needs of the COVID-19 crisis to the broader demands of nationwide immunization campaigns demonstrates its value as a long-term public health asset. Moving forward, Kuwait's model could serve as a blueprint for other countries looking to modernize their vaccination systems and improve their readiness for future health crises.

In future pandemics, community surveys will remain a valuable tool for early detection, real-time monitoring, and evaluating the effectiveness of public health measures. By offering insights beyond conventional reporting-based surveillance methods, community surveys proved to be a robust method for infectious disease surveillance in Kuwait, particularly during the COVID-19 pandemic. A dedicated Public Health Emergency Operations Centre (PHEOC) is in urgent need, and this has been already in implementation in Kuwait healthcare system.

References

European Centre for Disease Prevention and Control (2020) Methodology for estimating point prevalence of SARS-CoV-2 infection by pooled RT-PCR testing. ECDC, Stockholm

Gandhi M, Yokoe DS, Havlir DV (2020) Asymptomatic transmission, the Achilles' heel of current strategies to control covid-19. N Engl J Med 28;382(22):2158–2160

Gudbjartsson DF, Helgason A, Jonsson H, Magnusson OT, Melsted P, Norddahl G, Saemundsdottir J, Sigurdsson A, Sulem P, Agustsdottir AB, Eiriksdottir B, Fridriksdottir R, Gardarsdottir EE, Georgsson G, Gretarsdottir OS, Gudmundsson KR, Gunnarsdottir TR, Gylfason A, Holm H, Jensson BO, Jonasdottir A, Jonsson F, Josefsdottir KS, Kristjansson T, Magnusdottir D, le Roux L, Sigmundsdottir G, Sveinbjornsson G, Sveinsdottir KE, Sveinsdottir M, Thorarensen EA, Thorbjornsson B, Löve A, Masson G, Jonsdottir I, Möller AD, Gudnason T, Kristinsson KG, Thorsteinsdottir U, Stefansson K (2020) Spread of SARS-CoV-2 in the Icelandic population. N Engl J Med 382(24):2302–2315

Lohse S, Pfuhl T, Berko-Gottel B, Rissland J, GeiBler T, Gartner B, Becker SL, Schneitler S, Smola S (2020) Pooling of samples for testing for SARS-CoV-2 in asymptomatic people. Lancet Infect Dis 20(11):1231–1232

Padula WV (2020) Why only test symptomatic patients? Consider random screening for COVID-19. Appl Health Econ Health Policy 18(3):333–334

World Health Organization (2020) Laboratory testing for coronavirus disease (COVID-19) in suspected human cases: interim guidance. WHO

Chapter 13
Impact of Air Pollution on COVID-19 Mortality and Morbidity; Pre- and Post-COVID-19 Global and Turkish Scenarios

Nur Konyalilar, Muge Akpinar-Elci, and Hasan Bayram

Abstract Studies suggest a positive association between increased levels of air pollutants, such as particulate matter (PM), ozone, and nitrogen dioxide, and Coronavirus Disease 2019 (COVID-19) mortality and morbidity worldwide, including Turkey. PM has been shown to carry SARS-CoV-2 virus. Air pollutants can induce SARS-CoV-2 infection in human beings by both stimulating the expression of proteins (i.e., ACE2 and TMPRSS2) required for viral entry and impairing host immune defence. The objectives of this chapter are to discuss the recent epidemiological evidence, as well as the in vivo and in vitro mechanistic studies involving the impact of air pollutants on COVID-19 mortality and morbidity together with the Turkish perspective on COVID-19 from an epidemiological standpoint, focusing on air pollution and health infrastructure during and after the pandemic.

Keywords Air pollution · COVID-19 · Respiratory · Epidemiology · Mechanisms · Turkey

---

N. Konyalilar · H. Bayram (✉)
Koc University Research Centre for Translational Medicine (KUTTAM), Istanbul, Turkey
e-mail: habayram@ku.edu.tr

N. Konyalilar
e-mail: nkonyalilar18@ku.edu.tr

M. Akpinar-Elci
School of Public Health, University of Nevada Reno, Reno, NV, USA

H. Bayram
Department of Pulmonary Medicine, School of Medicine, Koc University, Istanbul, Turkey

R. Akhtar (ed.), *COVID-19: Impacts on Health and Health Systems*,
https://doi.org/10.1007/978-3-031-90255-0_13

13.1 Introduction

Air pollution is one of the world's most significant health and environmental concerns (Fuller et al. 2022). Based on Global Burden of Disease (GBD) data, 9% of deaths worldwide are attributed to air pollution, corresponding to approximately 5 million deaths each year (Stanaway et al. 2018). Air pollution originates from indoor and outdoor air pollutants such as particulate matter (PM) and harmful gases including ozone (O_3), oxides of nitrogen (NOx), sulfur oxides (SOx), carbon monoxide (CO), mainly deriving from industrial facilities, energy production, transportation, household combustion, desert dust, and forest fires (Abbasi-Kangevari et al. 2023).

Air pollution not only causes deterioration of human health but also has a negative impact on the prognosis of diseases (Amnuaylojaroen and Parasin 2021). These pollutants accelerate the risk for cardiopulmonary pathologies such as respiratory infections, COPD, asthma, lung cancer, and metabolic diseases including diabetes (Bayram et al. 2017; Boğan et al. 2022; Comunian et al. 2020; Ritchie 2024). Epidemiological studies reported a close association between air pollution and mortality and morbidity due to Coronavirus Disease 2019 (COVID-19) (Aykaç and Etiler 2022; Bowe et al. 2021; Fattorini and Regoli 2020; Filippini et al. 2020; Wu et al. 2020; Zhu et al. 2020). Elevated levels of $PM_{2.5}$, PM_{10} and gases like NO_2 were linked to higher infection rates and worse outcomes, with long-term exposure associated with increased mortality. Recent studies across various regions, including China, Italy, the USA, and Latin America, further confirmed that air pollution exacerbated COVID-19 severity (Bonilla et al. 2023; Conticini et al. 2020; Wu et al. 2020; Yao et al. 2020; Zhu et al. 2020). These findings underscore the importance of addressing environmental pollution as a key factor in public health responses to the pandemic.

Parallel results linking air pollution to higher COVID-19 cases and deaths were obtained in Turkey, too. Tanis and Karakaya (2021) and Aykaç and Etiler (2022) found that pollutants such as $PM_{2.5}$, PM_{10}, NO_2, and SO_2 were associated with increased COVID-19 cases and the disease's mortality, particularly in metropolitan cities like Istanbul (Aykaç and Etiler 2022; Tanis and Karakaya 2021). Broader studies confirmed similar trends with regional variations (Akan 2022; Yıldırım 2023). Besides that, the COVID-19 pandemic in Turkey highlighted significant challenges in public health preparedness, including limitations in healthcare infrastructure, data transparency, and logistical support. While early measures such as border closures and lockdowns were implemented, the healthcare system faced strains due to under-resourced facilities and inefficient management (Çapar et al. 2024; Kartoglu and Pala 2023). To enhance future pandemic response, Turkey must strengthen healthcare infrastructure, improve data transparency, and refine crisis management strategies while also addressing environmental health factors and ensuring more robust preparedness planning (Gouglas et al. 2023; Gul and Yucesan 2021).

This chapter reviews global published data regarding epidemiological and mechanistic studies, exploring the association between air pollution and COVID-19 incidence, hospitalization, and mortality. It also provides an updated perspective on

Turkey's COVID-19 management capacity and the current situation, focusing on air pollution and health infrastructure during and the post-pandemic period.

13.1.1 Air Pollution and Respiratory Effects

PM is a combination of solid and liquid particles suspended in the atmosphere with varying diameters and chemical compositions, containing organic and biological compounds, elemental and organic carbon, nitrates, sulfates, and metals (Rai 2016). O_3 is a potent oxidant that can be detrimental at increased levels. It is formed through a photochemical reaction between nitrogen oxides and volatile organic compounds (VOCs), which are emitted from both natural sources and anthropogenic activities. (Kim et al. 2020; Manisalidis et al. 2020). NOx are mostly produced by motor engines, a traffic-related pollutant, and during electricity generation by thermal power plants. SOx are harmful gases mainly emitted from fossil fuel combustion or industrial activities (Manisalidis et al. 2020). CO is a byproduct of incomplete combustion of carbonaceous fuels like wood, tobacco, gasoline, and coal (Levy 2015).

Unlike larger ones, small particles such as PM $\leq$ 2.5 μm ($PM_{2.5}$) can reach the alveolar region and enter the circulation, potentially affecting other organs (Pu et al. 2022). It has been shown that PM can cause pulmonary inflammation due to increased oxidative stress so that free radicals. It damages the immune defense by reducing the host resistance due to epithelial barrier integrity disruption, with elevated release of key inflammatory mediators such as interleukin (IL)-1β, IL-6, and IL-8 (Cooper and Loxham 2019; Donaldson et al. 2001). Elevated O_3 can also cause inflammatory changes and airway epithelial damage, disruption in mucociliary clearance, and epithelial barrier integrity (Akdis 2021; Bayram et al. 2002). Moreover, it can cause deterioration in respiratory functions (Gryparis et al. 2004). NOx can cause chronic lung diseases and sense of smell impairment in response to long-term exposure. SOx can lead to respiratory irritation, bronchospasm, and bronchitis. (Manisalidis et al. 2020). CO poses significant risks due to its high affinity for hemoglobin, leading to hypoxia and various respiratory and cardiovascular issues in high concentrations (Levy 2015). Numerous studies have established a link between excessive exposure to air pollutants and respiratory diseases worldwide. Lung cancer, COPD, and respiratory infections are the most affected pathologies (Bălă et al. 2021). In a comprehensive review of 799 studies across Asia, Europe, and North America, Sun et al. showed that mortality is the most common outcome related to outdoor air pollution, followed by respiratory diseases (Sun and Zhu 2019).

13.1.2 Air Pollutants and Respiratory Infections

Studies demonstrated an association between air pollution and respiratory infections. For instance, a study of 573,950 children from 35 countries found a correlation between mean annual $PM_{2.5}$ concentrations and acute respiratory infection (ARI) in children under 5 years of age in socioeconomically disadvantaged countries. A 10 $\mu g/m^3$ rise in $PM_{2.5}$ led to an increased risk for ARIs, especially in younger ones, boys, and rural residents (Odo et al. 2022). In China, Chen et al. found that ambient $PM_{2.5}$ may raise the chance of being infected with influenza, particularly on cooler days. It has been suggested that ambient $PM_{2.5}$ level may be the cause of about ten percent of influenza cases (Chen et al. 2017). Moreover, a strong association between respiratory syncytial virus (RSV)-related bronchiolitis in infants and air pollutants including NO_x, SO_2, PM_{10}, and $PM_{2.5}$ was reported in Rome, Italy (Nenna et al. 2017).

13.2 Impact of Air Pollution on COVID-19 Mortality and Morbidity; Epidemiological Evidence

COVID-19 is a systemic infection caused by Severe Acute Respiratory Syndrome-related Coronavirus 2 (SARS-CoV-2) that may affect all the systems, in particular the respiratory system, where it can lead to deadly pneumonia and acute respiratory distress syndrome (ARDS) (Sigrist et al. 2020). World Health Organization (WHO) declared it officially as a global pandemic on the 11th of March 2020 (*WHO, COVID-19 pandemic overview*). Globally, the number of reported cases reached 776,281,230 with 7,065,880 deaths as of September 15th, 2024 (*WHO, COVID-19 dashboard*). Besides that, in 2020 and 2021, the escalation of all-cause mortality rates was determined as 21.9% for males and 16.6% for females aged 15 years and above compared to the pre-pandemic data of 2019 (Strong et al. 2024).

Epidemiological studies indicate a potential linkage between SARS-CoV-2 infection, COVID-19 cases, and mortality with particulate and gaseous pollutants (Aykaç and Etiler 2022; Bowe et al. 2021; Fattorini and Regoli 2020; Filippini et al. 2020; Wu et al. 2020; Zhu et al. 2020). At the beginning of the pandemic, most of the papers showing the relationship between air quality and COVID-19 incidence and fatality were published in countries where the first outbreaks occurred, such as China, Italy, and the USA. For instance, Zhu et al. (2020) reported a 1.76% and 6.94% increase in daily confirmed cases with a 10 $\mu g/m^3$ rise in levels of $PM_{2.5}$ and PM_{10}, respectively, with the data collected from 120 cities in China (Zhu et al. 2020). Additionally, the same rise in $PM_{2.5}$ and PM_{10} enhanced the case fatality rate by 0.24% and 0.26%, respectively (Yao et al. 2020). In a study conducted in Wuhan and XiaoGan, a correlation was shown between the prevalence of COVID-19 and $PM_{2.5}$ concentration (Li et al. 2020). Furthermore, the highest death ratio was detected in Lombardy and Emilia Romagna, Italian provinces where air pollution is relatively higher than in other parts of the country (Conticini et al. 2020). There was a three-fold increase in the risk of infection by the rise of $PM_{2.5}$ concentration from 10 to 25 $\mu g/m^3$ while case fatality risk doubled with the increased $PM_{2.5}$ level from 10 to 22 $\mu g/m^3$ in Italy (Borro et al. 2020). Research conducted in the USA reported that a rise of 1 $\mu g/m^3$

in long-term average $PM_{2.5}$ concentration was associated with an 11% augmentation in deaths due to COVID-19 (Wu et al. 2020).

In the following years, as the number of COVID-19 cases increased, data on this subject began to emerge worldwide. Some of the most recent studies are summarized below. Recently, a systematic review, including 43 studies, demonstrated an increased risk of COVID-19 mortality with long-term exposure to $PM_{2.5}$, NO_2 (nitrogen dioxide), and O_3, while $PM_{2.5}$, PM_{10}, and NO_2 were positively correlated with short-term exposure (Yu et al. 2024). Sheppard and colleagues showed a 66% higher risk of COVID-19 infection with a 10 μg/m^3 increase in $PM_{2.5}$ concentration (Sheppard et al. 2023). In Southern California, research consisted of 21.415 individuals who had died within 90 days after positive test results revealed a greater risk of death due to COVID-19 with augmented $PM_{2.5}$ and $PM_{0.1}$ (diameter $\leq$ 0.1 μm) concentrations (Jerrett et al. 2023). In Spain, Bronte et al. presented an increased systemic inflammatory response and enhanced probability of death in COVID-19 pneumonia patients with enhanced exposure to PM_{10}, NO_2, and NO (nitric oxide) by examining 1548 patients data (Bronte et al. 2023). Another large population-based cohort study, which included 340.608 COVID-19 cases in Catalonia in 2020, demonstrated a significant association between higher annual exposure to $PM_{2.5}$ and NO_2 and greater hazard of COVID-19-related problems (Ranzani et al. 2023).

Furthermore, Bonilla and colleagues examined the relationship between long-term exposure to pollutants and COVID-19 mortality in Latin America, where just 4 out of 57 cities fulfill the WHO criteria for $PM_{2.5}$ concentration. Research conducted in Brazil, Chile, Colombia, and Mexico, countries facing the greatest fatality rates of the region with more than a million deaths by early 2022, showed that 1 μg/m^3 elevation of $PM_{2.5}$ was associated with a 2.7% rise in the death rate due to COVID-19 (Bonilla et al. 2023). In the Netherlands, individual-level data was evaluated for 8.3 million citizens to assess links between long-term exposure to ambient air pollution and SARS-CoV-2 incidence together with COVID-19 hospitalization risks. Based on the analysis, infection risk was amplified by 7%, while hospitalization risk was elevated by 16% with PM_{10}. $PM_{2.5}$ was a higher risk factor, with 8% increased infection risk and 18% increased hospitalization risk. Although not as much as particles, NO_2 raised the risk of infection (3%) and hospitalization (11%) (Zorn et al. 2024). Danish nationwide research containing 3.721.810 individuals over 29 years old, including 138.742 individuals who tested positive for SARS-CoV-2, represented a 23% and 18% higher risk of fatality with 0.53 and 3.59 $\mu g/m^3$ increase in $PM_{2.5}$ and NO_2 respectively. Moreover, a weaker but still significant positive association was found for PM_{10} in COVID-19 incidence and hospitalization too (Zhang et al. 2023). In addition, Feng et al. revealed that greater short-term exposure to air pollutants, which were $PM_{2.5}$, PM_{10}, NO_2, and CO, was correlated with increased risk of severe COVID-19 with an odds ratio (OR) of 1.89, 2.35, 2.87, and 2.01 respectively by using data obtained from 569 patients with confirmed SARS-CoV-2 test in Zhejiang province, China (Feng et al. 2024). Lately, in the UK, a cohort study involving 10,489 participants showed a 10 μg/m^3 escalation in 5-year average NO_2 and $PM_{2.5}$ positive correlation with a rising risk of seropositivity by 1.092 and 1.65, respectively (Hajmohammadi et al. 2024). In Sweden, a case-crossover study containing 425 young adults (mean age: 25.6) who have confirmed positive results for SARS-CoV-2 revealed an association between short-term exposure to $PM_{2.5}$ PM_{10} and BC with increased risk of SARS-CoV-2 infection too (Yu et al. 2022). Moreover, Nobile et al. presented a strong association between interquartile range (IQR) increments in $PM_{2.5}$ and NO_2 with COVID-19 mortality with a hazard ratio of 1.08 and 1.09, respectively, which was a higher ratio compared to non-COVID-19 mortality rates in Rome, Italy (Nobile et al. 2022). A prospective cohort study assessed a significant association between long-term (4 years) exposure to $PM_{2.5}$ and NO_2 and the risk of COVID-19-related hospital and intensive care unit (ICU) admission, including 151.105 people with confirmed SARS-CoV-2 infection in Ontario, Canada (Chen et al. 2022).

The observed correlation between higher COVID-19 cases and mortality rates in areas with poor air quality has heightened the significance of investigating the relationship between air pollution and COVID-19. Research has increasingly focused on how outdoor air pollution influences viral transmission and the severity of infections (Domingo and Rovira 2020). Emerging evidence suggests that airborne PM may play a role in the transport of viral material. Indeed, previous studies assessed that the influenza virus could be transported by ambient particles (Chen et al. 2010). Setti et al. isolated SARS-CoV-2 RNA from PMs in Bergamo, an industrial region of Northern Italy. They demonstrated for the first time that PM could act as a carrier in SARS-CoV-2 transmission (Setti et al. 2020). In a broader study, Kayalar and colleagues detected the presence of SARS-CoV-2 in PM samples collected from 10 different provinces and 13 different locations in Turkey. RNA of the virus was detected in 9.8% of the collected PMs. In particular, the genetic material of SARS-CoV-2 was encountered in samples taken from hospital gardens and crowded places in cities, where the infection was common (Kayalar et al. 2021). Recently, it was suggested that glycans of spike proteins mediate the direct contact with the carbon core surface of $PM_{2.5,}$ at the same time, other organic and inorganic components of particles surround the glycoprotein, as demonstrated by molecular dynamic simulation (Romeo et al. 2023). Surface retention of the virus on airborne dust and PM contributes to the long-range transport of the virus (Qu et al. 2020). However, further research is needed to determine whether these viral materials are infectious.

13.3 A Turkish Epidemiological Perspective on COVID-19: Air Pollution and Health Infrastructure During and After COVID-19

Turkey was faced significant challenges during the COVID-19 pandemic, ranking 12th globally in terms of case numbers (*WHO, COVID-19 dashboard*). The first COVID-19 case was announced by Turkish Ministry of Health (MoH) on March 11, 2020 (Güner et al. 2021). Unfortunately, the lack of updated epidemiological data from the MoH made it difficult to assess the ongoing situation. The total amount of cases and deaths, which were last reported in March 2023, were 17.232.066

and 102.174 cases, respectively (*Republic of Turkey Ministry of Health, COVID-19 Dashboard*). Besides that, according to the 2023 World Air Quality Report, Turkey ranked 44th among 134 countries with the most polluted air based on the $PM_{2.5}$ measurements. It ranked 5th out of 43 countries in Europe, containing 7 cities in the top 15 most polluted regional cities list. The annual average $PM_{2.5}$ concentration was 20.3 $\mu g/m^3$, approximately 4 times higher than the WHO standards indicating a pressing need for environmental health improvements (IQAir 2023 World Air Quality Report 2024).

Despite the high number of COVID-19 cases and excessive air pollution, the number of studies investigating the relationship between the two has been relatively low. Tanis and Karakaya investigated the association of air quality indexes and COVID-19 incidence and severity in İstanbul between July 15, 2020, and September 30, 2020. They found a positive correlation between increased $PM_{2.5}$, PM_{10}, NO_2, SO_2, CO, and the number of daily confirmed COVID-19 cases (Tanis and Karakaya 2021). Aykaç and Etiler also found a moderate correlation between the COVID-19 death rate per 100,000 individuals in Istanbul and PM_{10}, SO_2, and NO_2 levels (Aykaç and Etiler 2022). A broader study across 15 provinces demonstrated a positive association between outdoor air pollution and COVID-19 cases between the 8th of February to the 8th of May 2021 (Akan 2022).The strongest correlation between air pollutants, particularly PM_{10}, O_3, and COVID-19 cases was detected in Kütahya and Sivas cities (Akan 2022). Shahzad and colleagues revealed that NO_2 enhanced the infection and fatality rate in İstanbul by using data collected between the 11th of March 2020 and the 19th of February 2021 (Shahzad et al. 2023). Another research conducted between June 29, 2020, and November 23, 2020, for all provinces in the 12 regions at the nomenclature of territorial units for statistics (NUTS) level 1 of Turkey determined a positive correlation between PM_{10} concentration, population density, and COVID-19 incidence (Yıldırım 2023). Furthermore, the effect of air pollutants on the number of COVID-19 cases was investigated in Western Anatolia and the Western Black Sea region of Turkey. As a result, SO_2, CO, NO_X, and O_3 were observed to be effective in Western Anatolia while ineffective in the Western Black Sea region, suggesting the noteworthy impact of meteorological factors and geography on the infection (Ulutaş et al. 2023). Also, air pollution and COVID-19 incidence correlation analysis conducted in 9 big cities of Turkey, including İstanbul, Izmir, Ankara, Konya, Kocaeli, Sakarya, Bursa, Adana, and Zonguldak accounting for 78% of confirmed COVID-19 cases of the country from March 21, 2020, to April 3, 2020. According to the analysis results, among PM_{10}, SO_2, CO, NO_2, NOx, NO, and O_3, only SO_2 was significantly correlated with the number of cases (Şahin 2020).

Although it has been about 4 years since the pandemic was declared and about 3 years since the effective vaccine was produced, COVID-19 is still affecting human's physical and mental health, economy, and social life in Turkey, as well as around the world. Due to the inadequate health system infrastructure and management strategies, Turkey has not responded strongly to the COVID-19 pandemic (Kartoglu and Pala 2023).

As a preventive measure, Turkey's first action was introducing a travel ban for China on February 3rd, 2020, which was followed by the closure of borders for all countries at the end of March 2020. Later, partial or total lockdowns started to be announced all around the country (Cakir 2020). When the borders were reopened, thermal screening was used in airports, which was only a 50% effective method for detection (Quilty et al. 2020). Turkey followed the Pandemic Influenza National Preparedness Plan, published by The Turkish Ministry of Health, and updated in 2019. Although the scientific advisory committee was established before most countries in January 2021, its effectiveness was questioned, since the necessary epidemic strategy and preparedness plan for COVID-19 could not be taken rapidly, and the existing plan was not fully implemented, either (Kartoglu and Pala 2023). Moreover, the removal of restrictions (called as normalization plan) complicated efforts to control the number of cases. Family Health Centers, which could have played a more proactive role in the outbreak management with their resources, were notably passive during the crisis. In contrast, District Health Directorates, despite having fewer resources, were more actively involved. Also, Family Health Centers staff could not be supplied with the necessary protective equipment to cope with the pandemic and this led to infection or death of primary healthcare service workers (Mardin 2021).

One of the most crucial challenges during the pandemic was data transparency. Only the number of cases diagnosed by molecular methods was disclosed, while epidemiologic and clinical diagnoses were not included. In addition, details such as age, gender, and residency were not shared regarding the number of cases and deaths. More importantly, the MoH intervened in COVID 19 research and subjected research on COVID 19 to the permission of the MoH. This has restricted independent research and greatly limited the availability of solid scientific data regarding COVID 19 (Bayram et al. 2020; Pala 2020). Çapar and colleagues conducted a questionnaire cross-sectional study with 656 people in Turkey. Based on their findings, despite the general healthcare system's strong performance, it was observed that the transparency of the information shared during COVID-19 received lower marks than other dimensions (Çapar et al. 2024).

To assess the spread of the pandemic, it was instructed to perform more than five tests per thousand individuals per day. However, this was not fulfilled in Turkey, either (Kartoglu and Pala 2023). Another problem was that hospitals were not prepared for this process, facing logistical problems and the lack of personal protective equipment (Sandal et al. 2021). Sülkü and colleagues assessed the efficiency of public hospitals in Turkey during the pandemic by examining pre-pandemic and pandemic periods. First, it was noted that the capacity and service of outpatient and inpatient services was significantly improved by the medical staff including physicians, nurses, and other medical professionals. However, despite an increase in overall service production during the pandemic, the effectiveness of inpatient services and inpatient efficiency in 2021 was worse than in 2020. The decline in efficiency extended to outpatient services as well, indicating that the healthcare system was under significant strain due to the demands of the pandemic (Sülkü et al. 2023). Specifically in Istanbul, inadequate medical care increased by more than 1.5 times compared to the pre-pandemic period (Tokatlioglu and Sulku 2023).

Erdem et al. evaluated the changing trends in micro, medium, and macro categories that have emerged or are likely to emerge in the Turkish health system in the post-COVID period. At the micro level, patients understood the value of health personnel, and hospital managers gained experience in a crisis management. At the medium level, it brought about changes in the delivery of remote and home health care services. At the macro level, the negative impact of COVID-19 on the world economy led to an increase in the share allocated to health, financial, and inflation problems and a shortage of access to medicines and medical supplies. (Erdem et al. 2023).

By the end of the twentieth century, significant public health measures such as vaccination, hygiene, better nutrition, and sanitation led to a noteworthy reduction in deaths due to infectious diseases. However, the human footprint on the globe shows that the threat of serious infections remains in the twenty-first century, with at least 13–15 million infectious deaths per year, driven by travel, regional inequalities, environmental and social changes (Şenol 2023). Planning resources in circumstances in which time and resources are scarce is one of the most crucial factors in combating pandemics (Gul and Yucesan 2021). Moreover, it was also understood that "Vaccine Hesitancy" defined by WHO as one of the ten major threats to public health, is a much more widespread problem than anticipated and a significant obstacle to vaccination. Based on the 2022 Global Pandemic Preparedness Summit, the 100 Days Mission report, international coordination and cooperation, road mapping, and global cooperation should be strengthened, and diagnosis, treatment, and vaccines for a disease should be available within the first 100 days. Priorities should include sustained and dedicated leadership, the creation and strengthening of independent institutions and organizations, cooperation with regions that may be weak links, monitoring zoonoses, and strengthening primary preventive health care (Gouglas et al. 2023).

Overall, the COVID-19 pandemic in Turkey highlights serious issues with environmental health, data accessibility, and public health management. The necessity of addressing environmental elements in health interventions is emphasized by the significant association found between air pollution and the severity of COVID-19. In addition, Turkey must prioritize a more robust health infrastructure, open data sharing, and efficient crisis management going forward. To successfully protect public health during future emergencies, environmental health, and pandemic preparedness must be incorporated.

13.4 Respiratory Effects of Air Pollution and COVID-19: Underlying Mechanisms

The entrance of SARS-CoV-2 into host cells is a vital stage in the infection pathway and has been the focus of drug development. It contains 4 steps, which are prebinding, receptor recognition, proteolytic cleavage, and membrane fusion (Le et al.

2023). Initially, the spike (S) protein located on the surface of the SARS-CoV-2, which is a structural protein, has some conformational changes to become ready for the receptor binding (Le et al. 2023; V'kovski et al. 2021). The ectodomain of spike protein contains 2 units, one of them is receptor-binding unit S1, which has a receptor-binding domain (RBD) to interact with specific cell surface receptors, and a membrane-fusion unit S2, which initiates the endocytosis of virion, fuses the host membrane and viral membrane so that the viral genome can get into the host cell (Chen et al. 2020). SARS-CoV-2 enters the target cell by binding through the angiotensin-converting enzyme II (ACE-2) receptor on the host (Frontera et al. 2020). Subsequently, the spike protein cleavage site is cleaved by a protease, majorly transmembrane serine protease 2 (TMPRSS2), which is also present in the host cell membrane (Hoffmann et al. 2020). It is known that single-strand RNA viruses like coronaviruses tend to have several receptors (Zhang et al. 2019). Besides the TMPRSS2, SARS-CoV-2 can utilize cysteine proteases Furin, Cathepsin B, and L for the S priming like SARS-CoV (Hoffmann et al. 2020). Excessive expression of these proteases is hypothesized to affect the SARS-CoV-2's cell and tissue tropism, while increasing the transmissibility and pathogenicity (Walls et al. 2020).

In physiological conditions, ACE2 takes part in the renin-angiotensin system (RAS), regulating blood pressure, electrolyte homeostasis, and inflammation, the alteration of which is associated with diverse pathologies, such as ARDS (Prato et al. 2024). This system proceeds on 2 opposing paths; the first one consists of the ACE/Angiotensin II peptide (Ang II)/Angiotensin II type 1 receptor (AT1R), leading to the release of pro-inflammatory cytokines such as interleukin (IL)-6 and tumor necrosis factor (TNF)-α. The second pathway is the ACE-2/ Angiotensin 1–7 peptide (Ang1-7) /Mas receptor, which has an anti-inflammatory role. ACE2 cleaves Ang II into Ang1-7, which inhibits the inflammatory axis. However, in the presence of SARS-CoV-2, ACE2 is occupied by spike protein that leads to downregulation of the receptor so that the pathway shifts to the Ang II axis, initiating the inflammatory response (Prato et al. 2024). $PM_{2.5}$ may facilitate the establishment of viral infections by upregulating the ACE2 level at first, the increased viral load in the cells later causes the depletion of ACE2, which leads to acute lung inflammation (Frontera et al. 2020). It is also suggested that high levels of air pollutant NO_2 in the same environment can lead to more severe disease by further increasing inflammation in the cells (Frontera et al. 2020).

For instance, in 2018, Lin and colleagues conducted a $PM_{2.5}$-induced acute lung injury model using wild-type and ACE2 gene knockout mice and established the amplified ACE-2 expression in the lungs of wild-types at 2- and 5-days post-instillation. Moreover, it was observed that the signs of recovery, including respiratory rate and levels of inflammatory cytokines, were improved significantly in wild-type mice following 5 days of exposure, while there was only a partial recovery in ACE2 knockout mice, suggesting that the repair process after acute $PM_{2.5}$ exposure could be insufficient in the absence of ACE2 (Lin et al. 2018). Another research demonstrated both ACE2 and TMPRSS2 protein expression elevation in alveolar macrophages and type 2 alveolar cells in a mouse model with the PM treatment by using immunohistochemistry analysis (Sagawa et al. 2021). A more recent study conferred similar results with a 40% increase in ACE2 protein expression with $PM_{2.5}$ treatment in mouse model (Botto et al. 2023). Additionally, Zhu et al. established increased ACE2 levels by immunofluorescence staining in mice that were exposed to $PM_{2.5}$ for 3 consecutive days (Zhu et al. 2022).

In vitro studies presented similar results. For instance, Marchetti et al. assessed increased expression of ACE2 after 72h of $PM_{2.5}$ exposure in the A549 cell line (adenocarcinoma human alveolar epithelial cells). Moreover, they found enhanced expression of Ras-related protein Rab-5A (RAB5), which is an endosomal marker indicating viral internalization, when the cells were treated with $PM_{2.5}$before the SARS-CoV-2 infection compared to only SARS-CoV-2 infected ones (Marchetti et al. 2024). Kesic and colleagues showed disrupted protease/antiprotease balance due to O_3 exposure in human nasal epithelial cells. Besides an increased expression of TMPRSS2, there was enhanced inhibition of antiproteases, which may facilitate the viral cleavage so that entry and replication (Kesic et al. 2012).

It is thought that exposure to pollutants triggers a pro-inflammatory state, together with augmented oxidative stress, due to elevated reactive oxygen species (ROS) leading to cell death, such as apoptosis. If the exposure is continuous for longer periods, it may cause immune dysregulation, leading to distinctive diseases (Wang et al. 2020a, b). Decreased antioxidant capacity due to the pollutants in the respiratory tract–lining fluid and tissues is one of the other reasons for enhanced susceptibility to oxidative damage and inflammation (Glencross et al. 2020). In China, Dai et al (2021) exposed mice to $PM_{2.5}$ for 8 h/day, 6 days/week, for 8 weeks. Compared to the control group, there was remarkable pulmonary inflammation in $PM_{2.5}$ treated ones, which was conferred with micro-CT (computed tomography). Histochemistry staining showed inflammatory cell infiltration and alveolar structure damage in the alveoli, while there was epithelial hyperplasia in the bronchiolar region of those mice. Moreover, mRNA sequencing analysis revealed an upregulated B-cell receptor signaling pathway while cell adhesion molecules (CAMs) were downregulated (Dai et al. 2021). A recent study confirmed enhanced IL-6 and IL-8 in peripheral blood mononuclear cells (PBMCs) of healthy individuals exposed to PM_{10} for 24h prior to SARS-CoV-2 inoculation (Marín-Palma et al. 2023). There was a rise in pro-inflammatory markers (nuclear factor-kappa B, NF-kB; and IL-8) when A549 cells were treated with $PM_{2.5}$ for 3 days before the SARS-CoV-2 infection compared to only the virus-infected group without $PM_{2.5}$ treatment (Marchetti et al. 2024). These results confirm that pre-exposure to PM can contribute to the imbalance of the inflammatory response that is typically seen in COVID-19 prognosis (Chen et al. 2021).

Altered immune defense makes individuals more susceptible to infections (Ciencewicki and Jaspers 2007). For instance, it was shown that the phagocytic activity of alveolar macrophages was decreased by PM exposure (Gonzalez et al. 2018). In addition, PM was found to have a negative effect on the promoter activity and subsequent transcription of type-I interferons (IFN), a crucial antiviral cytokines (Wang et al. 2020a, b). Animal models showed that PM could cause alveolar collapse by changing the structure of the surfactant on the alveolar epithelial surface (F. Wang et al. 2020a, b). It is known that surfactant function is impaired in ARDS (Dushianthan et al. 2023). Therefore, people residing in locations with advanced PM concentrations had an augmented risk of severe COVID-19 with ARDS (Wang et al. 2020a, b; B; Wang et al. 2020a, b). Furthermore, PM can interfere with antimicrobial peptides, which have a regulatory role in the immune response by stimulating cytokine release, immune cell chemotaxis of immune cells, and degranulation of neutrophils and mast cells. Particles can bind to antimicrobial peptides and alter their efficiency, or they can directly inhibit the production or secretion of these peptides (Woodby et al. 2021).

Impaired epithelial barrier integrity by the virus is another mechanism inducing the SARS-CoV-2 infection. Downregulation of CAMs, together with a reduction of tight junction proteins like occludin, internalization of claudin from the plasma membrane to endosomal compartments and dissociation of Zona occludens (ZO)-1, due to pollutants facilitate the pathogen access to the basolateral side of the epithelium (Caraballo et al. 2011; Kayalar et al. 2024; Liu et al. 2019). Bayram and colleagues reported increased epithelial barrier permeability of human primary bronchial epithelial cell cultures exposed to O_3 and NO_2 (Bayram et al. 2002). Moreover, O_3 exposure resulted in neutrophile infiltration into the lungs (Morrison et al. 2006; Mudway and Kelly 2004; Pino et al. 1992). These pollutants can deteriorate the innate immune response by disrupting mucociliary clearance (Fiorito et al. 2022). PM exposure can cause mucus hypersecretion, which can contribute to the hypoxia in COVID-19 patients (Liu et al. 2020; Yang et al. 2020). It has been established that diesel exhaust particles (DEP) could reduce ciliary beating frequency (CBF) and stimulate the release of inflammatory mediators, altering the cell cycle and leading to apoptosis in human airway epithelial cells (Bayram et al. 1998, 2013, 2006). A study conducted in primary bronchial epithelial cells showed ciliary loss 4 days after the SARS-CoV-2 inoculation (Robinot et al. 2021). Moreover, electron microscopy images demonstrated ultrastructurally abnormal cilia including misshapen cilia, axoneme loss, and accumulation of mis localized basal bodies. There was an inhibition of ciliogenesis regulator *Foxj1* gene expression due to the infection. Lastly, immobile or randomly oriented limited movement of cilia was observed 7 days after the SARS-CoV-2 inoculation by beat tracking method, as mucociliary clearance assay, representing an impairment of the mucociliary clearance (Robinot et al. 2021). Figure 13.1 (Bayram et al. 2024) summarizes the interaction between air pollutants and SARS-CoV-2 infection and underlying mechanisms at the cellular level.

To sum up, poor air quality can increase the risk of SARS-CoV-2 infection, COVID-19 severity, and mortality by increasing the expression of viral entry-related genes, altering inflammatory cytokine release and immune cell responses, augmenting epithelial permeability and impairing mucociliary clearance.

13.5 Conclusion and Research Needs

In conclusion, a significant amount of research suggests an association between air pollution and mortality and morbidity due to COVID-19 both locally in Turkey, and globally worldwide. These findings underscore the impact of improving air quality

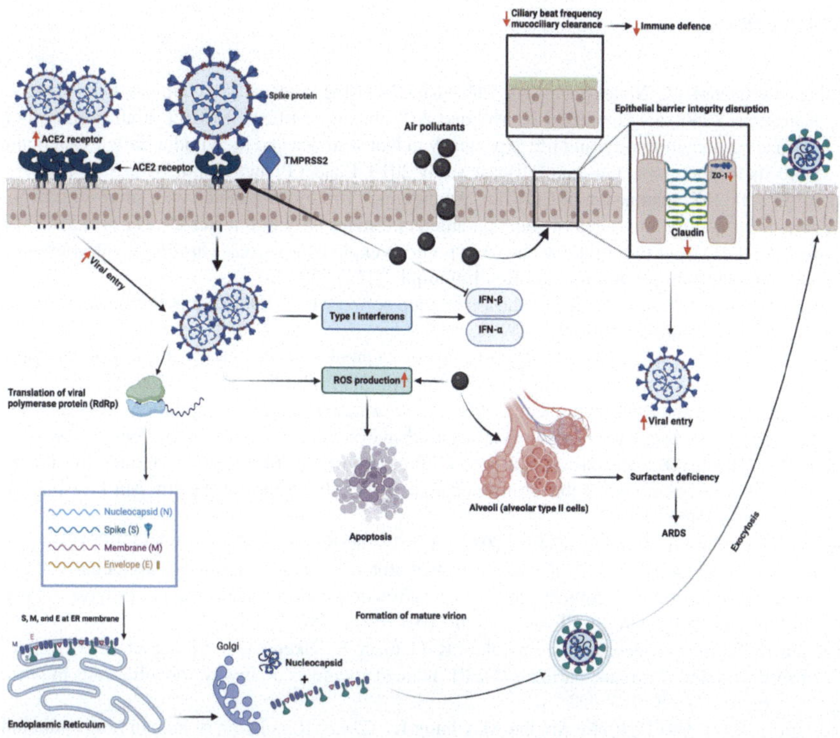

Fig. 13.1 Mechanisms underlying the interaction between air pollution and SARS-CoV-2 infection at cellular level. Air pollutants compromise the integrity of the epithelial barrier and reduce the ciliary beat frequency, impairing mucociliary clearance and facilitating viral entry. Additionally, they enhance SARS-CoV-2 entry by upregulating the expression of the angiotensin-converting enzyme (ACE)2 receptor. Air pollutants also stimulate the production of reactive oxygen species (ROS) in host cells, which can result in cellular damage and apoptosis. The combined effect of viral replication and elevated ROS levels leads to cellular injury, potentially causing severe clinical outcomes such as acute respiratory distress syndrome (ARDS). IFN, interferon; TMPRSS2, Transmembrane protease, serine 2 (Bayram et al. 2024)

to protect public health and reduce risks during pandemics. To effectively tackle the ongoing challenges of COVID-19 and other infections, some key actions are critical. Foremost, a better understanding is needed on how air pollution affects viral infections. Strengthening public health infrastructure and training for healthcare workers will improve response times. Data transparency is essential for guiding decisions. Stricter air quality regulations and public awareness about air pollutants' health impacts are required. Future health interventions should integrate environmental considerations, recognizing the link between air quality and health. Focusing on these needs will better protect public health against possible future pandemics.

References

Abbasi-Kangevari M, Malekpour M-R, Masinaei M, Moghaddam SS, Ghamari S-H, Abbasi-Kangevari Z, Rezaei N, Rezaei N, Mokdad AH, Naghavi M (2023) Effect of air pollution on disease burden, mortality, and life expectancy in North Africa and the Middle East: a systematic analysis for the global burden of disease study 2019. Lancet Planet Health 7(5):e358–e369

Akan AP (2022) Transmission of COVID-19 pandemic (Turkey) associated with short-term exposure of air quality and climatological parameters. Environ Sci Pollut Res 29(27):41695–41712

Akdis CA (2021) Does the epithelial barrier hypothesis explain the increase in allergy, autoimmunity and other chronic conditions? Nat Rev Immunol 21(11):739–751

Amnuaylojaroen T, Parasin N (2021) The association between COVID-19, air pollution, and climate change. Front Public Health 9:662499

Aykaç N, Etiler N (2022) COVID-19 mortality in Istanbul in association with air pollution and socioeconomic status: an ecological study. Environ Sci Pollut Res 29(9):13700–13708

Bălă G-P, Râjnoveanu R-M, Tudorache E, Motişan R, Oancea C (2021) Air pollution exposure—the (in) visible risk factor for respiratory diseases. Environ Sci Pollut Res 28(16):19615–19628

Bayram H, Bauer AK, Abdalati W, Carlsten C, Pinkerton KE, Thurston GD, Balmes JR, Takaro TK (2017) Environment, global climate change, and cardiopulmonary health. Am J Respir Crit Care Med 195(6):718–724

Bayram H, Devalia JL, Khair OA, Abdelaziz MM, Sapsford RJ, Sagai M, Davies RJ (1998) Comparison of ciliary activity and inflammatory mediator release from bronchial epithelial cells of nonatopic nonasthmatic subjects and atopic asthmatic patients and the effect of diesel exhaust particles in vitro. J Allergy Clinical Immun 102(5):771–782

Bayram H, Fakili F, Gögebakan B, Bayraktar R, Öztuzcu S, Dikensoy Ö, Chung KF (2013) Effect of serum on diesel exhaust particles (DEP)-induced apoptosis of airway epithelial cells in vitro. Toxicol Lett 218(3):215–223

Bayram H, Ito K, Issa R, Ito M, Sukkar M, Chung KF (2006) Regulation of human lung epithelial cell numbers by diesel exhaust particles. Eur Respir J 27(4):705–713

Bayram H, Konyalilar N, Elci MA, Rajabi H, Aksoy GT, Mortazavi D, Kayalar Ö, Dikensoy Ö, Taborda-Barata L, Viegi G (2024) Issue 4—Impact of air pollution on COVID-19 mortality and morbidity: an epidemiological and mechanistic review. Pulmonology S2531–0437(24):00051–00055

Bayram H, Köktürk N, Elbek O, Kilinc O, Sayiner A, Dagli E (2020) Interference in scientific research on COVID-19 in Turkey. Lancet, 396, 10249

Bayram H, Rusznak C, Khair O, Sapsford R, Abdelaziz M (2002) Effect of ozone and nitrogen dioxide on the permeability of bronchial epithelial cell cultures of non-asthmatic and asthmatic subjects. Clin Exp Allergy 32(9):1285–1292

Boğan M, Kul S, Al B, Oktay MM, Akpinar Elçi M, Pinkerton KE, Bayram H (2022) Effect of desert dust storms and meteorological factors on respiratory diseases. Allergy, 77(7)

Bonilla JA, Lopez-Feldman A, Pereda PC, Rivera NM, Ruiz-Tagle JC (2023) Association between long-term air pollution exposure and COVID-19 mortality in Latin America. PLoS ONE 18(1):e0280355

Borro M, Di Girolamo P, Gentile G, De Luca O, Preissner R, Marcolongo A, Ferracuti S, Simmaco M (2020) Evidence-based considerations exploring relations between SARS-CoV-2 pandemic and air pollution: involvement of PM2. 5-mediated up-regulation of the viral receptor ACE-2. Int J Environ Res Public Health 17(15):5573

Botto L, Lonati E, Russo S, Cazzaniga E, Bulbarelli A, Palestini P (2023) Effects of PM2. 5 exposure on the ACE/ACE2 pathway: possible implication in COVID-19 pandemic. Int J Environ Res Public Health, 20(5):4393

Bowe B, Xie Y, Gibson AK, Cai M, van Donkelaar A, Martin RV, Burnett R, Al-Aly Z (2021) Ambient fine particulate matter air pollution and the risk of hospitalization among COVID-19 positive individuals: cohort study. Environ Int 154:106564

Bronte O, García-García F, Lee D-J, Urrutia I, Uranga A, Nieves M, Martínez-Minaya J, Quintana J, Arostegui I, Zalacain R (2023) Impact of outdoor air pollution on severity and mortality in COVID-19 pneumonia. Sci Total Environ 894:164877
Cakir B (2020) COVID-19 in Turkey: lessons learned. J Epidemiol Global Health 10(2):115–117
Caraballo JC, Yshii C, Westphal W, Moninger T, Comellas AP (2011) Ambient particulate matter affects occludin distribution and increases alveolar transepithelial electrical conductance. Respirology 16(2):340–349
Chen C, Wang J, Kwong J, Kim J, van Donkelaar A, Martin RV, Hystad P, Su Y, Lavigne E, Kirby-McGregor M (2022) Association between long-term exposure to ambient air pollution and COVID-19 severity: a prospective cohort study. CMAJ 194(20):E693–E700
Chen G, Zhang W, Li S, Zhang Y, Williams G, Huxley R, Ren H, Cao W, Guo Y (2017) The impact of ambient fine particles on influenza transmission and the modification effects of temperature in China: a multi-city study. Environ Int 98:82–88
Chen H, Liu W, Wang Y, Liu D, Zhao L, Yu J (2021) SARS-CoV-2 activates lung epithelial cell proinflammatory signaling and leads to immune dysregulation in COVID-19 patients. EBioMedicine 70
Chen P-S, Tsai FT, Lin CK, Yang C-Y, Chan C-C, Young C-Y, Lee C-H (2010) Ambient influenza and avian influenza virus during dust storm days and background days. Environ Health Perspect 118(9):1211–1216
Chen Y, Guo Y, Pan Y, Zhao ZJ (2020) Structure analysis of the receptor binding of 2019-nCoV. Biochem Biophys Res Commun 525(1):135–140
Ciencewicki J, Jaspers I (2007) Air pollution and respiratory viral infection. Inhalation Toxicol 19(14):1135–1146
Comunian S, Dongo D, Milani C, Palestini P (2020) Air pollution and COVID-19: the role of particulate matter in the spread and increase of COVID-19's morbidity and mortality. Int J Environ Res Public Health 17(12):4487
Conticini E, Frediani B, Caro D (2020) Can atmospheric pollution be considered a co-factor in extremely high level of SARS-CoV-2 lethality in Northern Italy? Environ Pollut 261:114465
Cooper DM, Loxham M (2019) Particulate matter and the airway epithelium: the special case of the underground? Eur Respir Rev 28(153)
Çapar H, Çınar F, Ekinci G (2024) Evaluation of Turkey's healthcare system performance in COVID-19 pandemic period. J Health Manage 09720634241246920
Dai S, Wang Z, Yang Y, Guo T, Li W (2021) Assessment on the lung injury of mice posed by airborne PM2. 5 collected from developing area in China and associated molecular mechanisms by integrated analysis of mRNA-seq and miRNA-seq. Ecotoxicol Environ Saf 224:112661
Domingo JL, Rovira J (2020) Effects of air pollutants on the transmission and severity of respiratory viral infections. Environ Res 187:109650
Donaldson K, Stone V, Seaton A, MacNee W (2001) Ambient particle inhalation and the cardiovascular system: potential mechanisms. Environ Health Perspect 109(suppl 4):523–527
Dushianthan A, Grocott MP, Murugan GS, Wilkinson TM, Postle AD (2023) Pulmonary surfactant in adult ARDS: current perspectives and future directions. Diagnostics 13(18):2964
Erdem R, Alkan A, Çam S (2023) Post-pandemic change trends in Turkish health system. Abant Sosyal Bilimler Dergisi 23(3):1446–1453
Fattorini D, Regoli F (2020) Role of the chronic air pollution levels in the Covid-19 outbreak risk in Italy. Environ Pollut 264:114732
Feng B, Lian J, Yu F, Zhang D, Chen W, Wang Q, Shen Y, Xie G, Wang R, Teng Y (2024) Impact of short-term ambient air pollution exposure on the risk of severe COVID-19. J Environ Sci 135:610–618
Filippini T, Rothman KJ, Goffi A, Ferrari F, Maffeis G, Orsini N, Vinceti M (2020) Satellite-detected tropospheric nitrogen dioxide and spread of SARS-CoV-2 infection in Northern Italy. Sci Total Environ 739:140278

Fiorito S, Soligo M, Gao Y, Ogulur I, Akdis CA, Bonini S (2022) Is the epithelial barrier hypothesis the key to understanding the higher incidence and excess mortality during COVID-19 pandemic? Case Northern Italy Allergy 77(5):1408–1417

Frontera A, Cianfanelli L, Vlachos K, Landoni G, Cremona G (2020) Severe air pollution links to higher mortality in COVID-19 patients: the "double-hit" hypothesis. J Infect 81(2):255–259

Fuller R, Landrigan PJ, Balakrishnan K, Bathan G, Bose-O'Reilly S, Brauer M, Caravanos J, Chiles T, Cohen A, Corra L (2022) Pollution and health: a progress update. Lancet Planet Health 6(6):e535–e547

Glencross DA, Ho T-R, Camina N, Hawrylowicz CM, Pfeffer PE (2020) Air pollution and its effects on the immune system. Free Radical Biol Med 151:56–68

Gonzalez Y, Carranza C, Iñiguez M, Torres M, Quintana R, Osornio A, Gardner C, Sarkar S, Schwander S (2018) Effect of inhaled air pollution particulate matter in alveolar macrophages on local pro-inflammatory cytokine and peripheral interferon γ production in response to mycobacterium tuberculosis. Lancet Glob Health 6:S29

Gouglas D, Christodoulou M, Hatchett R (2023) The 100 days mission—2022 global pandemic preparedness summit. Emerg Infect Dis 29(3)

Gryparis A, Forsberg B, Katsouyanni K, Analitis A, Touloumi G, Schwartz J, Samoli E, Medina S, Anderson HR, Niciu EM (2004) Acute effects of ozone on mortality from the "air pollution and health: a European approach" project. Am J Respir Crit Care Med 170(10):1080–1087

Gul M, Yucesan M (2021) Hospital preparedness assessment against COVID-19 pandemic: a case study in Turkish tertiary healthcare services. Math Probl Eng 2021(1):2931219

Güner AE, Sürmeli A, Kural K, Şahin E, Alkan P, Kocayiğit E, Hatipoğlu M, Birinci Ş, Memişoğlu K, MARAL I (2021) First known COVID-19 case and contact tracing efforts in İstanbul, Turkey. Turk J Med Sci 51(4):1653–1658

Hajmohammadi H, Talaei M, Fecht D, Wang W, Vivaldi G, Faustini SE, Richter AG, Shaheen SO, Martineau AR, Sheikh A (2024) Long-term air pollution exposure and risk of SARS-CoV-2 infection: a UK-wide cohort study. Respir Med 224:107567

Hoffmann M, Kleine-Weber H, Schroeder S, Krüger N, Herrler T, Erichsen S, Schiergens TS, Herrler G, Wu N-H, Nitsche A (2020) SARS-CoV-2 cell entry depends on ACE2 and TMPRSS2 and is blocked by a clinically proven protease inhibitor. Cell 181(2):271–280. e278

IQAir, 2023 World air quality report (2024). https://www.iqair.com/dl/2023_World_Air_Quality_Report.pdf

Jerrett M, Nau CL, Young DR, Butler RK, Batteate CM, Su J, Burnett RT, Kleeman MJ (2023) Air pollution and meteorology as risk factors for COVID-19 death in a cohort from Southern California. Environ Int 171:107675

Kartoglu U, Pala K (2023) Evaluation of COVID-19 pandemic management in Türkiye. Front Public Health 11:1142471

Kayalar Ö, Arı A, Babuççu G, Konyalılar N, Doğan Ö, Can F, Şahin ÜA, Gaga EO, Kuzu SL, Arı PE (2021) Existence of SARS-CoV-2 RNA on ambient particulate matter samples: a nationwide study in Turkey. Sci Total Environ 789:147976

Kayalar Ö, Rajabi H, Konyalilar N, Mortazavi D, Aksoy GT, Wang J, Bayram H (2024) Impact of particulate air pollution on airway injury and epithelial plasticity; underlying mechanisms. Front Immunol 15:1324552

Kesic MJ, Meyer M, Bauer R, Jaspers I (2012) Exposure to ozone modulates human airway protease/antiprotease balance contributing to increased influenza A infection. PLoS ONE 7(4):e35108

Kim S-Y, Kim E, Kim WJ (2020) Health effects of ozone on respiratory diseases. Tuberc Respir Dis 83(Supple 1):S6

Le K, Kannappan S, Kim T, Lee JH, Lee H-R, Kim KK (2023) Structural understanding of SARS-CoV-2 virus entry to host cells. Front Mol Biosci 10:1288686

Levy RJ (2015) Carbon monoxide pollution and neurodevelopment: a public health concern. Neurotoxicol Teratol 49:31–40

Li H, Xu X-L, Dai D-W, Huang Z-Y, Ma Z, Guan Y-J (2020) Air pollution and temperature are associated with increased COVID-19 incidence: a time series study. Int J Infect Dis 97:278–282

Lin C-I, Tsai C-H, Sun Y-L, Hsieh W-Y, Lin Y-C, Chen C-Y, Lin C-S (2018) Instillation of particulate matter 2.5 induced acute lung injury and attenuated the injury recovery in ACE2 knockout mice. Int J Bio Sci 14(3):253

Liu J, Chen X, Dou M, He H, Ju M, Ji S, Zhou J, Chen C, Zhang D, Miao C (2019) Particulate matter disrupts airway epithelial barrier via oxidative stress to promote Pseudomonas aeruginosa infection. J Thorac Dis 11(6):2617

Liu Y, Lv J, Liu J, Li M, Xie J, Lv Q, Deng W, Zhou N, Zhou Y, Song J (2020) Mucus production stimulated by IFN-AhR signaling triggers hypoxia of COVID-19. Cell Res 30(12):1078–1087

Manisalidis I, Stavropoulou E, Stavropoulos A, Bezirtzoglou E (2020) Environmental and health impacts of air pollution: a review. Front Public Health 8:14

Marchetti S, Colombo A, Saibene M, Bragato C, La Torretta T, Rizzi C, Gualtieri M, Mantecca P (2024) Shedding light on the cellular mechanisms involved in the combined adverse effects of fine particulate matter and SARS-CoV-2 on human lung cells. Sci Total Environ 952:175979

Mardin FD (2021) Turkey's healthcare policies and the Covid-19 pandemic. Analyses pluridisciplinaires sur la crise sanitaire Covid-19 en Turquie. Istanbul: Institut français d'études anatoliennes 10

Marín-Palma D, Tabares-Guevara JH, Zapata-Cardona MI, Zapata-Builes W, Taborda N, Rugeles MT, Hernandez JC (2023) PM10 promotes an inflammatory cytokine response that may impact SARS-CoV-2 replication in vitro. Front Immunol 14:1161135

Morrison D, Rahman I, MacNee W (2006) Permeability, inflammation and oxidant status in airspace epithelium exposed to ozone. Respir Med 100(12):2227–2234

Mudway IS, Kelly FJ (2004) An investigation of inhaled ozone dose and the magnitude of airway inflammation in healthy adults. Am J Respir Crit Care Med 169(10):1089–1095

Nenna R, Evangelisti M, Frassanito A, Scagnolari C, Pierangeli A, Antonelli G, Nicolai A, Arima S, Moretti C, Papoff P (2017) Respiratory syncytial virus bronchiolitis, weather conditions and air pollution in an Italian urban area: an observational study. Environ Res 158:188–193

Nobile F, Michelozzi P, Ancona C, Cappai G, Cesaroni G, Davoli M, Di Martino M, Nicastri E, Girardi E, Beccacece A (2022) Air pollution, SARS-CoV-2 incidence and COVID-19 mortality in Rome: a longitudinal study. Eur Respir J 60(3)

Odo DB, Yang IA, Dey S, Hammer MS, van Donkelaar A, Martin RV, Dong G-H, Yang B-Y, Hystad P, Knibbs LD (2022) Ambient air pollution and acute respiratory infection in children aged under 5 years living in 35 developing countries. Environ Int 159:107019

Pala K (2020) COVID-19: case and death notification problems in Turkey. Turkish Thoracic Journal 21(3):213

Pino MV, Levin JR, Stovall MY, Hyde DM (1992) Pulmonary inflammation and epithelial injury in response to acute ozone exposure in the rat. Toxicol Appl Pharmacol 112(1):64–72

Prato M, Tiberti N, Mazzi C, Gobbi F, Piubelli C, Longoni SS (2024) The renin-angiotensin system (RAS) in COVID-19 disease: where we are 3 years after the beginning of the pandemic. Microorganisms 12(3):583

Pu S, Peng S, Zhu J, Liu Z, Long D, Lim T (2022) Characteristics of PM2. 5 and its correlation with feed, manure and NH3 in a pig-fattening house. Toxics 10(3):145

Qu G, Li X, Hu L, Jiang G (2020) An imperative need for research on the role of environmental factors in transmission of novel coronavirus (COVID-19). Environ Sci Technol 54:(3730–3732)

Quilty BJ, Clifford S, Flasche S, Eggo RM (2020) Effectiveness of airport screening at detecting travellers infected with novel coronavirus (2019-nCoV). Eurosurveillance 25(5):2000080

Rai PK (2016) Particulate matter and its size fractionation. Bio Monitor Part Matter 1:1–13

Ranzani O, Alari A, Olmos S, Milà C, Rico A, Ballester J, Basagaña X, Chaccour C, Dadvand P, Duarte-Salles T (2023) Long-term exposure to air pollution and severe COVID-19 in Catalonia: a population-based cohort study. Nat Commun 14(1):2916

Republic of Turkey ministry of health, COVID-19 Dashboard. Retrieved 18.Oct 24. https://covid19.saglik.gov.tr/

Ritchie HR, Max (2024) Air pollution. Our overview of indoor and outdoor air pollution. Retrieved on 02 02 2024. https://ourworldindata.org/air-pollution

Robinot R, Hubert M, de Melo GD, Lazarini F, Bruel T, Smith N, Levallois S, Larrous F, Fernandes J, Gellenoncourt S (2021) SARS-CoV-2 infection induces the dedifferentiation of multiciliated cells and impairs mucociliary clearance. Nat Commun 12(1):4354
Romeo A, Pellegrini R, Gualtieri M, Benassi B, Santoro M, Iacovelli F, Stracquadanio M, Falconi M, Marino C, Zanini G (2023) Experimental and in silico evaluations of the possible molecular interaction between airborne particulate matter and SARS-CoV-2. Sci Total Environ 895:165059
Sagawa T, Tsujikawa T, Honda A, Miyasaka N, Tanaka M, Kida T, Hasegawa K, Okuda T, Kawahito Y, Takano H (2021) Exposure to particulate matter upregulates ACE2 and TMPRSS2 expression in the murine lung. Environ Res 195:110722
Sandal A, Töreyin ZN, Saltürk C, Arbak PM (2021) The perspective of the Turkish thoracic society members on institutional preparedness during the COVID-19 pandemic in Turkey. Turkish Thoracic J 22(4):317
Setti L, Passarini F, De Gennaro G, Barbieri P, Perrone MG, Borelli M, Palmisani J, Di Gilio A, Torboli V, Fontana F (2020) SARS-Cov-2RNA found on particulate matter of Bergamo in Northern Italy: first evidence. Environ Res 188:109754
Shahzad K, Farooq TH, Doğan B, Zhong Hu L, Shahzad U (2023) Does environmental quality and weather induce COVID-19: case study of Istanbul, Turkey. Environ For 24(1–2):9–20
Sheppard N, Carroll M, Gao C, Lane T (2023) Particulate matter air pollution and COVID-19 infection, severity, and mortality: a systematic review and meta-analysis. Sci Total Environ 880:163272
Sigrist CJ, Bridge A, Le Mercier P (2020) A potential role for integrins in host cell entry by SARS-CoV-2. Antiviral Res 177:104759
Stanaway JD, Afshin A, Gakidou E, Lim SS, Abate D, Abate KH, Abbafati C, Abbasi N, Abbastabar H, Abd-Allah F (2018) Global, regional, and national comparative risk assessment of 84 behavioural, environmental and occupational, and metabolic risks or clusters of risks for 195 countries and territories, 1990–2017: a systematic analysis for the global burden of disease study 2017. Lancet 392(10159):1923–1994
Strong K, You D, Banerjee A, Azevedo JP (2024) Global health estimates should be more responsive to country needs. Lancet
Sun Z, Zhu D (2019) Exposure to outdoor air pollution and its human health outcomes: a scoping review. PLoS ONE 14(5):e0216550
Sülkü SN, Mortaş A, Küçük A (2023) Measuring efficiency of public hospitals under the impact of Covid-19: the case of Türkiye. Cost Effective Res Allocat 21(1):1–12
Şahin M (2020) The association between air quality parameters and COVID-19 in Turkey. Pharm Bio Res 6:49–58
Şenol E (2023) Pandeminin 3. yılına girerken COVID-19'a güncel yaklaşım. In PDEÖ, Ulubay G, Köktürk N (ed.) Gelecek Endemi/Pandemilere Hazır mıyız?. 38:255–261. Turkish Thoracic Society
Tanis C, Karakaya K (2021) Investigating the impacts of air quality and weather indicators on the spread of SARS-Cov-2 in Istanbul, Turkey. Thermal Sci 25(4 Part A):2755–2765
Tokatlioglu Y, Sulku SN (2023) The impact of the covid-19 outbreak on unmet health care needs in istanbul. Prev Med Reports 36:102400
Ulutaş K, Abujayyab SK, Abu Amr SS, Alkarkhi AF, Duman S (2023) The effect of air quality parameters on new COVID-19 cases between two different climatic and geographical regions in Turkey. Theoret Appl Climatol 152(1):801–812
V'kovski P, Kratzel A, Steiner S, Stalder H, Thiel V (2021) Coronavirus biology and replication: implications for SARS-CoV-2. Natv Rev Microbiol 19(3):155–170
Walls AC, Park Y-J, Tortorici MA, Wall A, McGuire AT, Veesler D (2020) Structure, function, and antigenicity of the SARS-CoV-2 spike glycoprotein. Cell 181(2):281–292. e286
Wang CH, Chan YL, Oliver BG (2020) Is there an association between the level of ambient air pollution and COVID-19? Am J Physiol Lung Cell Mol Physiol 319(3):L416–L421
Wang B, Chen H, Chan YL, Oliver BG (2020a) Is there an association between the level of ambient air pollution and COVID-19? Am J Phys Lung Cell Mol Phys 319(3):L416–L421

Wang F, Liu J, Zeng H (2020b) Interactions of particulate matter and pulmonary surfactant: implications for human health. Adv Coll Interface Sci 284:102244
WHO, COVID-19 dashboard. Retrieved 15 Sept 2024. https://data.who.int/dashboards/covid19/deaths?n=o
WHO, COVID-19 pandemic overview. Retrieved 08 10 2024. https://www.who.int/europe/emergencies/situations/covid-19
Woodby B, Arnold MM, Valacchi G (2021) SARS-CoV-2 infection, COVID-19 pathogenesis, and exposure to air pollution: what is the connection? Ann N Y Acad Sci 1486(1):15–38
Wu X, Nethery RC, Sabath MB, Braun D, Dominici F (2020) Air pollution and COVID-19 mortality in the United States: strengths and limitations of an ecological regression analysis. Sci Adv 6(45):eabd4049
Yang L, Li C, Tang X (2020) The impact of PM2. 5 on the host defense of respiratory system. Front Cell Develop Bio 8:91
Yao Y, Pan J, Wang W, Liu Z, Kan H, Qiu Y, Meng X, Wang W (2020) Association of particulate matter pollution and case fatality rate of COVID-19 in 49 Chinese cities. Sci Total Environ 741:140396
Yıldırım E (2023) The relationship between PM10 and SO2 exposure and Covid-19 infection rates in Turkey using nomenclature of territorial units for statistics level 1 regions. Heliyon 9(11)
Yu K, Zhang Q, Wei Y, Chen R, Kan H (2024) Global association between air pollution and COVID-19 mortality: a systematic review and meta-analysis. Sci Total Environ 906:167542
Yu Z, Bellander T, Bergström A, Dillner J, Eneroth K, Engardt M, Georgelis A, Kull I, Ljungman P, Pershagen G (2022) Association of short-term air pollution exposure with SARS-CoV-2 infection among young adults in Sweden. JAMA Netw Open 5(4):e228109–e228109
Zhang J, Lim Y-H, So R, Jørgensen JT, Mortensen LH, Napolitano GM, Cole-Hunter T, Loft S, Bhatt S, Hoek G (2023) Long-term exposure to air pollution and risk of SARS-CoV-2 infection and COVID-19 hospitalisation or death: danish nationwide cohort study. Europ Res J 62(1)
Zhang Z, Zhu Z, Chen W, Cai Z, Xu B, Tan Z, Wu A, Ge X, Guo X, Tan Z (2019) Cell membrane proteins with high N-glycosylation, high expression and multiple interaction partners are preferred by mammalian viruses as receptors. Bioinformatics 35(5):723–728
Zhu P, Zhang W, Feng F, Qin L, Ji W, Li D, Liang R, Zhang Y, Wang Y, Li M (2022) Role of angiotensin-converting enzyme 2 in fine particulate matter-induced acute lung injury. Sci Total Environ 825:153964
Zhu Y, Xie J, Huang F, Cao L (2020) Association between short-term exposure to air pollution and COVID-19 infection: evidence from China. Sci Total Environ 727:138704
Zorn J, Simões M, Velders GJ, Gerlofs-Nijland M, Strak M, Jacobs J, Dijkema MB, Hagenaars TJ, Smit LA, Vermeulen R (2024) Effects of long-term exposure to outdoor air pollution on COVID-19 incidence: a population-based cohort study accounting for SARS-CoV-2 exposure levels in the Netherlands. Environ Res 252:118812

Chapter 14
The Impact of COVID-19 in Mozambique: Health and Health System

Maria dos Anjos Ernesto Hauengue and Rais Akhtar

Abstract This chapter summarizes the impact of pandemic COVID 19 in Mozambique, particularly in the Health System, the risk factors associated to the disease and the country epidemiological profile in the post-COVID-19 period. COVID-19 was primarily discovered in China and until February 23rd, 2020, more than 77,000 cases of COVID-19 infection and 60% of confirmed cases were reported in the city of Wuhan. In Mozambique, the first case of COVID-19 infection was confirmed on March 22. Since then, the disease infected a total number of 137,413 people and caused 1,690 official deaths (MISAU, 25/03/2020—http://www.misau.gov.mz/index.php/informacao-sobrecprpnavirus-covid-19; (MISAU, 20 May 2020). The negative impact caused by this disease was enormous, affecting different areas of development. However, for the interest of this document, we underline here the consequences brought to National Health System in terms of provision of services; demand of services by patients especially in the area of maternal –child health care. It also analyses the implication of the disease on the epidemiological features of the country; measures implemented by the Government to overcome the situation. At the end, the reports recommend the need for more investment either in financial resources or material and specific equipment to increase the capacity of public health hospitals to prevent and minimize the prevalence and incidence of the disease, and revert the epidemiological profile of the country.

Keywords COVID-19 · Risk analysis · Epidemiological feature · Mozambique

---

M. dos Anjos Ernesto Hauengue
Master On Public Health, Maputo, Mozambique

R. Akhtar (✉)
Formerly Professor of Geography, University of Kashmir, Srinagar, Jammu and Kashmir, India
e-mail: raisakhtar@gmail.com

R. Akhtar (ed.), *COVID-19: Impacts on Health and Health Systems*,
https://doi.org/10.1007/978-3-031-90255-0_14

14.1 Introduction

The COVID-19 pandemic, a disease caused by the SARS-Cov 2 virus in humans, has rapidly spread out to about 210 countries around the world. The first case emerged in the city of Wuhan, Hubei, China and was first reported by the World Health Organization (WHO) on December 31, 2019 (WHO, 25/03/2020)- https://www.afro.who.int/health-topics/coronavirus-covid-19. Due to the dramatic impacts that this disease has caused in the different social and economic areas of countries, the World Health Organization (WHO, 2020) has classified the outbreak of this COVID 19 pandemic as an international public health emergency, due to the high rates of incidence, prevalence and related lethality, given the high virulence and infective capacity of the viral pathogen, the fragility of the human population's immunity to the virus, and the absence of a specific vaccine to prevent the disease (OMS, 25/03/2020—https://www.int/emergencies/diseases/novel-coronavirus-2019).

According to the WHO (WHO, 21 January; 20 May 2020) in the affected countries, SARS-Cov-2 has caused about 4,789,205 infections and approximately 318,789 deaths until May 20, 2020, most of them adults.

Apart from morbidity and mortality, COVID 19 has also highlighted the precariousness of the existing health network and the fragility of the social protection network, especially in countries in the sub-Saharan African region for the implementation of prevention measures and quarantine enforcement (www.pscm-ps.org.mz, 2022). Other parallel services provided by the sector to patients and patients with other illnesses was also limited and with reduced qualified medical staff on duty in hospitals, having further affected the quality of life of communities, especially the most vulnerable segments of the population, such as children, the elderly who are unable to work permanently, people with disabilities with recognized incapacity for work, chronically ill people unable to work, families headed by children, the poorest, among others, whose immune response differs, and is deficient for some diseases. (WHO, 21 January, 20 May 2020).

The same concern regarding all the aspects narrated above is raised by the other member states of the African Union which, despite already facing some of the most serious threats to public health (HIV/AIDS, malaria, tuberculosis), have high rates of maternal mortality and mortality of children under 5 years of age, in addition to noncommunicable diseases (Guidance for the Continuation of Essential Health ENG_final.pdf.).

14.2 COVID 19 in Mozambique

14.2.1 The Impact of Disease

Mozambique was highly affected by the disease COVID-19 due mostly to the following risk factors: (i) weakness of health services; (ii) low capacity of Government in terms of humans resources and equipment to face emergencies situation within public health; (iii) low access to water and sanitation facilities to the public; (iv) lack of equipment protection for the health workers; (v) population poverty which affects the compliance to the governmental rules such as staying at home (WaterAID Mozambique 2024). The first case of COVID-19 infection in the country was confirmed on March 22, had a total of 137,413 cases of confirmed infections, and 1690 official deaths have been recorded since then. (MISAU, 25/03/2020- http://www.misau.gov.mz/index.php/informacao-sobrecprpnavirus-covid-19; (MISAU, 20 May 2020),

Overall, all services within the National Health System in Mozambique have faced challenges caused by this pandemic. It was seen a high and sudden demand for hospitalization, screening and treatment of this disease, combined with the restrictions that preventive measures imposed from the reduction of the labor mass, confinement, reinforcement of individual and collective hygiene measures and social distancing to reduce the spread of COVID-19 (https://www.wateraid.org/mz/covid-19. Moreover, the reduction in the flow of users in health units such as hospitals and health centers was notorious, with a drop of between 10 and 50% in demand being reported, according to the type of service; a reduction in the number of outpatient consultations in the order of 30% and the decline in full vaccination activities estimated at 10% in children under 1 year old in the second quarter of 2020 when compared to the same period of 2019 (Centro de Integridade Publica 2020). Family planning consultations have seen a 46% reduction due to the pandemic. (MISAU, 25/03/2020- http://www.misau.gov.mz/index.php/informacao-sobrecprpnavirus-covid-19; (MISAU, 20 May 2020). Consequently, there has been a decline in the provision of basic health services such as vaccination of children under five years of age, prenatal consultation (pregnant woman) and family planning, healthy child consultation (child growth assessment), screening (general consultations) and other specific consultations, as well as changes in the epidemiological profile in the country observing a resurgence of vaccine-preventable diseases such as diphtheria, measles and poliomyelitis in children under one year of age due to the fact that mothers were not complying with the vaccination schedule during the pandemic period (WHO, GAVI, UNICEF, COVID-19' disrupts routine vaccination efforts, May 2020).

The poor sanitation conditions and the lack of Personal Protective Equipment (PPE) were also seen as conducive to the rapid spread of the disease among communities and health professionals due to the higher rate of exposure to the virus and lead to the reflection that so far frontline employees against the new coronavirus did not have the complete PPE for their protection and continue to be a vulnerable population for contamination of the disease. INS (Parpia et al 2016).

Negative impacts also occurred on the country's economy, specifically in the pharmaceutical and manufacturing industry and others, related to business and economic productivity, and consequently more poverty for the country (UNDP & Ministry of Financ 2021). The disease also brought out to light the weak investment that has been made in the Mozambican public health system, especially in infrastructure and human capacities and equipment. Data from the SARA report showed that only 34% of health facilities (557) had an infrastructure that was poorly prepared for this pandemic scenario, compared to an acceptable average of 66% (1090) (Final report, SARA, 2018. Web.pdf.pdf.). Consequently, at the time of COVID 19, there was a greater demand for the diagnosis and treatment of the disease in services in the private sector, where the costs of treatment were enormous and a large part of the population could not afford it. In view of this fact, the impact of the third wave of the pandemic was quickly felt on the health of Mozambicans. (Barreto, 2021).

14.2.2 Implemented Measures in the Country

In the face of this disease, the Mozambican Government has implemented various measures to prevent and treat the disease, and the National Health System has been strengthened at the expense of investments in various components of medical care (MISAU, 2020). Among the actions carried out in this context, the following stand out:

(i) The installation in several provinces of laboratory capacity for the diagnosis of the pathogen using scientific techniques of molecular biology;
(ii) The establishment of isolation wards for infectious diseases;
(iii) Installation of intensive care units with mechanical ventilation capacity in hospitals;
(iv) Vaccination for increased immunization against COVID-19;
(v) Strengthening multisectoral actions to implement preventive measures against the spread of the disease;
(vi) The modernization of public health legislation regarding individual and collective hygiene measures, hygiene and safety in public places and dissemination through the media, to raise awareness among communities for the adoption of preventive measures against the disease.
(vii) Strengthening joint multisectoral actions to address this public health problem

14.2.3 Epidemiological Picture Post-COVID-19 in Mozambique

Data from the National Report on Chronic and Non-Communicable Diseases show that risk factors for these diseases are increasing, and the prevalence of hypertension in individuals aged between 25 and 64 years is 39.0%. The prevalence of diabetes is 7.4% and was more frequent in men (9.2% compared to women 6.0%) and in older individuals (10.1% compared to young people 5.7%). Cervical and breast cancers together accounted for about 50.0% of all cancers in women, while in men the most common cancer is Kaposi's sarcoma 33.0% ("National Report—2018", "language":"Portuguese", "author":[{ "family": Instituto Nacional de Saúde (INS, 2019)

Of the individuals with hypertension, 85.2% are not aware of the disease, 48.1% do not take any type of treatment; consequently, they have the disease not controlled. Hence, the numerous complications that arise from this lack of control, including cerebrovascular accident (CVA) and chronic kidney disease, which has been shown to be responsible for 53.7% of nephrology consultations at the Maputo Central Hospital as a consequence of untreated arterial hypertension (Damasceno et al. 2017).

14.3 Conclusion and Recommendations

COVID-19 has shown the world the need and importance of ensuring the provision of a more resilient health system in the face of pandemics that may arise, because the longer coverage reductions continue, the more lives will be lost. And once care-seeking patterns in hospitals are broken, it can be difficult to restore.

If the demand for health care by users continues to worsen over the subsequent years, the care supply gains gained over the years could slow down if the situation is not immediately reversed. Added here is the fact that the country is also prone to various climatic effects such as floods that in the absence of a pandemic the health system is unable to offer services to all who seek them.

These results underline the need to continue investing in the maintenance and improvement of essential health services during emergencies, such as the COVID-19 pandemic, and this evidence allows the Ministry of Health to advocate for the mobilization of funds for the most vulnerable target groups.

References

Barreto A (2021) COVID-19 exposes lack of investment on health Sector in Mozambique. Maputo

Centro de Integridade Publica (2020) Reduction on demand for health services during COVID-19

Containertitle (n.d.) JMIR Public Health and Surveillance 6(2):e19927. Source: DOI.org (Crossref.)

Damasceno A, Cubula B, Lunet N (2017) STEPS Mozambique, 2014/2015

Global Health Institute (2020) Impact assessment use of preventive measures against SARS-Cov-2 in Mozambique 2020

Guidance for the Continuation of Essential Health ENG_final.pdf

Children under five years old penalized by the impact of HIV—19. https://oPais.co.mz

INS (2019) National Report—2018 "language": "Portuguese", "author" "family": "Instituto Nacional de Saúde" (INS, 2019)

MISAU (2020) 25 Mar 2020. https://www.misau.gov.mz/index.php/informacao-sobre-coronavirus - COVID 19

MISAU (2020) 25 Mar 2020. http://www.misau.gov.mz/index.php/informacao-sobrecprpnavirus-covid-19

Parpia AS, Ndeffo-Mbah ML, Wenzel NS, Galvani AP (2016) Effects of response to 2014-2015 Ebola outbreak on deaths from malaria from malaria, HIV-AIDS and tuberculosis, West Africa. Emergence Infect Dis 22(3):433-41

Relatorio _Final_SARA_2018_Web.pdf.pdf.

UNDP (2021) The Socioeconomic impact of COVID 19 in the urban informal economy in Mozambique. Rangel Logistic Solution. Impact of COVID 19 in Mozambique, 2024

WaterAid Mozambique (2024) COVID 19 and hand wash

WHO 25 Mar (2020). https://www.who.int/emergencies/diseases/novel-coronavirus-2019

WHO, GAVI, UNICEF (2020) COVID-19 disrupts routine vaccination efforts

Chapter 15
The Impact of the COVID-19 Pandemic on the Incidence Dynamics of Main Disease Categories in the Russian Federation

Svetlana M. Malkhazova, Fedor I. Korennoy, Tamara V. Vatlina, and Dmitry S. Orlov

Abstract To assess the impact of the COVID-19 pandemic on the dynamics of other nosological forms in the Russian Federation (RF), the incidence of the main categories, such as Diseases of Circulatory System (DCS) and Malignant Neoplasms (MN), was analyzed. Time series of standardized incidence for each category were examined for the period 2007–2019 (pre-COVID-19), and 2020–2023 (COVID-19 and post-COVID-19). The post-COVID trends were compared to those hypothetically expected with no COVID impact. For the majority of the RF regions, upward trends of incidence were detected both in pre-COVID and post-COVID years. In the first year of the pandemic, a decrease in incidence was observed for all categories. The median incidence rates of MN in the post-COVID period were lower than expected in most of the country, while those of DCS demonstrated heterogeneous distribution with no clear spatial patterns.

Keywords COVID-19 · Russian Federation · ICD disease categories · Time series · Incidence trends · Trend change point

---

S. M. Malkhazova · D. S. Orlov (✉)
Lomonosov Moscow State University, Moscow, Russian Federation
e-mail: orlovds@list.ru

F. I. Korennoy
Federal Center for Animal Health (FGBI ARRIAH), Vladimir, Russian Federation

T. V. Vatlina
Smolensk State University, Smolensk, Russian Federation

R. Akhtar (ed.), *COVID-19: Impacts on Health and Health Systems*,
https://doi.org/10.1007/978-3-031-90255-0_15

15.1 Introduction

COVID-19, an infectious disease caused by the severe acute respiratory syndrome coronavirus, has spread rapidly throughout the world, affecting millions of people in various countries.

The pandemic has caused profound environmental, economic, and socio-cultural changes in the world. Lockdown measures and a sharp reduction in industrial production have resulted in a decrease in anthropogenic emissions into the atmosphere, resulting in a decrease in the concentration of air pollutants and greenhouse gas levels in many regions, especially in large urban agglomerations (Ginzburg et al. 2020). These effects varied widely depending on the characteristics of the natural variability of atmospheric impurities, as well as differences in the severity of restrictions (Baklanov et al. 2021).

The COVID-19 pandemic has had many consequences for global health in addition to those caused by the disease itself. There has been a decrease in the number of hospital visits, which has led to an increase in mortality from other causes.

The study of the impact of COVID-19 on health and the healthcare system at the local, regional, and global levels is highly important since the development of effective solutions to improve public health and increase the resilience of healthcare systems to global challenges is only possible through multi-scale experience (Coronavirus… 2021).

The study of the impact of COVID-19 on public health and the healthcare system in the Russian Federation (Malkhazova et al. 2021) is still highly relevant at present due to the scale of the pandemic which has had a significant impact on Russian society. This led to the need for an in-depth analysis of the consequences of the pandemic for the health of individual population groups, as well as for the healthcare system as a whole.

The assessment of the impact of the pandemic on the health of the Russian population, as well as the healthcare sector, continues as evidenced by a significant number of publications according to the Web of Science, Scopus, PubMed, Google School, and eLibrary databases.

Journals have been devoting special editions to assessments of the scale of the coronavirus pandemic, analysis of the main mechanisms of its impact on the economy and socio-economic consequences in different geographical conditions, medical, and social problems as well as the impact of the pandemic on the environment and air quality (Geography, Environment, Sustainability 2021).

The high contagiousness of COVID-19 and the variability of clinical manifestations have led to an increase in the number of hospitalizations which required significant resources from healthcare institutions. In addition, prolonged exposure to the virus can lead to post-COVID syndrome which manifests itself in various chronic symptoms. The situation is especially complicated in the case of the development of combined pathology in patients under observation for chronic non-communicable diseases. The pandemic has caused significant changes in the lifestyle of the population due to the transition to remote working, the introduction of restrictions on movement and social interaction, the deterioration of the financial situation, and all these factors have adversely affected the mental health of the population. Many studies note the vulnerability of the healthcare system in the context of a massive increase in COVID-19 cases due to an overload of medical institutions and a shortage of personnel, equipment, and medications (Krivenko 2021; Ulumbekova and Ginoyan 2022).

According to some authors, the so-called optimization of the healthcare system implemented in Russia in recent years has provoked the emergence of several problems that have manifested itself during the COVID-19 pandemic (Starodubov et al. 2020; Chernyshev et al. 2021). The reduction of infectious disease beds before the pandemic, the mergers of medical institutions, and the reduction of government funds for healthcare served as prerequisites for the emergence of difficulties during the pandemic (Perkhov and Gridnev 2020; Goroshko et al. 2022).

Many medical institutions were forced to adapt to changed working conditions, redistribute resources, and introduce new treatment protocols (Starodubov et al. 2020). There is a need to develop new strategies for resource management and optimization of healthcare delivery processes (Minakir 2020; Pesennikova and Perkhov 2020; Ulumbekova 2020).

The purpose of this article is to analyze the impact of the COVID-19 pandemic on the dynamics of the incidence of main categories of diseases in the Russian Federation, to identify patterns of distribution and geographic differentiation of the incidence of malignant neoplasms (MN) and diseases of the circulatory system (DCS) in the pre-COVID, COVID, and post-COVID periods.

15.2 Materials and Methods

15.2.1 Study Area

The Russian Federation is a country located in both Europe and Asia, ranking first in the world in terms of area (more than 17 million km^2). The country's population is more than 146 million people with an average density of 8.53 people per km^2. Administratively, the country is divided into 89 first-level units (subjects or regions), represented by oblasts, republics, autonomous Okrugs and three cities of federal significance. The research was conducted on the regional level of 85 subjects, for which incidence data are available for the entire analyzed period.

15.2.2 COVID-19 in Russia

The first cases of COVID-19 were registered in Russia in the first ten days of March 2020. By April 16, COVID-19 had been recorded in all regions of the country, and by the end of the month, the number of cases in Russia exceeded 100,000. A sharp rise occurred in late April–early May when the number of infections increased from 6000 to 11,000 reaching a maximum on May 11, 2020 (11,656 cases). Then, in the summer months of 2020, the incidence rate began to decline (Malkhazova et al. 2021).

After the number of cases decreased to less than 5000 per day by the end of August 2020, a new increase in cases ("second wave") emerged in the fall and was considered most intense in the second half of September 2020. In November, more than 20,000 infected people were detected each day. The second wave of the disease peaked in late December 2020 when the number of daily COVID-19 cases was almost 30,000, more than 2.5 times higher than the corresponding figure during the first wave. According to Rosstat (Federal Service for State Statistics) (Federal Service... 2024), 144,691 people with coronavirus infection died in Russia in 2020 which accounted for 6.8% of the total number of deaths in 2020. Despite significantly higher incidence and mortality rates, the country did not introduce a general lockdown. Essentially, only the universal mask regime and a ban on mass events were maintained on the federal level. The decision to apply additional restrictive measures was made by regional authorities. In December 2020, vaccination against COVID-19 began in Russia amid an increase in the number of infections. From January to June 2021, the number of infections decreased. For example, 27,039 cases of COVID-19 were confirmed on January 1, while the number was only 9500 on June 1.

A new surge in COVID-19 cases (the "third wave") was recorded in Russia in the last ten days of June 2021. The number of new cases exceeded 20,000 on June 24. Thus, the growth in registered cases of infection since the beginning of the month was 100%. During August, as well as the first and second ten days of September, the daily number of infected people was between 17,813 and 19,905.

Another surge in cases called the "fourth wave" was recorded on September 23, 2021. A new maximum since the beginning of the pandemic was recorded on November 6, 2021, when the daily number of cases reached 41,335.

Thus, despite the measures taken, Russia turned out to be one of the most affected countries.

In the following years, the COVID-19 situation gradually stabilized. The coronavirus has ceased to be a source of anxiety and a serious burden on the healthcare system as it was in the first year of the pandemic but its strains are still circulating and continuing to mutate.

15.2.3 Disease Data

Two categories of diseases according to the international classification ICD-10 were considered: II (Neoplasms) and IX (Diseases of the Circulatory System). These categories are the leading causes of death in the Russian Federation for the period from 2007 to 2023 accounting for about 70% of all causes of death in the population. Data on relative incidence (per 100 thousand people) annually for the period from 2007 to 2023 were obtained from official reports of the Federal State Statistics Service (Federal Service … 2024).

15.2.4 Analysis Methods

To assess changes in the dynamics of incidence in the population before and after the emergence of COVID-19, the median incidence values were considered for two time periods: 2007–2019 (pre-COVID-19) and 2020–2023 (COVID-19 and post-COVID). For each of the periods, incidence trends were also determined by fitting a linear regression model using the least squares method. The reliability of the trend was assessed using the determination coefficient R^2 (the proportion of variation explained by regression). A trend was considered reliable if $R^2 \geq 0.5$. Additionally, the change in incidence in the first year of the pandemic (2020) was assessed compared to the previous year. Furthermore, expected incidence values for the period from 2020 to 2023 were modeled in the absence of COVID-19 and compared with the observed values for this period. The simulation was carried out using a linear trend for the period from 2007 to 2019.

15.3 Results and Discussion

15.3.1 Literature Review

15.3.1.1 The Impact of COVID-19 on Mortality

COVID-19 has caused excess mortality in Russia. The natural population decline exceeded 1 million people in 2021 (Federal Service… 2024). It should be emphasized that the spread of COVID-19 affected mortality not only directly (leading to the death of those infected) but also indirectly since it made it difficult for other patients to receive medical care, and caused complications of diseases. Several publications characterizing the situation in Russia during the pandemic are devoted to the high level of excess mortality. This indicator is considered a representation of the impact of the pandemic including direct (deaths from COVID-19) and indirect (impact on the healthcare system and socio-economic factors) effects. The presence of excess mortality in all regions of the Russian Federation is confirmed by all authors who have studied this topic (Danilova 2020; Kotov et al. 2022; Kvasha and Kharkova 2022, etc.), the differences are only in the quantitative parameters of these estimates. According to some authors, excess mortality in 2020 amounted to 288,000 people and COVID-19 caused approximately 60% of those deaths (Goroshko and Patsala 2022). However, another study says that official data on mortality from COVID-19 in Russia are underestimated by more than two times (Druzhinin and Molchanova 2021; Livshits and Neklyudova 2020).

An assessment of excess mortality in the regions of Russia from April 2020 to February 2021 compared to the five-year average based on regression analysis showed (Kotov et al. 2022) that the most significant factors increasing excess mortality are the share of the elderly population and the employment structure represented by the share of workers in the manufacturing industry (economic activity C, European Skills, Competences, and Occupations classification). An assessment of excess mortality rates for the Russian Federation and its regions, as well as equivalent rates for 36 countries, showed that Russia had the highest excess mortality rate among all the countries considered. Most excess mortality cases were registered in the fourth quarter of 2020, and the level and trajectory of excess mortality in Russia and most Eastern European countries differed from those in Western countries. While the cumulative number of COVID-19 cases and deaths was positively correlated with excess mortality across countries, in Russia an inverse negative relationship was observed between excess mortality and cumulative incidence at the regional level which may be due to common underreporting of COVID-19 cases (Timonin et al. 2021). Rosstat published monthly statistics on deaths from coronavirus infection distinguishing four groups of deaths of patients diagnosed with COVID-19: (1) cases in which coronavirus was the main cause of death; (2) cases requiring additional research, when it is assumed that death occurred due to coronavirus; (3) cases in which coronavirus as a concomitant disease affected the underlying disease that caused death; (4) cases in which coronavirus did not cause complications of the diseases that caused death (Federal Service… 2024).

As a result of a comparative analysis of COVID-19-related mortality data, significant territorial heterogeneity was revealed across different regions of the Russian Federation (Goldshteyn 2020).

15.3.1.2 The Impact of COVID-19 on Chronic Non-communicable Diseases

Special attention was paid by the experts to the analysis of the influence of COVID-19 on the occurrence of new cases of chronic noncommunicable diseases and the aggravation of their course (Shelgunov et al. 2023). The literature review conducted by the authors is based on the results of 64 papers published between April 1, 2020 and January 31, 2023, exploring the impact of COVID-19 on various organs and systems as well as the aggravation of chronic non-communicable diseases.

The search for publications was carried out in the PubMed, Google Scholar, and eLibrary databases. The analysis showed that the infection not only affected various organs and systems and increased the risk of complications in the course of chronic noncommunicable diseases but also contributed to the emergence of new cases.

It is understood that COVID-19 is the primary cause of new cases of chronic non-communicable diseases in healthy people, also contributing to the development of combined pathology in patients undergoing dispensary observation for another reason. At the same time, patients with comorbidities are at higher risk of severe COVID-19 with an unfavorable outcome. Given the scale of the spread of COVID-19 and its consequences, it is important for healthcare workers to have a clearer understanding of the possible consequences for patients, including in terms of the development and exacerbation of chronic non-communicable diseases that this infection can cause. According to the authors of the review (Shelgunov et al. 2023), it is important not only to understand the extent of the impact of COVID-19 on the development of chronic pathology in patients but also to keep in mind the ever-increasing volume of medical care for such patients.

15.3.1.3 Impact on Diseases of the Circulatory System

The COVID-19 virus has an adverse effect on the cardiovascular system. Manifestations can be varied: myocardial damage, myocarditis, acute heart failure, acute myocardial infarction, cardiac arrhythmia, cardiac arrest, venous thromboembolism, and others (Driggin et al. 2020; Gushchina and Lozhkina 2021; Sugraliev 2021; Romanov 2022; Vorobyeva and Romanova 2022; Huseynov 2023; Pogosova et al. 2023; Yusov and Alpidovskaya 2023, etc.).

It has been established that patients with COVID-19 have a high prevalence of cardiovascular manifestations due to a systemic inflammatory response and immune system disorders as the disease progresses. According to the conducted studies, the most common complication from the cardiovascular system is the progression of forms of ischemic disease (41.9%). COVID-19 can also cause both primary cardiac pathology and aggravate existing cardiovascular diseases (Kravtsiva et al. 2021).

Although some studies have suggested a link between cardiovascular diseases and severe COVID-19, the efforts to accurately assess the prevalence of cardiovascular diseases in patients with COVID-19 are complicated by the lack of widespread testing, national surveillance, or standardized data collection in the Russian Federation.

There are concerns that people with cardiovascular diseases do not seek timely treatment during the pandemic and die at home. The shortage of essential medications is also important.

15.3.1.4 The Impact of COVID-19 on Oncologic Incidence

Cancer patients are at higher risk of contracting infectious diseases or developing serious complications caused by the virus due to the immunodeficiency caused by the underlying disease. Many anticancer drugs are known to have an immunosuppressive effect. Surgery is another risk factor for developing infections. Cancer patients are a special group since their treatment cannot be stopped. The need to find a balance between the risk of cancer patients and healthcare workers becoming infected with coronavirus and the need to continue treatment changes the work of oncology institutions. The high incidence of COVID-19 among healthcare workers significantly reduces the number of personnel to provide specialized care. The global healthcare system is faced with an atypical situation that requires, on the one hand, an urgent revision of the standards of treatment for cancer patients, and on the other hand, maintaining the quality of care provided. The available experience is extremely limited, heterogeneous and, in most cases, based on isolated observations. Therefore, it is important to organize the provision of care to patients during a pandemic as well as to search for an answer to the question of how coronavirus infection affects the prognosis and methods of treatment of malignant neoplasms. The leading oncology institute in the Russian Federation, the Hertsen Moscow Oncology Research Institute, conducted a detailed, systematic review of world literature data, including all aspects that reflect the impact of the COVID-19 pandemic on oncological practice (Kaprin et al. 2020). The search for information sources was carried out in the PubMed, MedLine, Scopus, Web of Science, and RSCI databases. Data from retrospective and prospective clinical studies were analyzed. The review reflects the experience of Russian and foreign researchers, presents data on the impact of COVID-19 on cancer patients, the mortality of cancer patients infected with COVID-19, the possibilities of treating COVID-19, etc. It is noted that, despite the pandemic, treatment of cancer patients must be continued since the presence of a tumor process does not allow a delay in therapy. The global oncology community continues to actively develop recommendations for the optimal management of cancer patients during the pandemic. High-risk groups for infection with the new coronavirus have been identified, including patients with cancer.

15.3.1.5 The Impact of COVID-19 on the Central Nervous System

As COVID-19 continues to be reported, recent literature on the effects of this new virus on the central nervous system may help guide clinical practice and identify potentially important research directions. In addition, understanding the potential mechanisms of neurological injury may focus efforts on more accurate detection and prevention of these complications (Belopasov et al. 2020; Simonenko et al. 2021; Ekusheva et al. 2022; Parfenov and Kulesh 2022). Reported nervous system manifestations range from relatively simple symptoms to cerebral hemorrhage and infarction. Postmortem studies of human brains indicate that human coronavirus variants and SARS-CoV-2 can infect neurons and glia. Furthermore, studies have demonstrated increased serum cytokine levels following SARS-CoV-2 infection, consistent with the notion that cytokine overproduction and toxicity may be an important potential mechanism of neurological injury parallel to the known lung injury pathway. Evidence that SARS-CoV-2 may be a vasculotropic and neurotropic virus is discussed. Early reports suggest that COVID-19 may be associated with severe neurological complications, and there are several plausible mechanisms to explain these observations. Detailed information on the potential neurological injury and further exploration of relevant pathophysiological interventions are needed to understand and ultimately mitigate SARS-CoV-2-associated neurological injury.

15.3.1.6 The Impact of COVID-19 on Infectious Diseases

Several countries have seen a notable decline in the spread of sexually transmitted infections, including HIV/AIDS, attributed to COVID-19 quarantines, social distancing measures, and advice against casual sex. Similarly, in some places, transmission rates of influenza and other respiratory viruses have dropped significantly during the pandemic. In addition, the influenza B/Yamagata lineage may have disappeared in 2020 or 2021 due to measures to control the COVID-19 pandemic, and no natural cases have been confirmed since March 2020. In 2023, the World Health Organization concluded that protection against the Yamagata lineage is no longer required in the seasonal influenza vaccine, reducing the number of lineages targeted by the vaccine from four to three. An assessment of the course of the COVID-19 pandemic and its impact on the epidemics of HIV infection, viral hepatitis C (HCV), tuberculosis, influenza, and acute respiratory infections was carried out in the Northwestern Federal District of the Russian Federation over two years of the spread of COVID-19 from the standpoint of the possible formation of a syndemic, or interference between the diseases (Belyakov et al. 2022). The pandemic did not significantly affect the patterns of the epidemic course of HIV, HCV and tuberculosis, which rejected the presence of a syndemic between these pathogens. Interference with respiratory infections was noted, with a clear predominance of SARS-CoV-2.

15.3.1.7 The Impact of COVID-19 on the Category "External Causes"

Experts paid special attention to the analysis of the influence of COVID-19 in terms of stress and anxiety that accompanied the pandemic's course which can be associated with an increase in mortality rates from injuries, poisoning, and conditions associated with alcohol and drug use (Briko et al. 2020; Nikiforov et al. 2020; Nikitina et al. 2021; Yakovleva et al. 2023).

According to experts, concerns about a possible sharp increase in suicide mortality were not confirmed; on the contrary, at the first stages, during the period with the strictest quarantine measures, the suicide rate decreased (Boyko et al. 2020; Rozanov 2020; Rozanov and Semenova 2022).

15.3.2 *Results*

15.3.2.1 Diseases of the Circulatory System (DCS)

The annual dynamics of the incidence of DCS on average in the subjects of the Russian Federation is presented in Fig. 15.1. The point of change in the trend in 2020 is obvious, followed by a continuation of the upward trend. The decline in incidence in 2020 compared to 2019 on average in the country was 6.2 ± 7.5%. At the same time, an increase in incidence was observed only in some regions in the southern part of the Russian Federation, such as Astrakhan and Kurgan oblasts, and the Republics of Chechnya, Kabardino-Balkaria, Adygea, Altai, and Buryatia.

The average incidence of DCS in the pre-COVID period in all regions of the Russian Federation was 22,177 ± 5,748 cases per 100,000 population. In most (66 out of 85) regions, an increasing trend was observed, and in 44 out of 66 subjects, the trend significance exceeded 0.5. In eight regions, a significant negative trend in the incidence of DCS was observed. These regions included the federal cities of Moscow, St. Petersburg and Sevastopol, the Republic of Crimea, Astrakhan Oblast, Altai Krai, Vladimir Oblast, and Nenets Autonomous Okrug (Fig. 15.2). During the COVID-19 period, the average incidence rate in all regions increased and amounted to 24,954 ± 6382 cases per 100,000 population. At the same time, in 70 out of 85 regions, a stable trend toward increasing incidence continues to form. Only four regions show a significant downward trend in the incidence of DCS during this period; these are the Tomsk and Novgorod oblasts as well as the Republics of Tyva and Khakassia. Compared with the hypothetical incidence for 2020–2023 (assuming the continuation of the pre-COVID trend), in 40 of the 85 regions studied, the actual median incidence was higher by up to 35%. In the remaining regions, a lower median incidence of up

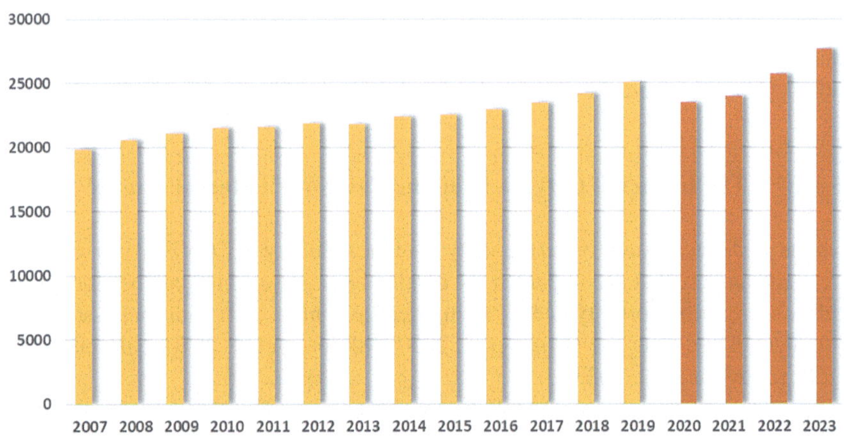

Fig. 15.1 Annual dynamics of the average incidence of diseases of the circulatory system in the Russian Federation for the period from 2007 to 2023

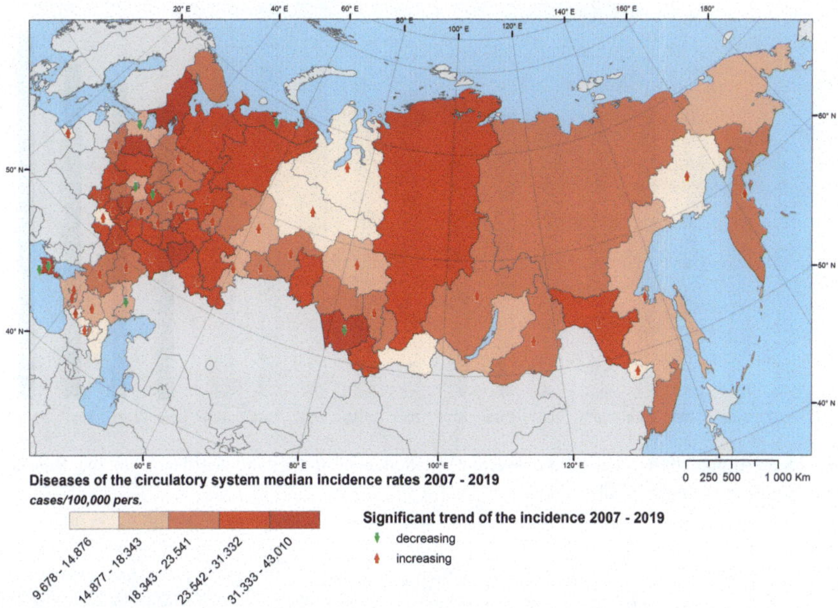

Fig. 15.2 Distribution of median incidence of diseases of the circulatory system in the Russian Federation in the period 2007–2019 and significant trends in its change

to 28% is observed. The spatial distribution of regions with an increase and decrease in the incidence of DCS is very heterogeneous and does not reveal any clear patterns.

15.3.2.2 Malignant Neoplasms (MN)

The annual dynamics of the incidence of MN on average in the regions of the Russian Federation is presented in Fig. 15.3. The point of change in the trend in 2020 with the subsequent continuation of the upward trend can be seen. The decline in incidence in 2020 compared to 2019 on average in the country was 11.8 ± 7.4%. At the same time, an increase in incidence was observed only in two regions, these being the Republic of Ingushetia and the Chukotka Autonomous Okrug.

The average incidence rate of malignant neoplasms in the pre-COVID period in all regions of Russia was 373.5 ± 81.3 cases per 100,000 population. In the overwhelming majority (84 out of 85) of regions, an upward trend was observed, and in 77 out of 84 regions, the trend significance exceeded 0.5. A significant negative incidence trend was observed only in the Chechen Republic (Fig. 15.4). During the COVID-19 period, the average incidence rate in all regions increased and amounted to 417.8 ± 97.7 cases per 100,000 population. The upward trend in incidence continued in 83 out of 85 regions studied (significant in 73 out of 83). An insignificant downward trend is observed in the Leningrad Oblast and the Republic of Ingushetia.

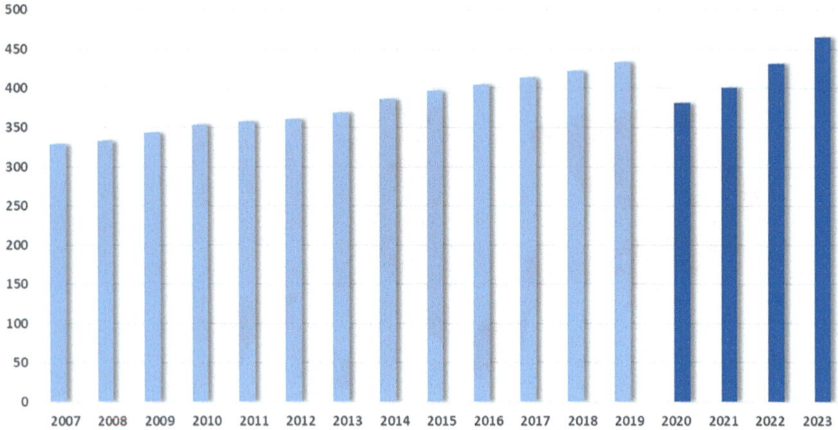

Fig. 15.3 Annual dynamics of the average incidence of malignant neoplasms in the Russian Federation for the period 2007–2023

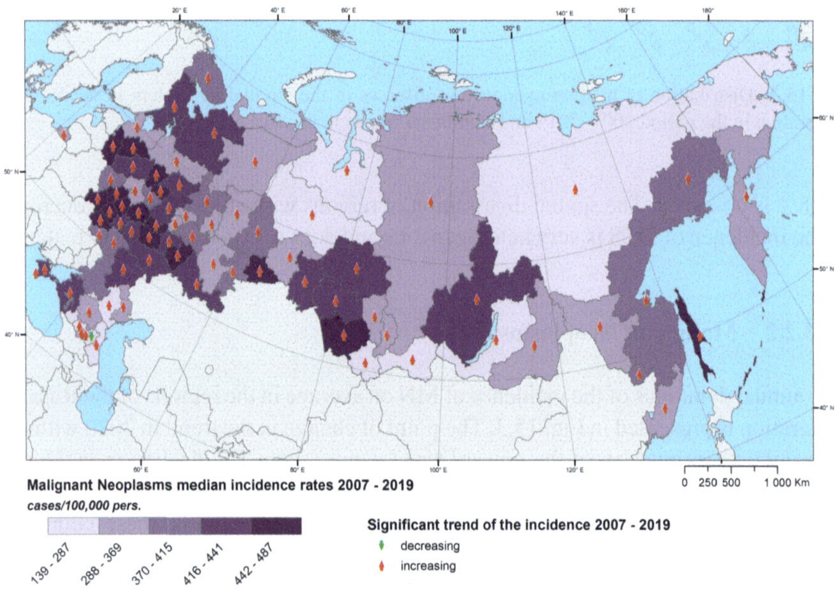

Fig. 15.4 Distribution of median incidence of malignant neoplasms in the Russian Federation in the period 2007–2019 and significant trends in its change

Compared with the hypothetical incidence rate for 2020–2023 (assuming the pre-COVID trend continues), the actual median incidence rate was higher in only nine out of 85 regions. Moreover, eight out of nine regions demonstrate a slight excess (up to 8%), while the Chechen Republic shows a 97% excess. Most regions have a lower median incidence rate of up to 30%.

The reduction in both DCS and MN morbidity during the COVID and post-COVID years may not have been due to an actual decline in cases, but rather to a significant reduction in the availability of health care and diagnostics, leading to a reduction in the detection of new cases. Similar patterns have been observed in other countries (Poucineau et al. 2024; Song et al. 2024).

15.3.3 Conclusion

The COVID-19 pandemic has dealt significant damage to healthcare systems around the world affecting thousands of people in different countries and is considered a serious threat to global health. Patients with underlying chronic diseases are at higher risk of a severe COVID-19 case with an unfavorable outcome. Moreover, as research results show, COVID-19 can cause new cases of chronic noncommunicable diseases in healthy people and the development of combined pathology in patients who are getting treatment for another reason.

An analysis of data on the incidence of diseases of the circulatory system and malignant neoplasms in the population for the period between 2007 and 2023 showed that in general; in Russia, there was a steady increase in the incidence even before COVID-19. At the onset of the COVID-19 pandemic, there was a decline in the incidence rate followed by a resumption of the increasing trend. At the same time, concerning malignant neoplasms, the decline in incidence was more pronounced: most regions of the country demonstrate lower average values of malignant neoplasm incidence after 2020 than would have been hypothetically expected without the impact of COVID-19.

Given the research data, it is important to have a good understanding of the extent of the impact of COVID-19 on the development of chronic pathology and, accordingly, on the likely increasing volume of medical care for such patients in order to plan medical care rationally within the framework of the free state medical care.

References

Baklanov AA, Chubarova NE, Kolosov VA, Malkhazova SM, Porfiriev BN (2021) Introduction to geography of Covid-19 pandemic: environmental issues, public health and socio-economic consequences. Geogr Environ Sustain 14(4):105–108. https://doi.org/10.24057/2071-9388-2021-044

Belopasov VV, Yashu Y, Samoilova EM, Baklaushev VP (2020) Damage to the nervous system in COVID-19. Klinicheskaya Praktika 11(2):60–80 (in Russian)

Belyakov NA, Boeva EV, Simakina OE, Svetlichnaya YuS, Ogurtsova SV, Serebryakova SL, Esaulenko EV, Zagdyn ZM, Yazenok AV, Lioznov DA, Stoma IO (2022) COVID-19 pandemic and its impact on other infections in Northwest Russia. HIV Infect Immunosuppressive Disord 14(1):7–24. https://doi.org/10.22328/2077-9828-2022-14-1-7-24 (in Russian)

Boyko OM, Medvedeva TI, Enikolopov SN, Vorontsova OYu, Kazmina OYu (2020) Psychological state of people during the COVID-19 pandemic and targets of psychological work. Psikhologicheskie issledovanija 13(70):1–12. https://doi.org/10.54359/ps.v13i70.196 (in Russian)
Briko NI, Kagramanjan IN, Nikiforov VV et al (2020) Pandemija COVID-19. COVID-19 pandemic. Measures to combat its spread in the Russian Federation. Epidemiologija i Vakcinoprofilaktika 19(2):4–12. https://doi.org/10.31631/2073-3046-2020-19-2-4-12 (in Russian)
Chernyshev VM, Voevoda MI, Strelchenko OV, Shalygina LS, Mingazov IF (2021) On the state of health of the population and healthcare in the Siberian Federal Okrug during the pandemic (some results of 2020). Sibirskij nauchnyj medicinskij zhurnal 41(6):101–109. https://doi.org/10.18699/SSMJ20210612 (in Russian)
Danilova IA (2020) Morbidity and mortality from COVID-19. The problem of data comparability. Demograficheskoe obozrenie 7(1):6–26 (in Russian)
Driggin E, Madhavan MV, Bikdeli B et al (2020) Cardiovascular considerations for patients, health care workers, and health systems during the COVID-19 pandemic. J Am Coll Cardiol 75(18):2352–2371. https://doi.org/10.1016/j.jacc.2020.03.031
Druzhinin PV, Molchanova EV (2021) Mortality of the population of Russian regions in the context of the COVID-19 pandemic. Regionologija 29(3):666–685. https://doi.org/10.15507/2413-1407.116.029.202103.666-685 (in Russian)
Ekusheva EV, Kovalchuk VV, Shchukin IA (2022) Neurological complications of COVID-19 and post-covid syndrome, 104 pp (in Russian)
Federal State Statistics Service. URL: https://eng.rosstat.gov.ru/. Accessed 09 Dec 2024
Geography, environment, sustainability (2021). Special issue: Geography of the COVID-19 pandemic: public health, economic and environmental consequences 14(4)
Ginzburg AS, Semenov VA, Semutnikova EG, Aleshina MA, Zakharova PV, Lezina EA (2020) The impact of COVID-19 restrictions on air quality in Moscow. Reports of the Russian Academy of Sciences Nauki o Zemle 1:74–79 (in Russian)
Goldshteyn EM (2020) Factors affecting mortality from a new coronavirus infection in different subjects of the Russian Federation. Zhurnal mikrobiologii, ehpidemiologii i immunobiologii 97(6):604–607. https://doi.org/10.36233/0372-9311-2020-97-6-11 (in Russian)
Goroshko NV, Patsala SV, Emelyanova EK (2022) Barriers to the fight against the COVID-19 pandemic in the healthcare system of Russia and its regions. Ekonomika. Informatika 49(2):217–233. https://doi.org/10.52575/2687-0932-2022-49-2-217-233 (in Russian)
Gushhina OI, Lozhkina NG (2021) Cardiovascular diseases combined with SARS-CoV-2 viral infection: course and prognosis. Ateroskleroz 17(3):97–105. https://doi.org/10.52727/2078-256X-2021-3-97-1055 (in Russian)
Huseynov A, Akin I, Duerschmied D, Scharf RE (2023) Cardiac arrhythmias in post-COVID syndrome: prevalence, pathology, diagnosis, and treatment. Viruses 15(2):389. https://doi.org/10.3390/v15020389
Kaprin AD, Gameeva EV, Polyakov AA, Korniyetskaya AL, Rubtsova NA, Fedenko AA (2020) The impact of the COVID-19 pandemic on oncology practice. Sibirskij onkologicheskij zhurnal 19(3):5–22. https://doi.org/10.21294/1814-4861-2020-19-3-5-22 (in Russian)
Kotov EA, Goncharov RV, Kulchitsky YV, Molodsova VA, Nikitin BV (2022) Spatial modelling of key regional level factors of Covid-19 mortality in Russia. Geogr Environ Sustain 2(15):71–83. https://doi.org/10.24057/2071-9388-2021-076
Kravtsiva AV, Gulyaeva AA, Golovanova ED, Airapetov KV (2021) Cardiovascular damage in COVID-19. Vestnik Smolenskoj Gosudarstvennoj Medicinskoj Akademii 20(4):59–65 (in Russian)
Krivenko NV (2021) Trends in economic development and aspects of healthcare reform in Russia in the context of the COVID-2019 pandemic. Uroven' zhizni naselenija regionov Rossii 17(4):503–513. https://doi.org/10.19181/lsprr.2021.17.4.7 (in Russian)
Kvasha EA, Kharkova TL (2022) The COVID-19 pandemic and mortality from the main causes of death in the regions of the Russian Federation in 2020. Regional'nye Issledovanija 4(78):61–75. https://doi.org/10.5922/1994-5280-2022-4-6. (in Russian)

Lifshits ML, Neklyudova NP (2020) COVID-19 mortality rate in Russian regions: forecasts and reality. R-Economy 6(3):171–182. https://doi.org/10.15826/recon.2020.6.3.015

Malkhazova SM, Korennoy FI, Shartova NV, Vatlina TV (2021) COVID-19 in the Russian Federation: Regional Differences and Public Health Response. In: Akhtar, R. (eds) Coronavirus (COVID-19) Outbreaks, Environment and Human Behaviour. Springer, Cham. https://doi.org/10.1007/978-3-030-68120-3_17

Minakir PA (2020) Pandemic economy: the Russian path. *Prostranstvennaja ekonomika* 16(2):7–18. https://doi.org/10.14530/se.2020.2.007-018 (in Russian)

Nikiforov VV, Suranova TG, Chernobrovkina TYa (2020) Novel coronavirus infection (COVID-19): clinical and epidemiological aspects. Arkhiv vnutrennej mediciny 10(2):87–93. https://doi.org/10.20514/2226-6704-2020-10-2-87-93 (in Russian)

Nikitina AYu, Chimagomedova ASh, Levin OS (2021) Neurological manifestations of COVID-19 in the elderly. Zhurnal nevrologii i psikhiatrii 121:10:2:5–15. https://doi.org/10.17116/jnevro20211211025 (in Russian)

Parfenov VA, Kulesh AA (2022) Acute and late neurological disorders in patients who have had coronavirus infection. Nevrologija, Nejjropsikhiatrija, Psikhosomatika 14(3):4–11 (in Russian)

Perkhov VI, Gridnev OV (2020) Lessons of the COVID-19 pandemic for public health policy. Sovremennye problemy zdravookhranenija i medicinskoj statistiki 2:206–222. https://doi.org/10.24411/2312-2935-2020-10043 (in Russian)

Pesennikova EV, Perkhov VI (2020) Directions for the development of medicine and healthcare in the post-pandemic world. Sovremennye problemy zdravookhranenija i medicinskoj statistiki 4:535–551. https://doi.org/10.24411/2312-2935-2020-00130 (in Russian)

Pogosova NV, Kuchiev DT, Popova AB (2023) Consequences of COVID-19 at a remote stage after hospitalization according to clinical, instrumental and laboratory research methods. Kardiologicheskij vestnik 18(4):56–66. https://doi.org/10.17116/Cardiobulletin20231804156 (in Russian)

Poucineau J, Khlat M, Lapidus N, Chouaïd C, Espagnacq M, Delory T, Le Cœur S (2024) Impact of the COVID-19 pandemic on lung cancer diagnoses and mortality: a nationwide study in France. Cancer Epidemiol 93:102679. https://doi.org/10.1016/j.canep.2024.102679

Romanov YuA (2022) SARS-CoV-2, COVID-19 and cardiovascular complications: a view from the perspective of vascular endothelium. Kardiologicheskij vestnik 17(1):21–28. https://doi.org/10.17116/Cardiobulletin20221701121 (in Russian)

Rozanov VA (2020) Urgent tasks in the sphere of suicide prevention in connection with the COVID-19 pandemic. Suicidologija 11(1):39–52. https://doi.org/10.32878/suiciderus.20-11-01(38)-39-52 (in Russian)

Rozanov VA, Semenova NV (2022) Suicidal behavior in the context of the COVID-19 pandemic. Psikhiatrija 20(3):74–84. https://doi.org/10.30629/2618-6667-2022-20-3-74-84 (in Russian)

Shelgunov VA, Zubko AV, Kungurcev OV, Zaporozhchenko VG (2023) The impact of a new coronavirus infection on the development of chronicnon-communicable diseases. Social'nye aspekty zdorov'ja naselenija 69(3):5. https://doi.org/10.21045/2071-5021-2023-69-3-5 (in Russian)

Simonenko VV, Vakal TN, Mikhalik DS, Zhukov GV, Nikolaenkova LI (2021) Neurological complications after coronavirus infection. Vestnik Smolenskoj Gosudarstvennoj Medicinskoj Akademii 20(2):59–64 (in Russian)

Song S, Guo C, Wu R, Zhao H, Li Q, Dou J-h, Guo F-s, Wei J (2024) Impact of the COVID-19 pandemic on cardiovascular mortality and contrast analysis within subgroups. Front. Cardiovasc. Med 11:1279890. https://doi.org/10.3389/fcvm.2024.1279890

Starodubov VI, Kadyrov FN, Obukhova OV, Bazarova IN, Endovitskaya YuV, Nesvetailo NY (2020) The impact of coronavirus COVID-19 on the situation in Russian healthcare. Menedzher Zdravookhranenija 4:58–71 (in Russian)

Sugraliev AB (2021) Heart damage in patients with COVID-19. Kardiologija 61(4):15–23. https://doi.org/10.18087/cardio.2021.4.n1408 (in Russian)

Timonin SA, Klimkin I, Shkolnikov VM, Andreev EM, Mckee M, Leon DA (2021) Excess mortality in Russia and its regions compared to high income countries: an analysis of monthly series of 2020. Population Health 17

Ulumbekova GE (2020) Proposals for the reform of healthcare in the Russian Federation after the peak of the COVID-19 pandemic. Vestnik VSHOUZ 6(2):9–26. https://doi.org/10.24411/2411-8621-2020-12001 (in Russian)

Ulumbekova GE, Ginoyan AB (2022) Lessons of the COVID-19 pandemic for healthcare in Russia. Nauchnye Trudy Vol'nogo ehkonomicheskogo obschestva Rossii 234(2):54–86. https://doi.org/10.38197/2072-2060-2022-234-2-54-86 (in Russian)

Vorobyeva OV, Romanova LP (2022) Clinical case of SARS-CoV-2 infection with the development of acute myocardial infarction, myomalacia and hemopericardium. Profilakticheskaja medicina 25(6):75–79. https://doi.org/10.17116/profmed20222506175 (in Russian)

Yakovleva OB, Safarova TP, Tsvetaeva DA (2023) The impact of the COVID-19 pandemic on the occurrence and course of mental illnesses in the elderly. Zhurnal nevrologii i psikhiatrii 123(5):131–138. https://doi.org/10.17116/jnevro2023123051131 (in Russian)

Yusov AA, Alpidovskaya OV (2023) Left ventricularthrombosis and a cutecardiovascular failure in patients after SARS-CoV-2 infection. Profilakticheskaja medicina 26(11):78–81. https://doi.org/10.17116/profmed20232611178 (in Russian)

Chapter 16
Lessons from COVID-19 Pandemic for Future Preparedness. Examples of GIS and Geocoding Applications in Rome

Cosimo Palagiano, Cristiano Pesaresi, Davide Pavia, Corrado De Vito, Alessandra di Masi, Omar Simonetti, and Enrico Di Rosa

Abstract The COVID-19 pandemic has highlighted the importance of dynamic spatial distribution systems, especially for nonpharmaceutical interventions and real-time monitoring of epidemiology for making predictions. A judicious use of these technologies could enable making decisions in advance (e.g., increasing the number of hospital beds and medical personnel, enhancing equipment for specific areas) and assessing the effectiveness of measures taken (e.g., border closures, vaccination). Additionally, it can be used to monitor the varied distribution of viral genetic variants of concern for epidemiological and therapeutic purposes. In this work, after a synthetic overview of some aspects concerning SARS-CoV-2 and COVID-19 pandemic, we provide an applied example of geocoding and data aggregation in GIS environment which may be used both in the first phase of emergencies, to identify spatial clusters and to hypothesize the evolution trend, and over the time to monitoring the disease patterns. For this example, also in the perspective of precision preparedness, we focus the GIS elaborations on a set of data regarding the period from 25 February to 26 September 2020 showing for the territory of Rome (Italy) some potentialities of geocoding and data aggregation.

C. Palagiano (✉)
Emeritus Professor, Sapienza University of Rome, Rome, Italy
e-mail: cosimo.palagiano@uniroma1.it

C. Pesaresi · D. Pavia · C. De Vito
Sapienza University of Rome, Rome, Italy

A. di Masi
Roma Tre University, Rome, Italy

Centro Linceo Interdisciplinare "Beniamino Segre", Accademia dei Lincei, Rome, Italy

O. Simonetti
Infectious Diseases Unit, University Hospital of Trieste, Trieste, Italy

E. Di Rosa
Local Health Unit Rome 1, Rome, Italy

R. Akhtar (ed.), *COVID-19: Impacts on Health and Health Systems*,
https://doi.org/10.1007/978-3-031-90255-0_16

Keywords COVID-19 · Geocoding · GIS applications · Space–time diffusion simulator · Therapeutic options · Virus-host interaction

16.1 COVID-19 Mechanisms in Humans

The COVID-19 pandemic had a devastating impact on human lives globally, with around 7 million deaths reported by April 13, 2024 according to WHO data (https://www.who.int/europe/emergencies/situations/covid-19). COVID-19 is caused by the severe acute respiratory syndrome coronavirus 2 (SARS-CoV-2), an enveloped, positive-sense, single-stranded RNA virus from the betacoronavirus genus. Other members of this genus, such as SARS-CoV-1 and the Middle East respiratory syndrome coronavirus (MERS-CoV), caused the SARS epidemic in 2002–2003 and the MERS epidemic in 2012, respectively.

Coronaviruses are diverse and can infect various animals, such as pigs, cows, dogs, cats, and chickens. Bats serve as a reservoir for coronaviruses, including SARS-CoV-2. In late 2019, when the first cases of pneumonia emerged in Hubei province, China, genome sequencing and phylogenetic analysis identified the pathogen as a novel coronavirus with genetic sequences from bats and pangolins, providing further evidence of the virus's zoonotic origin (Lau et al. 2020; Acosta et al. 2022).

The initial lack of knowledge about the interaction between the new SARS-CoV-2 coronavirus strain and the human host significantly hindered preventive and therapeutic efforts. However, extensive global research has rapidly provided substantial scientific evidence on various aspects of COVID-19. This emerging knowledge about the disease's pathogenic mechanisms has facilitated the development of vaccines and improved therapeutic strategies, helping to control the pandemic's spread. The emergence of multiple SARS-CoV-2 variants with increased transmissibility, virulence, and immune escape capabilities has complicated the pandemic's trajectory in 2020–2021. Emerging evidence is shedding light on these variants' enhanced virus-host interactions and immune evasion (Kumar et al., 2021; Lamers and Haagmans 2022).

16.1.1 Virus-Host Interaction

Infection occurs when viral particles are inhaled, enter the airways, and bind to receptors on the surface of host cells. SARS-CoV-2 host cell entry factors are widely expressed across various human tissue types. For sure, the entry of SARS-CoV-2 into human cells is facilitated by the angiotensin-converting enzyme 2 (ACE2) receptor, which is expressed on the cell surface of human cells, especially lung alveolar epithelium and enterocytes of the small intestine (Hoffmann et al. 2020). While ACE2 was quickly recognized as the primary receptor for the spike protein, other cell surface molecules have been proposed as alternative receptors or co-receptors, such as KIM1, AXL, L-SIGN, DC-SIGN, and SR-B1 (Wang et al. 2020, 2021; Wei et al. 2020; Amraei et al. 2021; Yang et al. 2021).

ACE2 interacts with the receptor-binding domain (RBD) of the SARS-CoV-2 spike (S) protein. The S protein consists of two subunits, i.e., S1 and S2. When the S1 subunit binds to the ACE2 receptor, serine proteases located in the host cell membrane such as transmembrane serine protease 2 (TMPRSS2), cathepsin B or L (CTS-B or -L), and furin, cleave the S protein to expose the S2 subunit. This causes significant conformational changes in the S protein, leading to the fusion of the viral membrane with the host cell. The viral particle is then engulfed by an endosome (see Kumar et al., 2021, and references therein). Once into the host cells, the virus starts the infective process by seizing the host cell machinery used for protein biosynthesis and energy production (Bouhaddou et al., 2020; Klann et al., 2020). Moreover, the virus induces a potent inflammatory response (see Acosta et al., 2022 and references therein) and the dysregulation of the renin–angiotensin–aldosterone system (RAAS) (Chung et al., 2020).

ACE2 is co-expressed with TMPRSS2/CTS-L in many tissues, which is crucial for SARS-CoV-2 infection (Sungnak et al., 2020; Kumar et al., 2021). Indeed, the extensive distribution of host cell entry factors for SARS-CoV-2 across the human body indicates that the virus has the potential to infect numerous organs and tissues. Indeed, while COVID-19 was initially viewed as primarily affecting the lungs, accumulating evidence has since revealed it to be a systemic illness with a wide range of manifestations (Gupta et al., 2020).

16.1.2 Brief Introduction to SARS-CoV-2 in Clinical Practice

COVID-19 is a syndrome that can be fatal. After an initial phase with mild symptoms, the mechanism that leads to further damage to the lungs and thus to respiratory failure is an immune-mediated acute respiratory distress syndrome (ARDS). ARDS is triggered by the so-called "cytokine storm," which is defined as the acute overproduction and uncontrolled release of pro-inflammatory markers, both locally and systemically (e.g., interleukin-6, interleukin-1, interleukin-17, and tumor necrosis factor-alpha) (Montazersaheb et al., 2022).

Since the emergence of the first wave of the pandemic caused by the wild-type strain of SARS-CoV-2 in 2020, a different spectrum of symptoms has been described. Interestingly, up to 74 different clinical manifestations have been associated with COVID-19. Fever, cough, muscle aches and fatigue are the most common manifestations since the onset of the illness (Luo et al., 2022). The current high immunity of the population, due to previous infections and/or vaccinations, has been associated with a significantly lower overall risk of ARDS and fatal outcomes. However, people with a weakened immune system are still at risk of disease progression and severe cases characterized by respiratory failure. Moreover, during the course of the pandemic, different variants of the original SARS-CoV-2 virus have emerged, which differ slightly in their symptom profile and mainly affect the upper respiratory tract (Meyerowitz et al., 2024).

In the years of the pandemic, clinicians have been confronted with a new syndrome related to SARS-CoV-2 that has been observed in convalescent patients. Long COVID is a debilitating subacute and/or chronic disease that occurs after acute SARS-CoV-2 infection in at least 10% of people. More than 200 multiorgan symptoms have been identified, and it is estimated that more than 65 million people worldwide suffer from long COVID (Davis et al., 2023).

16.1.3 SARS-CoV-2 and the Panorama of Variants

The availability of vaccines and previous infections have limited the severe illnesses and deaths caused by COVID-19, but SARS-CoV-2 is not expected to be eradicated due to the virus' ability to mutate and evade the immune system. In the course of SARS-CoV-2 evolutionary process, several variants such as Alpha, Beta, Gamma, Delta, and Omicron have emerged, leading to widespread outbreaks and pandemics (Sun et al. 2023).

Omicron lineages had fully adapted to the human host and targeted predominantly the upper respiratory tract. Moreover, the infection was characterized by a high viral load, increased transmissibility, and lower pathogenicity, which facilitated the spread of the virus (Sun et al. 2023). Consequently, the emergence of the Omicron viral variant in November 2021, characterized also by a pronounced humoral immune escape, has stressed the importance of global monitoring of SARS-CoV-2 evolution (Pagani et al. 2023). Indeed, the evolution of Omicron lineages and sub lineages has changed the assumption that a more fit variant replaces the previous one. In fact, in 2023, more circulating variants of concern generated the so-called COVID-19 "variant soup", making it difficult for clinicians to predict the best available therapies, especially against mild COVID-19 in patients at risk of severe disease (Callaway et al. 2022; Mykytyn et al. 2023).

In 2024, the variant of interest JN.1 accounts for more than a fifth of all COVID-19 cases in some scenarios, signaling a significant shift in the viral landscape and potentially reversing the trend of the "variant soup" (Altamini et al. 2024). As for the future evolution of SARS-CoV-2, it is very likely that the virus will undergo a multilineage evolution leading to intermittent small epidemic waves, relapses, and recurrences. Seasonal outbreaks of severe disease activity are unlikely (Sun et al. 2023).

16.1.4 Therapeutic Options for COVID-19

Since the advent of the COVID-19 pandemic, various therapeutic approaches have been developed to alleviate the consequences of the infection and reduce SARS-CoV-2 viral shedding. The two approaches that performed better in terms of reducing hospitalizations and reducing the progression to severe disease were direct antiviral therapies and monoclonal antibodies (mAbs) targeting the S protein (Ciccimarra et al. 2023). The antiviral drugs remdesivir (only available as an IV infusion) and nirmatrevir/ritonavir (available only as an oral formulation) are both used for mild COVID-19, although only remdesivir is approved for patients with respiratory failure (Godwin et al., 2024). On the other hand, various mAbs have also been marketed for mild COVID-19 treatment, with increasing evidence of in vitro failure of viral neutralization against new variants and concerns about their in vivo efficacy (Takashita et al. 2022). Quantifying the benefit of antiviral therapy and mAbs in SARS-CoV-2 early infection in vaccinated standard-risk patients after the end of the pandemic period is still a major knowledge gap (Meyerowitz et al. 2024). In addition, antiviral and antiviral/mAbs combination therapies or convalescent plasma have been used in immunocompromised patients and are still being studied to improve their place in therapy (Filippatos et al. 2023; Mikulska et al. 2023; Perico et al. 2023).

16.1.5 Brief Considerations About the Causes of Death and COVID-19 Impact in Italy

Mortality data are not always sufficiently reliable, because often the final cause is not indicative of the disease suffered. This is even more true for deaths that occurred during the COVID-19 pandemic that provoked an impressive impact at national and international scale also causing considerable consequences on the other causes of death. For example, if evaluated in the perspective of future years, COVID-19 pandemic could determine an increase in the deaths due to malignant tumors because the ordinary screening decreased in the years of pandemic, while a reduced number of deaths due to car accidents was recorded during the lockdown. Furthermore, an important if not excessive weight is attributed to the pandemic, when—especially in the first phase of the disease—there was uncertainly in the defining the prognosis.

An interesting element in the analysis of causes of death can be provided by the concept of avoidable mortality which derives from the uncertainly. According to Buzzi (2023, p. 85) "The 2021 mortality picture will be conditioned both by the incisive pandemic wave during the winter 2020–2021 and by the vaccination campaign started at the end of that winter and carried out gradually involving specific categories of workers, fragile subjects, elderly, up to the coverage of the population and continued with subsequent boosters from the second half of 2021, all set referred to social, economic, organizational contexts, etc., that in 2021 (and in following years)

Table 16.1 Deaths by cause and age in Italy (2020). Absolute values in thousands (Buzzi, 2023, p. 75; elaboration by study center Nebo on ISTAT data)

Causes of death	Death < 75 years	Of which avoidable	Death > 74 years	Total deaths
Malignant tumors	70.0	44.0	97.5	167.5
Circulatory system	31.8	25.6	194.4	226.3
Respiratory system	8.3	5.9	48.7	56.9
Endocrine, nutritional and metabolic	7.3	5.2	26.2	33.5
Nervous system and sense organs	6.6	0.6	26.5	33.1
Undefined morbid states	5.1	-	19.7	24.8
Other causes	27.7	17.5	94.7	122.4
Total excl. COVID-19	156.7	98.8	507.7	664.4
COVID-19	18.4	-	60.0	78.4
Total for all causes	175.1	98.8	567.7	742.8

are further changed if compared with 2020 and in any case notably different from the pre-pandemic period".

Table 16.1, extracted by Buzzi (2023, p. 74), shows deaths by cause and age (absolute values in thousands) in Italy in 2020 resuming the deaths in people aged less than 75 years, the avoidable estimated deaths, deaths in people aged more than 74 years, and total deaths. The table makes it possible to compare causes of death and the value of avoidable mortality; moreover, it also shows the deaths due to COVID-19 in people aged less than 75 years (18,400) and in people aged more than 74 years (60,000). The total amount of deaths due to COVID-19 in 2020 is equal to about 78,400 (the value of avoidable mortality is not expressed) while the total for all causes is equal to 742,800.

16.2 Geospatial Analysis Perspectives in Public Health Opened by Tests Made Within COVID-19

The lesson learned from the recent pandemic suggests that monitoring the spread of infectious diseases involves a complex process of collecting and analyzing vast amounts of data from various sources, such as healthcare facilities, laboratories, and field reports. These data, which include information on the number and location of cases and demographic and environmental factors, must be integrated and analyzed using advanced data management and analysis tools. Public health officials can take proactive measures to prevent further spread by detecting early signs of outbreaks and predicting their trajectory. GIS technology plays a pivotal role in this process by enabling the visualization and analysis of spatial data related to disease spread, clusters of infections, potential hotspots, and transmission patterns. This spatial awareness is critical for implementing targeted interventions and allocating resources efficiently.

During the COVID-19 pandemic, GIS technology was extensively used to track the spread of the virus, identify hotspots, and manage public health responses (Pesaresi et al. 2021, 2023). By overlaying infection data with demographic, environmental, and mobility data, health authorities could gain insights into factors driving the spread and implement measures to curb transmission (Atek et al. 2022, 2023). This approach, made possible by the advanced spatial analysis capabilities of GIS technology, not only helped manage the immediate crisis but also provided valuable data for future preparedness. The lessons learned from this pandemic highlight the importance of integrating GIS technology and geospatial analysis into public health strategies for effective disease surveillance and control (Hertelendy et al. 2021).

Moreover, predictive models incorporating geospatial data are invaluable for anticipating the spread of infectious diseases. These models, which are based on complex algorithms and vast amounts of geospatial data, can simulate various scenarios according to different variables, such as population density, mobility patterns, and public health interventions (Atek et al. 2022, 2023). By predicting the potential impact of different interventions on the spread of a disease, these models can help governments and health organizations promote specific actions and measures. For instance, models can predict the potential impact of lockdowns, travel restrictions, or vaccination campaigns on the spread of a disease. Such models were instrumental during the COVID-19 pandemic, helping governments and health organizations to make informed decisions about containment strategies (Fry et al. 2021).

Geospatial analysis empowers public health officials by identifying social and environmental factors contributing to disease spread (Nicodemo et al. 2020). For instance, areas with high population density or limited access to health care are often more vulnerable to infectious diseases. Public health officials can prioritize interventions and allocate resources where they are most needed by identifying these high-risk areas. This targeted approach improves the efficiency of public health responses and helps reduce disparities in healthcare access and outcomes (Logan et al. 2016).

In conclusion, as we continue to face new and emerging infectious diseases, the importance of GIS technologies will only grow, underscoring the urgent need for ongoing investment in geospatial infrastructure and expertise.

16.3 An Applied Example of Geocoding and Data Aggregation in GIS Environment Toward a Future Preparedness

In order to provide an applied contribution to the geospatial analysis of COVID-19 diffusion on a large geographical scale, also useful in the case of future emergencies due to infectious diseases, we have elaborated detailed data provided by the UOC Hygiene and Public Health Service—Local Health Unit Rome 1, for the period from February 25th to September 26th 2020, referred to the territory of Rome.

In particular, for this exemplification, we have first of all extracted two elaborations by the Dynamic Space–Time Diffusion Simulator in a GIS Environment developed to analyze the COVID-19 spread in Rome, produced after an accurate phase of data cleaning and optimization and configured to represent the evolution of the phenomenon in video mode, as described in previous works (Pesaresi et al. 2021, pp. 87–89; Pesaresi et al. 2023, p. 59). We then aggregated these data geolocalized on the territory according to the census sections (CS)[1] and census areas (CEA),[2] providing two meaningful "cartographic photos" on diffusion and critical situations, based on the experience conducted with research aimed at analyzing emergency department overcrowding through data concerning the geocoding of code white attendances at the Policlinico Umberto I in Rome (Pesaresi et al. 2020).

The goal of this kind of GIS elaborations is to support the near real-time monitor of the emergency, providing specific inputs for the decision-making regarding medical supply, the implementation of ad hoc measures in certain zones and the management of some hospitals and their intensive care departments. In a perspectival view, hypothesizing needs arising due to other epidemics or pandemics, having tested similar tools and functionalities makes it possible to seek to anticipate fast and extensive spread and constitutes a geospatial starting point to predict clusters, patterns and trends based on continuative data that must feed the model.

16.3.1 Elaborations, Discussion, and Results

Figure 16.1 shows the cases recorded of COVID-19 according to the geocoding process on the basis of an extract of the Simulator. This map makes it possible to observe the diffusion of COVID-19 cases on 26 March 2020 during the period of the Italian lockdown, declared with the DPCM dated 9 March 2020. Each case is represented with a red dot and on 26 March 2020 in particular 692 cases are geolocalized on a Dark Gray Canvas basemap where the submunicipal areas have been overlaid to provide additional geographical information on spatial diffusion.

---

[1] The census sections are the minimum administrative units of the municipality based on which the census survey is organized (Istat 2016, p. 30).

[2] The census areas are groups of contiguous census sections (Istat 2016, p. 28).

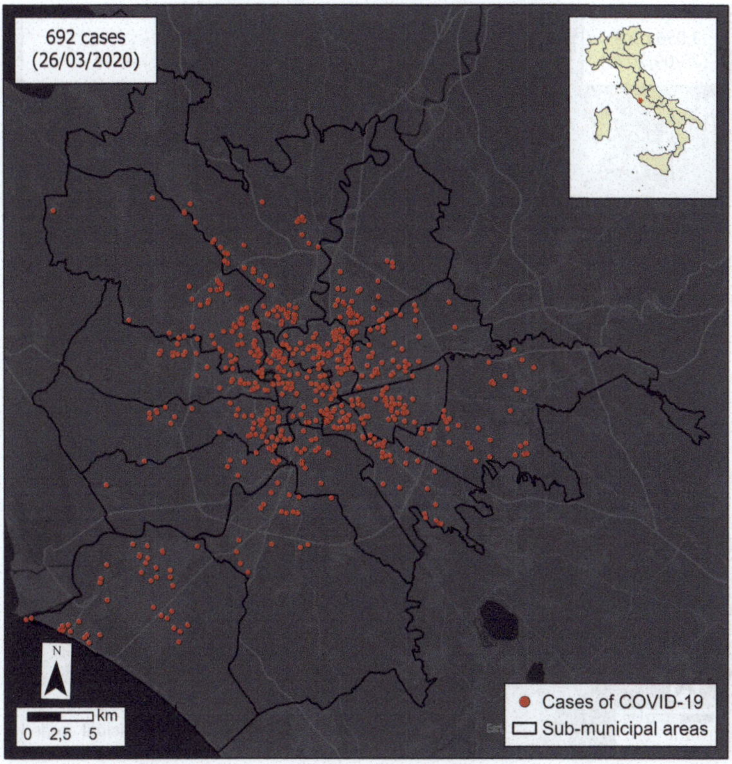

Fig. 16.1 The cases of COVID-19 (according to the geocoding process) extracted by the simulator on 26 march 2020. Authors' elaboration on data by UOC hygiene and public health service—local health unit Rome 1

Figure 16.2—created with the same characteristics as Fig. 16.1—highlights the cases of COVID-19 recorded according to an extract of the Simulator dated 26 September 2020, the last day available according to the data provided by the UOC of the Local Health Unit Rome 1. This map shows a considerably greater problematic nature with 3,056 cases, and passing dynamically from the previous elaboration to this it is possible to note a certain tendential rise in cases in proximity to the areas and axis already affected. Framed in a temporal key, with the support of the Simulator which enables the monitoring of the day by day evolution, it is possible for example to observe and understand that the restrictive measures of the lockdown were efficacious but with a certain phase shift, and how in the months of August and September there was a relevant increase in cases, with a marked accelerated rhythm, being perceived as a psychological release but destined to have a sort of tail effect.

Elaborations like these can play an important role above all in the first phases of an emergency when the cases are still minimal, and it is possible to intervene with the hope of limiting the spread with ad hoc measures and actions. Obviously, to reach a similar goal, it is necessary to promote a process of data sharing which makes it

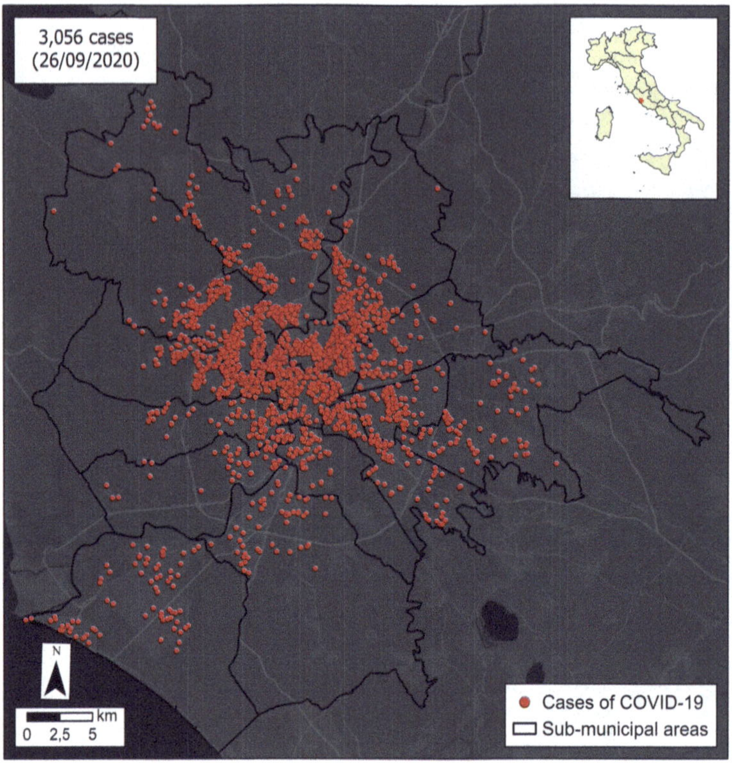

Fig. 16.2 The cases of COVID-19 (according to the geocoding process) extracted by the simulator on 26 september 2020. Authors' elaboration on data by UOC hygiene and Public Health service—local health unit Rome 1

possible to continuously feed the system and model where the data are geolocated on the basis of the domicile or residence addresses.

On the other hand, these maps can be useful also to understand if there are particular elements, structures, healthcare residences, zones with conditions of socio-economic hardship or high concentrations of elderly people where the contagion or lethality tend to be higher. These maps can therefore support detailed analysis of the characteristics and possible criticality at territorial level.

Figure 16.3 is built through aggregation for census sections of the data represented in Fig. 16.2. It is produced by considering three classes defined on the basis of the quantile criterion with a little adjustment and a pink-red chromatic scale. In white are then indicated the numerous census sections where no cases are recorded, while the above-mentioned three classes appear progressively darker with the increase in the case numbers. This first digital choropleth map, based on quantitative data derived by the aggregation of geocoding per census section, provides new information particularly useful during a series of zoom in. In fact, it is built through an overlay on satellite imagery and a territorial screening for macrozones is possible, as is evident thanks to a specific example (top right) where all the defined classes are present and the buildings below are visible. First indications in terms of spatial diffusion begin to appear, with some relevant aggregations of high numbers of cases in census sections, above all in the north and west sectors.

Further and even more communicative data, in terms of geovisualization for higher order aggregations, can be obtained from Fig. 16.4 which shows the data per census areas on a digital street map. Figure 16.4 directly derives from Fig. 16.3 and also in this map the quantile is the criterion used, with a proposal of four classes represented according to a pink-red chromatic scale. Now the presence of a major concentration of cases in the north and west sectors is evident with also a ramification in the north-east, and the general lower cases of the south section are also clear. The enormous census area known as ACE 0 is represented in gray, with its specific value, because it is anomalous in dimensions if compared with all the other CEAs and it could provoke misunderstanding without a specific symbolism.

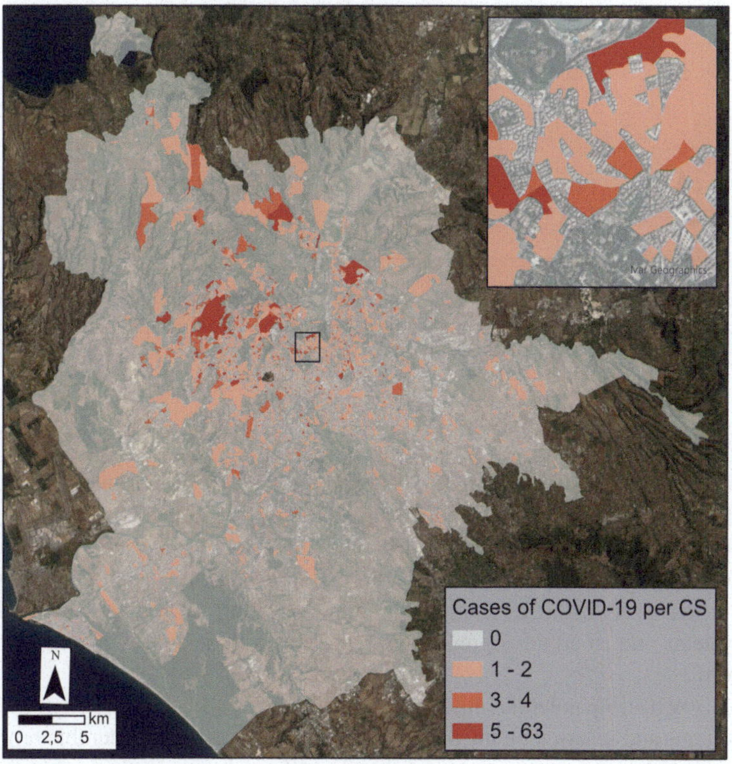

Fig. 16.3 The cases of COVID-19 on 26 september 2020 aggregated in census sections. Authors' elaboration on data by UOC hygiene and public health service—local health unit Rome 1

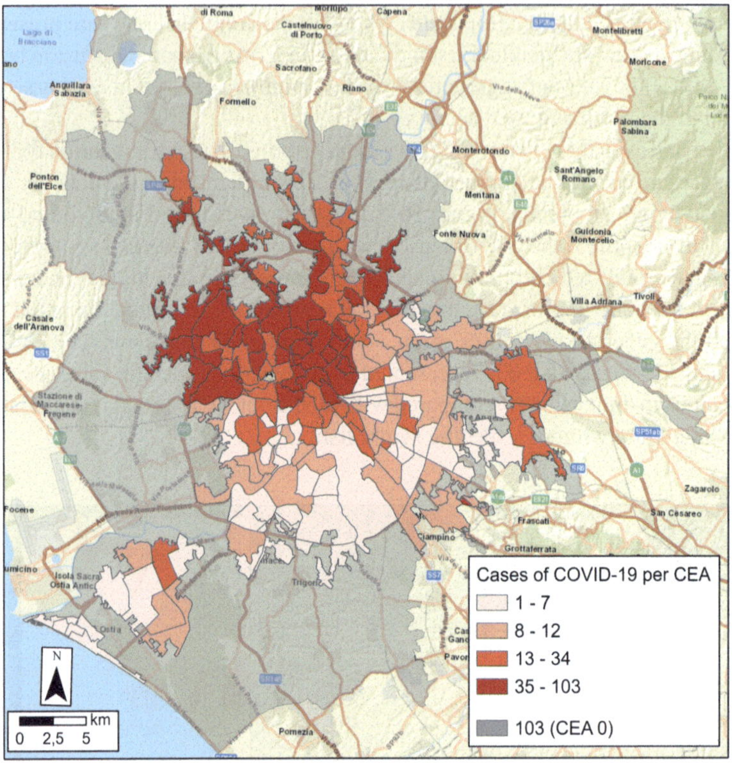

Fig. 16.4 The cases of COVID-19 on 26 September 2020 aggregated in census areas. Authors' elaboration on data by UOC hygiene and public health service—local health unit Rome 1

16.4 Conclusions

Even if the main causes of deaths in developed countries are chronic diseases, the COVID-19 pandemic suddenly and overwhelmingly reminded us of the terrible impact that infectious diseases can cause on a global and local scale, both owing to direct causes and to delay or suspend the normal screening and therapy. The COVID-19 pandemic upset balances and programming, showing the weaknesses of healthcare systems and a certain degree of unpreparedness in dealing with extraordinary situations.

The COVID-19 pandemic imposed relevant changes upon the capacity of healthcare systems and reaction on the part of the institutions, with considerable repercussions on healthcare provision, above all end of life and with particular attention to the elderly (Costeira et al. 2024, p. 1). In fact, they were particularly susceptible due to contemporary frailty and comorbidities, often affected by immune deficiency and severe illnesses (Chehrehgosha 2020, p. 973).

Therefore, since the pandemic underlined the extent of global unpreparedness to deal with the fast-arising necessities of impactful threats to health, several works have summarized some key practices and measures required to better tackle future possible emergencies and health scares (Maccaro et al. 2023, pp. 1, 15–16). Moreover, various international initiatives have been planned to improve the global framework for pandemic preparedness, in order to learn the lesson because the international and national organizations "should not repeat the mistakes of past health security efforts that ultimately contributed to the rapid spread of the COVID-19 pandemic and disproportionately affected vulnerable and marginalised populations" (Lal et al. 2022, p. 1675). In this perspective, the COVID-19 pandemic, happened as a global shock, has also given "an opportunity to be a catalyst for action and provide momentum to […] improve health systems resilience" (Sagan et al. 2022, p. 8).

Among the different measures, principles, and actions, a crucial role must be played by geotechnologies and innovations, rapid data sharing, interdisciplinary collaboration, and active involvement of citizenship, with a virtuous circle among institutions, universities and research centers, hospitals and population.

Mitigating local transmission, coordinating a correct communication, and promoting a conscious sensitization are cornerstones of an efficacious emergency preparedness plan able to make health systems operational to assure critically important medical assistance and care (Kuy et al., 2020).

In order to increase resilience through geospatial analysis, in previous studies, various examples have been given of the use of retrospective data and the improvement of algorithms to support predictive hypothesis for future perspectives in a geotechnological environment where Earth observation, GIS and AI can be harmonized to support decision-making, availing of simulations regarding, i.e., the possible number of emergency department admissions, purchase of medicines, the number of daily contagions, the changes in the region color coding classification for COVID-19 (Atek et al. 2022, 2023; Pesaresi et al. 2022).

In this work, we present some extracts from a Dynamic Space–Time Diffusion Simulator in a GIS Environment developed to analyze the COVID-19 spread in Rome because similar applications can prepare the way to replicable exemplifications in case of future needs. Moreover, some maps deriving from the previous ones are elaborated and discussed to foster the importance of geospatial and evolutionary analysis, looking for the identification of clusters, patterns and trends, and for other investigations able to ascertain the possible presence of specific acceleration and amplification factors on the territory.

Furthermore, as showed in the first phases of the COVID-19 pandemic, webGIS applications, dashboards, and story maps can carry out relevant purposes of education and awareness, spreading data (Kamel Boulos and Geraghty 2020; Dangermond et al. 2020), guidelines, and best practices based on official sources, applied models, and shared documentary evidence, and it is very important to benefit of the past experiences.

Acknowledgments Even if the paper was devised together by the Authors: Cosimo Palagiano coordinated the research and wrote sub-paragraph 16.1.5; Cristiano Pesaresi wrote paragraph 16.4

and is co-author of paragraph 16.3 and sub-paragraph 16.3.1; Davide Pavia is co-author of paragraph 16.3 and sub-paragraph 16.3.1; Alessandra di Masi and Omar Simonetti wrote paragraph 16.1 and sub-paragraphs 16.1.1-16.1.4; Corrado De Vito and Enrico Di Rosa wrote paragraph 16.2.

References

Acosta HRA, Garrigos ZE, Marcelin JR, Vijayvargiya P (2022) COVID-19 pathogenesis and clinical manifestations. Infect Dis Clin North Am 36(2):231–249

Altamimi I, Alabdulkarim IM, Alhumimidi AS, Albabtain MA, Temsah MH (2024) Navigating novel uncertainties of COVID-19: the rise of the JN.1 variant. Cureus 16(1):e51497

Amraei R, Yin W, Napoleon MA, Suder EL, Berrigan J, Zhao Q, Olejnik J, Chandler KB, Xia C, Feldman J, Hauser BM, Caradonna TM, Schmidt AG, Gummuluru S, Mühlberger E, Chitalia V, Costello CE, Rahimi N (2021) CD209L/L-SIGN and CD209/DC-SIGN act as receptors for SARS-CoV-2. ACS Cent Sci 7(7):1156–1165

Atek S, Bianchini F, De Vito C, Cardinale V, Novelli S, Pesaresi C, Eugeni M, Mecella M, Rescio A, Petronzio L, Vincenzi A, Pistillo P, Giusto G, Pasquali G, Alvaro D, Villari P, Mancini M, Gaudenzi P (2023) A predictive decision support system for coronavirus disease 2019 response management and medical logistic planning. Digit Health 9:1–20

Atek S, Pesaresi C, Eugeni M, De Vito C, Cardinale V, Mecella M, Rescio A, Petronzio L, Vincenzi A, Pistillo P, Bianchini F, Giusto G, Pasquali G, Gaudenzi P (2022) A geospatial artificial Intelligence and satellite-based earth observation cognitive system in response to COVID-19. Acta Astronaut 197:323–335

Bouhaddou M, Memon D, Swaney DL, Beltrao P et al (2020) The global phosphorylation landscape of SARS-CoV-2 infection. Cell 182(3):685–712

Buzzi N (2023) Covid-19 e mortalità evitabile: il primo anno della pandemia. In: Spandonaro F, d'Angela D, Polistena B (eds) 19° Rapporto Sanità. Il futuro (incerto) del SSN, fra compatibilità macro-economiche e urgenze di riprogrammazione, pp 74–85. CREA Sanità, Rome

Callaway E (2022) COVID 'variant soup' is making winter surges hard to predict. Nature 611(7935):213–214

Chehrehgosha M (2020) The unpreparedness of the healthcare system for the management of COVID-19 pandemic leading to the mistreatment of the elderly: a newly emerging moral dilemma. J Nutr Health Aging 24(9):973–974

Chung MK, Karnik S, Saef J, Bergmann C, Barnard J, Lederman MM, Tilton J, Cheng F, Harding CV, Young JB, Mehta N, Cameron SJ, McCrae KR, Schmaier AH, Smith JD, Kalra A, Gebreselassie SK, Thomas G, Hawkins ES, Svensson LG (2020) SARS-CoV-2 and ACE2: the biology and clinical data settling the ARB and ACEI controversy. EBioMedicine 58:102907

Ciccimarra F, Luxi N, Bellitto C, L'Abbate L, De Nardo P, Savoldi A, Yeomans A, Molokhia M, Tacconelli E, Trifirò G (2023) Exploring the use of monoclonal antibodies and antiviral therapies for early treatment of COVID-19 outpatients in a real-world setting: a nationwide study from England and Italy. BioDrugs 37(5):675–684

Costeira C, Dixe MA, Querido A, Rocha A, Vitorino J, Santos C, Laranjeira C (2024) Death unpreparedness due to the COVID-19 pandemic: a concept analysis. Healthcare 12(188):1–17

Dangermond J, De Vito C, Pesaresi C (2020) Using GIS in the time of the COVID-19 crisis, casting a glance at the future a joint discussion. J-READING (J Res Didact Geogr) 1(9):195–205

Davis HE, McCorkell L, Vogel JM, Topol EJ (2023) Long COVID: major findings, mechanisms and recommendations. Nat Rev Microbiol 21(3):133–146

Filippatos C, Ntanasis-Stathopoulos I, Sekeri K, Ntanasis-Stathopoulos A, Gavriatopoulou M, Psaltopoulou T, Dounias G, Sergentanis TN, Terpos E (2023) Convalescent plasma therapy for COVID-19: a systematic review and meta-analysis on randomized controlled trials. Viruses 15(3):765

Fry R, Hollinghurst J, Stagg HR, Thompson DA, Fronterre C, Orton C, Lyons RA, Ford DV, Sheikh A, Diggle PJ (2021) Real-time spatial health surveillance: mapping the UK COVID-19 epidemic. Int J Med Inform 149:104400

Godwin PO, Polsonetti B, Caron MF, Oppelt TF (2024) Remdesivir for the Treatment of COVID-19: A Narrative Review. Infect Dis Ther 13(1):1–19

Gupta A, Madhavan MV, Sehgal K, Nair N, Mahajan S, Sehrawat TS, Bikdeli B, Ahluwalia N, Ausiello JC, Wan EY, Freedberg DE, Kirtane AJ, Parikh SA, Maurer MS, Nordvig AS, Accili D, Bathon JM, Mohan S, Bauer KA, Leon MB, Krumholz HM, Uriel N, Mehra MR, Elkind MSV, Stone GW, Schwartz A, Ho DD, Bilezikian JP, Landry DW (2020) Extrapulmonary manifestations of COVID-19. Nat Med 26(7):1017–1032

Hertelendy AJ, Goniewicz K, Khorram-Manesh A (2021) The COVID-19 pandemic: How predictive analysis, artificial intelligence and GIS can be integrated into a clinical command system to improve disaster response and preparedness. Am J Emerg Med 45:671–672

Hoffmann M, Kleine-Weber H, Schroeder S, Krüger N, Herrler T, Erichsen S, Schiergens TS, Herrler G, Wu NH, Nitsche A, Müller MA, Drosten C, Pöhlmann S (2020) SARS-CoV-2 cell entry depends on ACE2 and TMPRSS2 and is blocked by a clinically proven protease inhibitor. Cell 181(2):271–280

Istat (2016). Descrizione dei Dati Geografici e Delle Variabili Censuarie Delle Basi Territoriali per i Censimenti: Anni 1991, 2011. Available online: https://www.istat.it/it/files//2013/11/Descrizione-dati-Pubblicazione-2016.03.09.pdf

Kamel Boulos MN, Geraghty EM (2020) Geographical tracking and mapping of coronavirus disease COVID-19/severe acute respiratory syndrome coronavirus 2 (sars-cov-2) epidemic and associated events around the world: how 21st century GIS technologies are supporting the global fight against outbreaks and epidemics. Int J Health Geogr 19(8)

Klann K, Bojkova D, Tascher G, Ciesek S, Münch C, Cinatl J (2020) Growth factor receptor signaling inhibition prevents SARS-CoV-2 replication. Mol Cell 80(1):164–174

Kumar A, Narayan RK, Prasoon P, Kumari C, Kaur G, Kumar S, Kulandhasamy M, Sesham K, Pareek V, Faiq MA, Pandey SN, Singh HN, Kant K, Shekhawat PS, Raza K, Kumar S (2021) COVID-19 mechanisms in the human body-what we know so far. Front Immunol 1(12):693938

Kuy SR, Gupta R, Correa R, Tsai R, Vohra S (2020) Best practices for a COVID-19 preparedness plan for health systems. NEJM Catal 1:2

Lal A, Abdalla SM, Chattu VK, Erondu NA, Lee TL, Singh S, Abou-Taleb H, Vega Morales J, Phelan A (2022) Pandemic preparedness and response: exploring the role of universal health coverage within the global health security architecture. Lancet Glob Health 10:1675–1683

Lamers MM, Haagmans BL (2022) SARS-CoV-2 pathogenesis. Nat Rev Microbiol 20(5):270–284

Lau SKP, Luk HKH, Wong ACP, Li KSM, Zhu L, He Z, Fung J, Chan TTY, Fung KSC, Woo PCY (2020) Possible Bat origin of severe acute respiratory syndrome coronavirus 2. Emerg Infect Dis 26(7):1542–1547

Logan JJ, Jolly AM, Blanford JI (2016) The sociospatial network: risk and the role of place in the transmission of infectious diseases. PLoS One 11(2):1–14

Luo X, Lv M, Zhang X, Estill J, Yang B, Lei R, Ren M, Liu Y, Wang L, Liu X, Wang Q, Meng M, Chen Y (2022) COVID-19 evidence and recommendations working group. Clinical manifestations of COVID-19: an overview of 102 systematic reviews with evidence mapping. J Evid Based Med 15(3):201–215

Maccaro A, Audia C, Stokes K, Masud H, Sekalala S, Pecchia L, Piaggio D (2023) Pandemic preparedness: a scoping review of best and worst practices from COVID-19. Healthcare 11(2572):1–20

Meyerowitz EA, Scott J, Richterman A, Male V, Cevik M (2024) Clinical course and management of COVID-19 in the era of widespread population immunity. Nat Rev Microbiol 22(2):75–88

Mikulska M, Sepulcri C, Dentone C, Magne F, Balletto E, Baldi F, Labate L, Russo C, Mirabella M, Magnasco L, Di Grazia C, Ghiggi C, Raiola AM, Giacobbe DR, Vena A, Beltramini S, Bruzzone B, Lemoli RM, Angelucci E, Bassetti M (2023) Triple combination therapy With 2 antivirals and

monoclonal antibodies for persistent or relapsed severe acute respiratory syndrome coronavirus 2 infection in immunocompromised patients. Clin Infect Dis 77(2):280–286
Montazersaheb S, Hosseiniyan Khatibi SM, Hejazi MS, Tarhriz V, Farjami A, Ghasemian Sorbeni F, Farahzadi R, Ghasemnejad T (2022) COVID-19 infection: an overview on cytokine storm and related interventions. Virol J 19(1):92
Mykytyn AZ, Rosu ME, Kok A, Rissmann M, van Amerongen G, Geurtsvankessel C, de Vries RD, Munnink BBO, Smith DJ, Koopmans MPG, Lamers MM, Fouchier RAM, Haagmans BL (2023) Antigenic mapping of emerging SARS-CoV-2 omicron variants BM.1.1.1, BQ.1.1, and XBB.1. Lancet Microbe 4(5):e294–e295
Nicodemo C, Barzin S, Cavalli N, Lasserson D, Moscone F, Redding S, Shaikh M (2020) BMJ Open 10(9):1–8
Pagani I, Ghezzi S, Alberti S, Poli G, Vicenzi E (2023) Origin and evolution of SARS-CoV-2. Eur Phys J Plus 138(2):157
Perico N, Cortinovis M, Suter F, Remuzzi G (2023) Home as the new frontier for the treatment of COVID-19: the case for anti-inflammatory agents. Lancet Infect Dis 23(1):e22–e33
Pesaresi C, Atek S, De Vito C, Cardinale V, Bianchini F, Novelli S, Eugeni M, Mecella M, Rescio A, Petronzio L, Vincenzi A, Pistillo P, Giusto G, Pasquali G, Alvaro D, Gaudenzi P, Mancini M, Villari P (2022) The ECO4CO project among retrospective data, improving of algorithms, predictive hypothesis and future perspectives to tackle health emergencies. J-READING (J Res Didact Geogr) 1(11):27–48
Pesaresi C, Migliara G, Pavia D, De Vito C (2020) Emergency department overcrowding: a retrospective spatial analysis and the geocoding of accesses. A pilot study in Rome. ISPRS Int J Geo Inf 9(579):1–26.
Pesaresi C, Pavia D, De Vito C, Barbara A, Cerabona V, Di Rosa E (2021) Dynamic space-time diffusion simulator in a GIS environment to tackle the covid-19 emergency. Testing a geotechnological application in Rome. Geogr Tech 16:82–99
Pesaresi C, Pavia D, De Vito C (2023) Upgrading of a dynamic space-time diffusion simulator in a GIS environment developed to analyse the COVID-19 spread in Rome. A replicable exemplification. Boll Della Assoc Ital di Cartogr 175:54–74
Sagan A, Greer SL, Webb E, McKee M, Muscat NA, Lessof S, de la Mata I, Figueras J (2022) Strengthening health system resilience in the COVID-19 era. Eurohealth 28(1):4–8
Sun Y, Wang M, Wei F, Huang S, Xu J (2023) COVID's future: viral multi-lineage evolution and the dynamics of small epidemic waves without seasonality in COVID-19. J Biosaf Biosecurity 5(3):96–99
Sungnak W, Huang N, Bécavin C, Berg M, Queen R, Litvinukova, M., Talavera-López C, Maatz H, Reichart D, Sampaziotis F, Worlock KB, Yoshida M, Barnes JL HCA Lung Biological Network. (2020) SARS-CoV-2 entry factors are highly expressed in nasal epithelial cells together with innate immune genes. Nat Med 26(5):1–7
Takashita E, Yamayoshi S, Simon V, van Bakel H, Sordillo EM, Pekosz A, Fukushi S, Suzuki T, Maeda K, Halfmann P, Sakai-Tagawa Y, Ito M, Watanabe S, Imai M, Hasegawa H, Kawaoka Y (2022) Efficacy of antibodies and antiviral drugs against omicron BA.2.12.1, BA.4, and BA.5 Subvariants. N Engl J Med 387(5):468–470
Wang K, Chen W, Zhang Z, Deng Y, Lian J-Q, Du P, Wei D, Zhang Y, Sun X-X, Gong L, Yang X, He L, Zhang L, Yang Z, Geng J-J, Chen R, Zhang H, Wang B, Zhu Y-M, Nan G, Jiang J-L, Li L., Wu J, Lin P, Huang W, Xie L, Zheng Z-H, Zhang K, Miao J-L, Cui H-Y, Huang M, Zhang J, Fu L, Yang X-M, Zhao Z, Sun S, Gu H, Wang Z, Wang C-F, Lu Y, Liu Y-Y, Wang Q-Y, Bian H, Zhu P, Chen Z-N (2020) CD147-spike protein is a novel route for SARS-CoV-2 infection to host cells. Signal Transduct Target Ther 5(1):1–10
Wang S, Qiu Z, Hou Y, Deng X, Xu W, Zheng T, Wu P, Xie S, Bian W, Zhang C, Sun Z, Liu K, Shan C, Lin A, Jiang S, Xie Y, Zhou Q, Lu L, Huang J, Li X (2021) AXL is a candidate receptor for SARS-CoV-2 that promotes infection of pulmonary and bronchial epithelial cells. Cell Res 31(2):126–140

Wei C, Wan L, Yan Q, Wang X, Zhang J, Yang X, Zhang Y, Fan C, Li D, Deng Y, Sun J, Gong J, Yang X, Wang Y, Wang X, Li J, Yang H, Li H, Zhang Z, Wang R, Du P, Zong Y, Yin F, Zhang W, Wang N, Peng Y, Lin H, Feng J, Qin C, Chen W, Gao Q, Zhang R, Cao Y, Zhing H (2020) HDL-scavenger receptor B type 1 facilitates SARS-CoV-2 entry. Nat Metab 2(12):1391–1400

Yang C, Zhang Y, Zeng X, Chen H, Chen Y, Yang D, Shen Z, Wang X, Liu X, Xiong M, Chen H, Huang K (2021) Kidney injury molecule-1 is a potential receptor for SARS-CoV-2. J Mol Cell Biol 13(3):185–196

Chapter 17
The Evolution of the Impact of SARS-Covid-19 on the Italian Healthcare System

Giovanni De Santis and Annamaria Bartolini

Abstract The concepts of universality, equality and globality are the basis of the Italian healthcare system established in 1978. However, political intervention in 1992 forced its corporatisation, the divestiture of Territorial Medicine and the closure of small hospitals. As a result, the healthcare system was unprepared for the Covid-19 pandemic, causing symptomatic patients to come to the emergency room, which accelerated the spread of the infection. In preparing the new pandemic plan, authorities in charge did not take into account the critical issues that also emerged in the private sector; they identified the necessary emergency measures, but did not prepare the steps to implement them.

Keywords Italy · Health · Pandemic · Covid-19 · Healthcare · Pandemic plan

17.1 Introduction

It is essential, in dealing with the evolution of healthcare and treatment in Italy—especially in the presence of a reality for which no one was absolutely prepared, such as the global explosion of the pandemic triggered by the Sars-CoV-2 viral strain causing the Covid-19 syndrome—to retrace the main stages of its legislative

---

All authors contributed to the conception and design of the study. Giovanni De Santis wrote Introduction (17.1) and Conclusions (17.5); Annamaria Bartolini wrote §§ 17.2, 17.3, 17.4.

G. De Santis (✉)
Faculty of Humanities—Ancient and Modern Languages, Literatures and Civilisations, University of Perugia, Perugia, Italy
e-mail: giovanni.desantis@unipg.it

A. Bartolini
Faculty of Philosophy, Social Sciences, Humanities and Education, University of Perugia, Perugia, Italy
e-mail: annamaria.bartolini@dottorandi.unipg.it

R. Akhtar (ed.), *COVID-19: Impacts on Health and Health Systems*,
https://doi.org/10.1007/978-3-031-90255-0_17

path,[1] as prodromes of the current reality of a country that had placed one of the basic principles enshrined in the current Constitution in the protection of health. In spite, in fact, of the inevitable negativities clearly highlighted during the pandemic, the Italian one, being founded on equality, universality and globality, still remains one of the most functional health systems in the world.[2]

The constant changes to the health management system, however, have undermined the efficiency of health care structures and the effectiveness of services, making their accessibility extremely complicated, especially in the presence of an unknown pandemic. In such a context, moreover, the consequent difficulties of an extremely conflictual relationship between the State and the Regions on the subject of health, as a concurrent matter,[3] have been deleterious, and have reached their acme precisely on this occasion.

This dichotomy has often led to conflicting and even opposing decisions, with serious repercussions on the population to the extent that, after the closure imposed by the government to stop the contagion, Regions were opposed Regions and advocated total openings to avoid economic collapse, generating strong friction between central and peripheral powers. The historical mismanagement by the political class of public spending, which hardly conforms to regulatory principles, and the continuous cuts that, over the past decades, the health sector has suffered at the hands of various Governments, highlighting, precisely during this emergency, its dramatic criticality, must also be considered.

Now, in the more specific geographical field, 'geography is more concerned with the health of whole human groups, that is, with the morbid forms that, by affecting

[1] Since the unification of Italy, there have been many successive legislative interventions in the health sector, culminating in Law 833/1978, which established the National Health Service (SSN) and the birth of the Local Health Units (USL). Unfortunately, to change radically their management and functioning, Legislative Decree 502/1992 transformed the USL into Local Health Authorities (ASL), with inevitable repercussions on the economic-financial front and, consequently, on the welfare front.

[2] The health emergency has inevitably entailed a drop in life expectancy, which in Italy—albeit with regional variables—has fallen from 83 (2019) to 82.7 (2023) years (OECD 2023), confirming the country's 20th place in the ranking on the quality of European health systems drawn up by the EHCI (2024). Nevertheless, in the annual ranking of the best hospitals in the world that the American magazine *Newsweek* (2024) has drawn up for 2024, no less than 14 of the top 250 nosocomas are Italian. The top three are Policlinico Universitario 'A. Gemelli' in Rome (35th), Grande Ospedale Metropolitano Niguarda in Milan (52nd) and IRCCS Ospedale San Raffaele—Gruppo San Donato in Milan (57th).

With regard to the situation of the various healthcare systems in Europe, but which can be taken as a model for the entire globe, it is useful to consult the thesis by Minardi (2008) and the study by Fasano et al. (2015).

[3] Constitutional Law No. 3 of 2001 redefined the powers of the State and the Regions in the field of health. The 'protection of health' (a much broader concept than the 'hospital care' provided for in the previous system) falls within the scope of matters subject to concurrent legislation between the State and the Regions. According to art. 117 (2) (m) of the Constitution, the State is entrusted with the task of defining the essential levels of care (LEA) and the services relating to civil and social rights throughout the national territory. Regions are assigned the function of government, strengthening their powers of planning and control, also in order to define new instruments of intervention aimed at rationalising health expenditure within the overall financial resources available.

several individuals in the same way, are capable of generating geographically significant phenomena, such as the modification of geographical relations in a given territory' (Palagiano 1996:41). It is therefore essential to study the relationships that arise and are determined in the relations between man as a whole (society) and the environment, whether natural or built (territory). In this context, it is essential to analyse the efficiency of health facilities, taking into account:

(a) Their presence and their more or less widespread distribution throughout the territory;
(b) Their accessibility (i.e. the possibility for the user to obtain the service he needs in a timely manner and at a sustainable cost);
(c) The infrastructure's equipment to make the service offered more or less adequate;
(d) The nature of the service itself.

However, the quality of the provided service is essential as it is inextricably linked to the idea of equity, which is closely linked to the concept of accessibility to care and facilities mentioned earlier. All of this is essential to avoid—as has unfortunately happened in the pandemic- the total collapse of health structures, not only in developing countries but also in industrialised ones, given the number of people affected. From the beginning of the pandemic until August 4th, the total number of confirmed cases reached 775,867,547 and 7,057,145 deaths (Who Data 2024a, b). The situation in Italy was no less catastrophic, with a total of 26,796,561 infections and 196,875 deaths (Civil Protection Department 2024) recorded by the same date (Fig. 17.1).

Similar data underline the need to establish essential services that guarantee human health, with the concepts of equity, efficiency and accessibility as cornerstones. However, this new Covid-19 has highlighted the huge inequalities between developed and developing countries in terms of access to appropriate health services, which are often non-existent (McCann et al. 2020). In fact, the incidence of any disease is closely linked to the socio-economic, political, environmental and therefore health conditions that characterise the areas where populations live. In the developed world, significant progress has been made in controlling diseases, particularly infectious and parasitic diseases, whose mortality rates had fallen to around 1–2% of the total before Covid-19, leading to a sharp increase in life expectancy. On the contrary, the situation is very different in developing countries, where 5.6 million children died in 2017, half of them in sub-Saharan Africa and 30% in South Asia, due to a lack of resources, not only for treatment, but above all for lack of prevention and vaccination. In the event of a pandemic, it is essential to have effective vaccines to ensure the health of all components of the 'Earth system'. Unfortunately, this is not the case, as evidenced by the disparity in vaccination rates around the world,

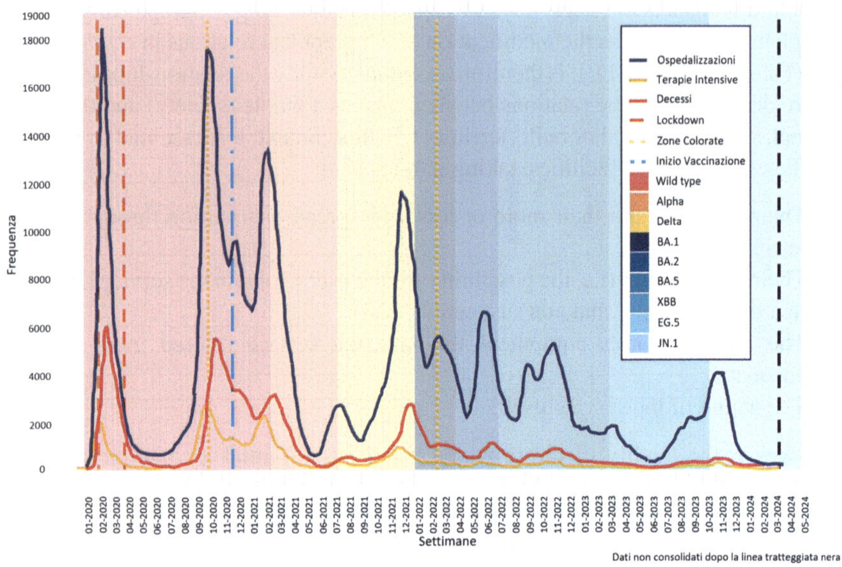

Fig. 17.1 Weekly number of hospitalisations, intensive care admissions and deaths per week of the event since the start of the pandemic in Italy. *Source* ISS, 2024:9

which are needed to prevent the uncontrolled spread of infection[4] and to restore the concept of 'good health'[5] (Figs. 17.2 and 17.3).

17.2 From China to Pandemic Escalation in Italy

The presence of abnormal pneumonias, not caused by known pathogens, has been reported around the world for some time. China recorded the outbreak in November 2019,[6] particularly in the densely populated eastern city of Wuhan, and its origin was linked in each case to the type of work carried out by workers in the local market. It was not until January 9th, 2020 that a new strain of coronavirus 2019-nCoV, later renamed Covid-19 or Sars-CoV-2, was isolated. On January 30th, after recognising

[4] In this context, unfortunately, there has been an ever-increasing push toward privatization of the public health service, devolving care to private entities or insurance underwriting. While the private, and not necessarily, could provide a better service, it should not be forgotten that care will always be better for the wealthier classes because they will be able to afford care coverage for every illness, while the less affluent classes, having less economic possibilities, will not even have the right to care. The importance of the public service lies precisely in the possibility of care, so that it guarantees everyone at least the right to be able to safeguard their health.

[5] See the study by Palagiano (1996), which looks at the dominant features of the 'health system', mainly in relational terms, taking into account the human–environment-disease relationship.

[6] For a reconstruction of the main facts in the history of Covid-19, see Lab24 (2021).

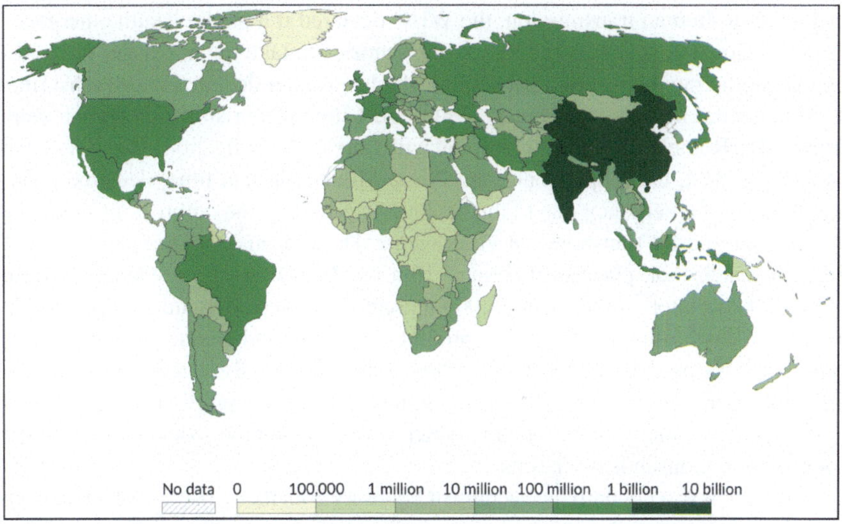

Fig. 17.2 Doses of anti-Covid-19 vaccine administered one year (December 8th 2021) after the first official worldwide administration (December 8th 2020). *Source* Our World in Data (2024)

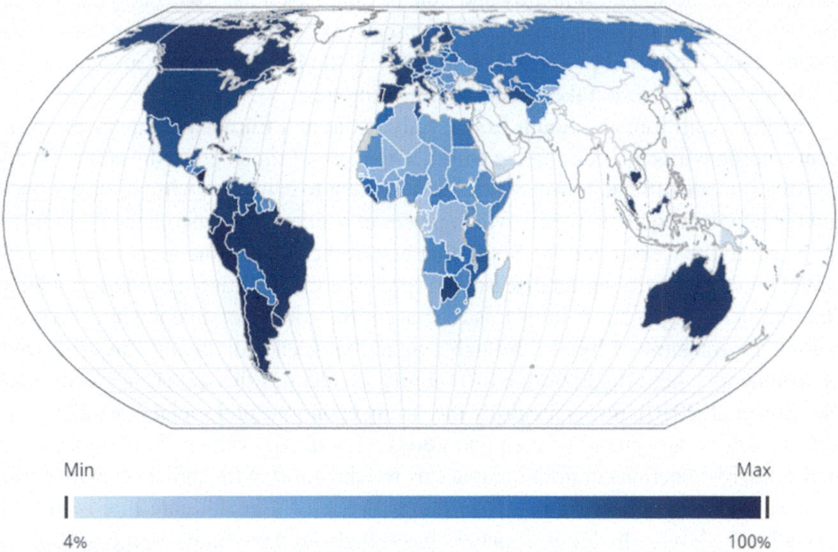

Fig. 17.3 Percentage of total population vaccinated with at least one dose of anti-Covid-19 vaccine by December 31st 2023. Official data are not available for Saudi Arabia, India and China. However, it is estimated that 73, 69 and 89% of the population have been vaccinated respectively (data as of November 26th 2023). *Source* Who Data (2023a)

its human-to-human transmission, the WHO declared it a public health emergency of international concern and called on all countries to take the necessary measures to prevent the rapid onset of a pandemic, which was later declared on March 11th.

Meanwhile, in Italy, on January 29th, two Chinese tourists from Wuhan were admitted to Rome's Spallanzani Hospital with Sars-CoV-2 infection. Two days later, on January 31st, Giuseppe Conte, the Prime Minister at that time, declared a six-month state of emergency due to the health risk posed by the outbreak of unknown diseases caused by transmissible viral agents. Despite this, the Atalanta-Valencia football match took place on February 19th at Milan's San Siro stadium in front of 50,000 spectators: more than 36,000 among them were Bergamo supporters. In the fifteen days following the event, around eight thousand claimed to have shown symptoms of Covid-19, and just under three thousand actually tested positive for the infection (BergamoNews 2020). It is thought that the event contributed significantly to the spread of the virus in the city, which soon became the worst affected in the country with thousands of victims.

However, it was not until the evening of February 20th that a 38-year-old man from Codogno (in the province of Lodi) was admitted to hospital with severe pneumonia. He was the first confirmed case of Covid-19 in Italy that could not be traced back to a return from China. The following day, with the death of the first victim, 11 municipalities between Lombardy and Veneto established the ‘red zone’, but people continued to hope that the virus would not spread. By March 4th, however, there were already 2700 confirmed cases, and on March 9th, the Prime Minister announced the lockdown, which was to take effect the following day.

Together with Iran and South Korea, Italy is the first European country to document a coronavirus infection in the world. The lapse of time before the containment measure is made official, however brief, is sufficient to generate an impressive movement of people who, try to return to their places of residence: a generalized flight from the North to the Center and the South that determines, due to the great rapidity and spatial spread capacity of the disease, the spread of the infection across the country. The timing of propagation, in fact, is crucial for the control of an epidemic—whose viruses are transmitted by air, through contact between individuals—in urbanized environments where the high rate of mobility accelerates the spread of the disease (Jackson et al. 2010): it was precisely the strong economic and social dynamics that, with a massive movement of men and goods, characterizes the peri-urban areas of the Lombard-Venetian conurbation that favored the onset of the infection in the area. In these realities, moreover, one is in the presence of a health care that is indeed very advanced, but which unfortunately has relied—and continues to rely—heavily on privatization, little able on the other hand to intervene in accessibility to services and care.

17.3 How the Pandemic Evolved in Italy

Italy imposed severe and restrictive measures on the country to contain the spread of infection. This marked the beginning of the most dramatic period of the pandemic. The confrontation between central and regional authorities escalated, and aroused opposition from sectors that feared the lockdown would jeopardise the economic structure of the territory. Schools, factories and offices shut down. Where possible, they continued to operate via telematics. Tourism came to a standstill and the general climate of concern affected the financial markets, which closed down by more than 10%. On March 17th, decree-law no. 18, the 'Cure Italy', came into force, with which the State tried to provide economic support to workers by preventing dismissals for 'justified objective reasons', by extending the network of payroll subsidies and by allowing, under certain conditions, the suspension of mortgage repayments.

The discovery that a large proportion of those infected were asymptomatic added to the tension. The government took new, tougher measures:

- Reducing the price of masks to encourage their widespread use;
- Extending testing to those who have not shown symptoms of flu;
- Closing parks;
- Banning sporting activities unless they take place close to home;
- Preventing people from moving to a different local authority area from where they live.

Barely two weeks after the proclamation of the lockdown, on March 27th, 2020, Italy registered 86,498 cases, with 3732 people admitted to intensive care; 43.1% of the cases were concentrated in Lombardy, among the provinces of Bergamo, Milan and Brescia, followed by Emilia Romagna with 13.4%. The figures exceed those of China: in a single day, there were 969 deaths,[7] a sad record for the country, distributed between the provinces of Piacenza and Reggio nell'Emilia. Most of the victims were elderly people, often hospitalised in Assisted Living Facilities (RSA), where the need for physical assistance to patients made it more difficult to comply with the rules of distancing and to adopt measures to contain the pandemic (Fig. 17.4).

From that date, a slow decline began: on April 5th, the number of patients in intensive care fell by 79, with 3994 hospitalisations. A month later, on May 4th, the Italian authorities announced the start of 'phase 2' of the pandemic and certain categories of workers—as restaurateurs—were allowed to return to work and, always respecting the distance and wearing a mask, to visit and meet their relatives. The decree of the President of the Council of Ministers (DPCM) of June 11th, 2020 allows further reopening. It also abolishes the obligation to wear masks in open spaces. Although in July the Council of Ministers extended the state of emergency until the following October 15th, and a new outbreak broke out in Veneto, 'Phase 3' allowed discos to reopen and various sporting activities to resume.

However, with the reopening of schools, the epidemic exploded again, reaching a peak on October 18th (414,241 cases, 11,705 more than the previous day): a week

[7] Cf. Ministero della Salute (2020b), https://tinyurl.com/2s4av5ux (last accessed 07/20/2024).

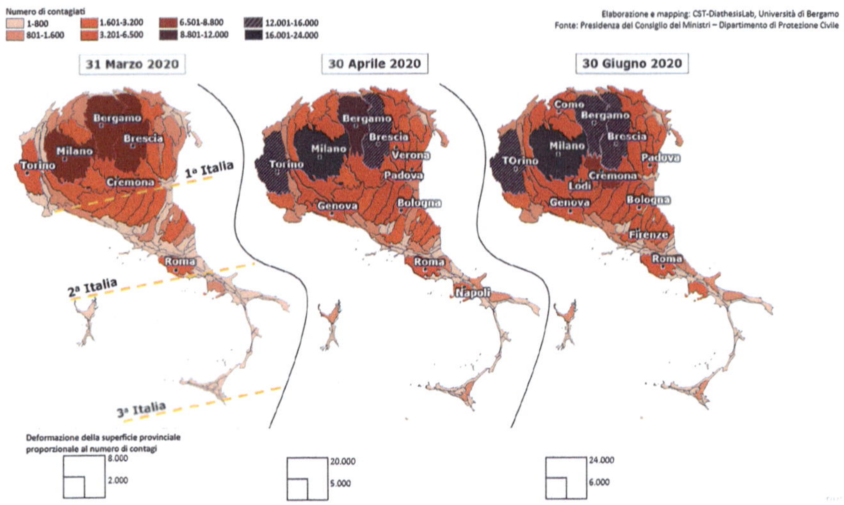

Fig. 17.4 Italy: quantification and evolution of Covid-19 infection (cumulative absolute data). *Source* Casti and Riggio (2022:11)

later, Prime Minister Giuseppe Conte will force the closure of gyms and swimming pools, promote the use of smart work, and distance learning for secondary schools. On November 4th, in response to the growing discontent among Italians, he signed a new prime ministerial decree dividing Italy into three different zones—yellow, orange and red—according to the severity of the outbreak, with corresponding containment measures.

17.3.1 The Vital Contribution of Vaccines

Just as Phase 3 was taking place in Italy, US President Donald Trump announced the arrival of an anti-Covid vaccine by 2020, but Soviet leader Vladimir Putin beat him to the punch by announcing the approval of Sputnik V on August 11th. On November 9th, 2020, pharmaceutical giants Pfizer and BioNTech declared the 90% effectiveness of a first vaccine against the coronavirus, followed on November 16th by US biotech company Moderna, which announced the 95% effectiveness of its serum.

On December 27th, the EMA approved the administration of Pfizer-BioNTech: in Italy, Nurse Claudia Alivernini, who works at Spallanzani Hospital in Rome, was the first person vaccinated. Meanwhile, in Israel, the first confirmations are coming in of the reduction in infections due to the effectiveness of the vaccine. At the same time, however, fears are beginning to spread about the different variants of the virus already circulating in England and in Italy itself, where, in Umbria, the South African and Brazilian variants are the cause of the new increase in infections.

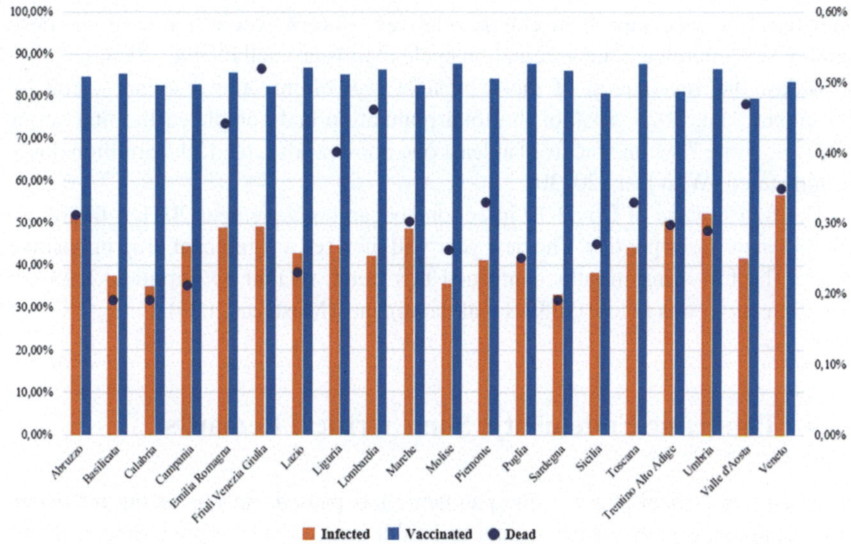

Fig. 17.5 Percentage of infected and vaccinated (left) and deaths (right) in relation to total population by region on September 24th 2023. *Source* Authors' elaboration on data from the Department of Civil Protection (2024) and WHO Data (2023b)

In fact, the third wave of the pandemic will explode in the country on February 26th, 2021. The vaccination campaign therefore continued at a rapid pace, reaching a record of 300,000 doses per day for more than a week on April 16th, a figure that was surpassed on June 10th with 631,817 doses administered in 24 h (Fig. 17.5).

The positive effects of the vaccine meant that the following June 30th Italy recorded the lowest number of infections during the second wave: 'only' 639 cases. The next day, the number of cases began to rise again, due to the Indian strain, later nicknamed 'Delta'.

On July 1st the 'Green Pass' is introducted, the green card that facilitates travel to and from all EU countries and within the Schengen area; it becomes compulsory on August 6th for access to restaurants and public places. In addition, on October 4th, the EMA authorised the third dose of the vaccine for the most vulnerable, while on October 15th the Green Pass was made compulsory for access to workplaces, exacerbating the resentment of the 'no-vaccine' population, who increased the number of swabs carried out (+ 56% compared to the previous day), overloading pharmacies and general practitioners.

Nevertheless, the efficient vaccination campaign in Italy continue, stopping only after a very high number of doses administered, which allowed the infection and spread of the disease to be contained. By September 24th, 145,134,032 doses had been given, with 90.25% of the population over 12 years of age having completed the vaccination cycle and 91.72% having received at least one dose. The data were also positive for children aged between 5 and 11 who, while attending school (distance learning was only recommended for secondary schools), could unintentionally infect their families, especially their elderly relatives: 38.60% received at least one dose and 35.37% completed the vaccination cycle (Ministero della Salute 2024a).

Today, the vaccination of those most at risk of infection continues, and by December 31st, 2023, 83% of the total population had completed the first series of doses, while 73% had received at least one booster dose, for 150.32 million doses administered (Who Data 2023b).

The fourth wave of Covid-19 infections began on November 2021 with around 180,000 new cases per day. The new year will then see a significant drop in positive cases. The following months confirmed this trend, so that by February 2023 the infection curve was below 0.03% on all continents (Mostacci 2024).

17.4 Pandemic Endemicity: Some Critical Remarks

Now that the critical phase of the pandemic has passed, Italy is facing the disastrous consequences of years of cuts in health care. With the borderline cases of the 'Forlanini'[8] hospital in Rome and 'Santa Maria della Misericordia'[9] hospital in Perugia, the number of beds has fallen from 6.79/1000 inh. census in 1991 to 3.5 beds per 1000 inh (Ministero della Salute 2019) . in 2019, (about 190,000 beds in total for ordinary inpatient care, of which 21.4% in accredited private facilities), distributed among 992 nursing institutions.

Consequently, in order to deal with the pandemic emergency, it was necessary to structurally increase hospital allocations (Law Decree 34/2020 and subsequent amendments) by 0.14 PL per 1000 inhabitants for intensive care units (ICUs) and 0.07 PL per 1000 inhabitants for semi-intensive care units. The divestment of public hospitals in favour of private ones has also led to a reduction in the number of medical and health personnel available. The government has therefore had to temporarily recruit 13,610 units, including doctors and nurses, and even to call in foreign medical personnel specialised in emergencies (such as the 38 operators of the Cuban Henry Reeve Brigade who arrived in Turin, or the Russian ones in Bergamo).

Entire wards, including oncology wards, have been converted to Covid wards. Thanks to the contribution of the Ministry of Defence, at least 32 field hospitals have

[8] Inaugurated in 1934 as a sanatorium for tuberculosis patients, it covered an area of 170,000 m^2 and had six pneumology wards, around 800 beds, laboratories, outpatient clinics and, since 2011, a crisis unit for intensive care for post-comatose patients. It was decommissioned in 2015 when some wards were merged with 'San Camillo' and thoracic surgery was transferred to 'Spallanzani'. The most recent Declaration signed on February 8th 2024 between the Italian State and the Holy See identified it as the new home of the private children's hospital 'Bambino Gesù' (https://tinyurl.com/bdhyy4vj).

[9] In 1991, in Umbria, there were 24 hospitals with a total of 5112 beds (6.28/1000 inhabitants); following the downsizing imposed by political decisions, there are now only 16 left, including the 'Santa Maria della Misericordia' hospital in Perugia, which at the beginning of the 1990s had more than 1200 beds, now reduced to 768. The alarming fact is that in order to cope with the pandemic, a field hospital was set up, which the regional administration chose to keep, the only one in Italy, to support the ER-OBS.

been set up throughout the country, with millions of euros invested in the various branches of the armed forces (Army, Navy, Air Force), in addition to the tensile structures erected outside the emergency rooms to facilitate the triage of patients.

The most shocking aspect, however, is how many of them—in Schiavonia (Padua), Marco di Rovereto (Trento), Avezzano (L'Aquila)—have remained unused, either because they arrived incomplete with the materials needed for assembly, or because of the lack of health workers to manage them (Bari). In some cases, once inaugurated, they were closed because they had received only one patient in three weeks out of the 84 beds available (Civitanova Marche, in the province of Macerata), or they were not used at all, except to receive patients from the emergency room evacuated after a fire (Genoa).[10]

Among other critical issues is the inadequacy of the preventive protocols in place in 2020. Designed for the management of a flu epidemic, these did not provide for the management of massive telephone calls to the single European emergency number (NUE) 112. Thus, the State also provided for the drafting of a new National Pandemic Risk Communication Plan drawn up according to the provisions of the then Pan-Flu 2021–2023 in force and the communication strategies suggested by the WHO. Considering also the rapidly evolving pandemic situation and the need to reshape epidemic response measures, a National Prevention Plan (PNP) 2020–2025 was also issued at the same time to support the Regional Prevention Plans (PRP).

Once the emergency was over, the critical issues that emerged at the management level led to an international acknowledgement of the need to revise the approach to pandemic risk. In Italy, this has led to the drafting of a new Pandemic Plan 2024–2028. Although still a draft, this incorporates the WHO's indications regarding the national update of the International Health Regulations (IHR) and the conclusion of a pandemic agreement. Wanting to focus here on the most incisive actions taken at the national level, and not being able to ignore the critical issues related to the impossibility of having all the knowledge about a possible new pathogen available in the immediate future, it is stressed that the new plan:

- Take into account the adoption of the RespiVirNet system suitable for 'monitoring the circulation of respiratory viruses in the community and among the hospitalised population, and can be extended to continuous surveillance throughout the year' (Ministero della Salute 2024b), enhancing the results of the Italian Network set up for the genomic surveillance of the SARS-CoV-2 virus and its variants;
- Take into account the estimates developed on the possible impact of a future pandemic caused by two influenza viruses and a coronavirus, according to three degrees (mild, moderate, severe) of pathogenicity ask the Regions to plan the response in the event of a pandemic on the basis of current allocations, providing for changes according to the progressive activation of new beds;
- Require that each hospital facility have a specific contingency plan to be implemented on the basis of evolving epidemiological needs and to be coordinated

[10] Cf. Pepino (2021), Rutigliano (2021), Del Frate (2020), Pellicani (2020) (last accessed 07/22/2024).

with the PRP and local operational plans to ensure the operational continuity of essential health services.

17.5 Conclusions

The WHO hopes that the experience of different countries in dealing with and managing the pandemic emergency caused by Covid-19 will enable them to understand the necessary measures to be taken when the next pandemic occurs (WHO 2020). However, doubts remain that Italy has not learned its lesson.

The vaccination campaign was undoubtedly proactive and contributed to a drastic reduction in the number of cases, especially where it was possible to administer the doses to the most vulnerable groups, who are at greater risk of contracting the virus and are most likely to spread it (Hunter and Kendall 2020).

However, the impact of the pandemic on the health system revealed the serious legacy of the cuts made to the Italian health system following Decree-Law 502/92, particularly in the public sector, which had to cope with a shortage of accredited intensive care units and which for months exceeded the critical saturation point. Suffice it to mention:

(1) The shortage of beds, at least 100,000 (https://tinyurl.com/2n67fx6r), to such an extent that dozens of field hospitals had to be built ex novo and wards converted into Covid hospitals;
(2) The now structural shortage of health workers needed to meet the needs of an increasingly elderly population, with a deficit of more than 3100 GPs, 30,000 hospital specialists and 70,000 nurses, to which 11,400 retirements will be added by 2026, which they have tried to cover by hiring only 13,610 temporary staff;
(3) The extreme length of waiting lists, to the point where they no longer even meet the standards set by the LEAs. The lists are still unable to catch up with operations postponed during the pandemic because they were not considered urgent or because they might require the use of intensive care units that should have been kept free for Covid-19 admissions.

The only response to all this is now decrees that hardly recognise as a priority the founding principles of the SSN, which was once able to guarantee health care for all thanks to the necessary economic investments. Today, however, repeated cuts in access to care mean that only those who can afford it can be cured, violating not only the right but also human dignity.

Thus, in the absence of wise political foresight and an adequate pandemic plan, it is difficult to see how we will be caught unprepared again in the event of the next pandemic, except for the purchase, which has already been made, of certain health equipment to counter the effects of pneumo-respiratory diseases. The new pandemic plan, however, for the time being only outlines the guidelines and objectives to be put in place to reduce the effects of a possible new pandemic, without, however, identifying the necessary funding to increase facilities, equipment and, above all, healthcare personnel in order to be in a position to fight the pandemic. And today, unfortunately, Italy is not in this situation.

References

AA VV (1994) La Geografia della salute, Sistema Terra, Laterza, Bari-Roma, I: 6-43

Akhtar R (1995) Changing map of malaria in India. In: Palagiano C, De Santis G, Cardillo C (eds) Le strutture sanitarie tra pubblico e privato, "Atti Quinto Seminario Internazionale di Geografia medica, Roma 1–3.12.1994", Ed. Rux., Perugia, pp 445–454

Arena G (1991) La qualità della vita in Italia. In: Palagiano C (ed) Geografia della salute in Italia, Angeli, Milano, pp 282–304

Armocida G (2000) La Geografia medica di ieri e di oggi, relazione presentata al 43° Convegno Nazionale dell'Associazione Italiana degli Insegnanti di Geografia, Varese 28-31.08.2000

Baldacci O (1974) Geografia generale. UTET, Torino, p 940

BergamoNews (2020) A Report i risultati dell'indagine su Atalanta-Valencia e la diffusione del Covid. https://tinyurl.com/3s3xy6zm

Bottura R (nd) Appunti per una storia della sanità italiana. http://www.societasalutediritti.com

Capua I (2022) Il dopo. Il virus che ci ha costretto a cambiare mappa mentale, Mondadori, Milano, p 144

Casti E, Riggio A (2022) Atlante Covid-19. Geografie del contagio in Italia, A.Ge.I., Roma. https://www.ageiweb.it/wp-content/uploads/2022/08/Atlante-COVID-19.pdf

Cremonese A (2008) Il Sistema Sanitario Italiano: una Configurazione Regionale del Modello Pubblico, Libreria Universitaria degli Studi Sociali (LUISS). Roma, Quaderno di Ricerca 152:11–15

Del Frate C (2020) Coronavirus, a Civitanova Marche chiude l'ospedale Covid da 12 milioni. https://tinyurl.com/4m768wv8

De Marco E (2020) Situazioni di emergenza sanitaria e sospensioni di diritti costituzionali. Considerazioni in tema di legittimità al tempo della pandemia da Coronavirus, Consulta online, fasc. II: 369–377

De Santis G (1995) Qualità della vita e malattie croniche in Italia. In: Brandis P, Scanu G (eds) La qualità della vita e la Geografia della salute, La Sardegna nel mondo mediterraneo, "Atti IV Convegno Internazionale Pianificazione territoriale e ambiente", Sassari-Alghero 15-17.04.1993, Pàtron, Bologna, 11: 225–251

De Santis G (1998) Il concetto di salute nel rapporto uomo-ambiente-malattia. Dalla percezione della salute alla geografia medica. In: Palagiano C, De Santis G, Castagnoli D (eds) op. cit., pp 375–384

De Santis G (2002) Dalla cura della malattia alla riconquista della salute. Percorsi di Geografia medica. In: Palagiano C (ed) Linee tematiche di ricerca geografica, Pàtron, Bologna, pp 143–163

De Santis C (2015) Gestione della salute: una prospettiva pedagogica. In: De Santis G (ed) Gestione della Salute, "Atti Undicesimo Seminario Internazionale di Geografia medica, Perugia 18-20.12.2014", Guerra Ed., Perugia, pp 133–146

De Santis G, Patella Scola LV (1995) Le strutture sanitarie in Umbria. In: Palagiano C, De Santis G, Cardillo MC (eds) Le strutture sanitarie tra pubblico e privato, "Atti Quinto Seminario Internazionale di Geografia medica, Roma 1-3.12.1994", Ed. Rux., Perugia, pp 285–306

Dinucci M (2000) Il sistema globale, Zanichelli, Bologna, p 360

Dipartimento della Protezione Civile (2024) Coronavirus. La situazione/dekstop. https://tinyurl.com/3ryya2j6

Donati P, Maspero G (2021) Dopo la pandemia: Rigenerare la società con le relazioni, Città Nuova, Roma, p 160

EHCI (2024) Euro health consumer index by country 2024. https://tinyurl.com/2jt7fwjz

Eyles J, Woods, JK (1983) The social geography of medicine and health. Croom Helm, London, p 272
Fasano A et al (2015) I sistemi sanitari nell'UE 28: una nuova tipologia multidimensionale. In De Santis G (ed) Gestione della salute, "Atti Undicesimo Seminario Internazionale di Geografia medica, Perugia 18-20.12.2014", Guerra Ed., Perugia, pp 147–151
Fondazione Gimbe (2024) Medici di famiglia: ne mancano oltre 3.100. Entro il 2026 oltre 11.400 pensionamenti: al Sud le nuove leve non basteranno a rimpiazzarli. Il 47,7% dei medici supera il limite di 1.500 assistiti: in forte crisi accessibilità e qualità dell'assistenza. https://tinyurl.com/3k3mencx
Gili Borghet AM (1991) Geografia, Ambiente, Salute, Ed. Libreria Cortina, Torino, p 350
Hunter K, Kendall D (2020) Equitable and efficient distribution of a covid-19 vaccine, Third Way. https://www.jstor.org/stable/resrep25616
Iossa Fasano A, Rizzi R (2002) Ospitare e curare, Angeli, Milano, p 240
IIS (2024) Aggiornamento nazionale relativo al periodo 29/04/2024 – 05/05/2024 dei dati della Sorveglianza Integrata COVID-19. https://www.epicentro.iss.it/coronavirus/bollettino/Bollettino-sorveglianza-integrata-COVID-19_8-maggio-2024.pdf
ISTAT (2001) Istituzioni non-profit in Italia. I risultati della prima rilevazione censuaria. Anno 1999, Roma. http://www.istat.it/dati/catalogo/20011212_00/
Jackson C et al (2010) Estimates of the transmissibility of the 1968 (Hong Kong) influenza pandemic: evidence of increased transmissibility between successive waves. Am J Epidemiol 171(4):465–478. https://doi.org/10.1093/aje/kwp394
Jones K, Moon G (1987) Health, disease and society: a critical medical geography. Routledge & Kegan Paul Ldt., New York, p 396
Lab24 (2021) Cose che noi umani. https://lab24.ilsole24ore.com/storia-coronavirus/
Levaggi R (2004) Sanità in Prima Linea. LaVoce.info
McCann G et al (2020) Covid-19 and global inequality. In: McCann G et al (eds) Covid-19 in the global south: impacts and responses. Bristol University Press, 1st ed., pp 161–172. https://doi.org/10.2307/j.ctv18gfz7c.22
Minardi F (2008) Sistemi sanitari: un'analisi comparata, Tesi di laurea, Facoltà di Economia dell'Università di Torino, relatore Dott.ssa Cinzia Di Novi
Ministero della Salute (2019) Servizio Sanitario Nazionale, i dati dell'Annuario 2019. https://www.salute.gov.it/portale/news/p3_2_1_1_1.jsp?lingua=italiano&menu=notizie&p=dalministero&id=5523
Ministero della Salute (2020a) Piano Nazionale della Prevenzione 2020–2025. https://www.salute.gov.it/imgs/C_17_pubblicazioni_2955_allegato.pdf
Ministero della Salute (2020b) Monitoraggio 27 marzo. Aggiornamento 27/03/2020 ore 17.00. https://tinyurl.com/2s4av5ux
Ministero della Salute (2022) Annuario Statistico del Servizio Sanitario Nazionale Assetto organizzativo. attività e fattori produttivi del SSN. Anno 2022. www.salute.gov.it/statistiche
Ministero della Salute (2023) Piano nazionale di comunicazione del rischio pandemico Quadro strategico, strutturale e procedurale 2023–2028 ad interim. https://www.salute.gov.it/imgs/C_17_pagineAree_722_0_file.pdf
Ministero della Salute (2024a) Report Vaccini anti-Covid-19. https://www.governo.it/it/cscovid19/report-vaccini/
Ministero della Salute (2024b) Piano strategico operativo di preparazione e risposta ad una pandemia da patogeni a trasmissione respiratoria a maggiore potenziale pandemico 2024–2028. https://www.quotidianosanita.it/allegati/allegato1705510003.pdf
Mostacci F (2024) Monitoraggio Coronavirus–medie mobili. http://www.francomostacci.it/?page_id=3424
Nava F (2020) Il focolaio: da Bergamo al contagio nazionale, Laterza, Roma-Bari, p 200
Newsweek (2024) World's best hospitals 2024. https://tinyurl.com/mpp2hr96
OECD (2023) Health at a glance 2023: OECD indicators, Paris, OECD Publishing. https://doi.org/10.1787/7a7afb35-en

Our World in Data (2024) Coronavirus (Covid-19) vaccinations. https://ourworldindata.org/covid-vaccinations
Palagiano C (1981) Geografia e Medicina, Il Bagatto, Roma, p 106
Palagiano C (1982) L'ambiente climatico e l'uomo, Il Bagatto, Roma, p 91
Palagiano C (1987) La geografia medica. In: Corna Pellegrini G (ed) Aspetti e problemi della geografia, Marzorati ed., Settimo Milanese, I: 465–507
Palagiano C (1989) Geografia della salute in Italia, Angeli, Milano, p 304
Palagiano C (1990) La salute artificiale. In: Meco G, Carunchio V (eds) Ambiente e salute, "Atti I Congresso Nazionale CIPSMS, Maratea 1988", Roma, pp 21–28
Palagiano C (1991) Salute e malattia nella prospettiva geoambientale, Scheda 2001 – Ecologia antropica, Schena ed., Fasano, III, n. 5–6: 95–115
Palagiano C (1996) Geografia medica, NIS, Roma, p 253
Palagiano C et al (eds) (1998) Metodi di raccolta dati e percezione della salute, "Atti Sesto Seminario Internazionale di Geografia medica, Roma 4-6.10.1997", Ed. Rux, Perugia
Palagiano C, Pesaresi C (2011) La salute nel mondo: Geografia medica e qualità della vita, Carocci, Roma, p 343
Pellicani N (2020) È vuoto l'ospedale da campo di Schiavonia. https://nicolapellicani.it/e-vuoto-lospedale-da-campo-di-schiavonia/
Pepino C (2021) Covid-19: Croce Rossa Italiana smantella l'ospedale da campo al San Martino. https://cri.it/liguria/2021/07/22/covid-19-croce-rossa-italiana-smantella-lospedale-da-campo-al-san-martino/
Phillips DR, Verhasselt Y (eds) (1994) Health and development. Routhedge, London, New York, p 348
Quammen D (2021) Perché non eravamo pronti, Adelphi, Milano, p 100
Redazione IMN (2020) Avezzano: "Risposte da Asl o pronti ad azioni incisive". https://infomedianews.com/avezzano-risposte-da-asl-o-pronti-ad-azioni-incisive/
Regione Umbria – Direzione Salute e Welfare (2020) Libro bianco al 31.12.2019, Perugia, p 92
Rezza G (2020) Epidemie. I perché di una minaccia globale, Carocci, Roma, p 142
Rezza G (2024) Piani pandemici: ieri, oggi e domani. https://careonline.it/2024/06/piani-pandemici-ieri-oggi-e-domani/
Romagnoli L (ed) (1999) Per una storia e una geografia della epidemiologia in Italia, "Atti del Convegno – Roma 26.03.1999", Geografia, XXII: 1–2
Rutigliano V (2021) Puglia: resta chiuso per assenza di personale il nuovo Covid Hospital di Bari. https://tinyurl.com/3d5tjx3j
Signorelli C (2009) Note didattiche di organizzazione e legislazione sanitaria. http://raffrag.files.wordpress.com/2009/10/note-didattiche-di-organizzazione-e-legislazione-sanitaria.pdf
Tedeschi M (2020) Il grande flagello. Covid-19 a Bergamo e Brescia, Scholé, Brescia, p 237
Who Covid-19 Dashboard (2020) Guidance on developing a national deployment and vaccination plan for COVID-19 vaccines: Interim Guidance. https://www.jstor.org/stable/resrep28002.18
Who Data (2023a) Covid-19 vaccination, World data. https://data.who.int/dashboards/covid19/vaccines?n=c
Who Data (2023b) Covid-19 vaccination, Italy data. https://data.who.int/dashboards/covid19/vaccines?m49=380&n=c
Who Data (2024a) Number of Covid-19 cases reported to WHO. https://data.who.int/dashboards/covid19/cases?n=c
Who Data (2024b) Number of COVID-19 deaths reported to WHO. https://data.who.int/dashboards/covid19/deaths?n=c

Chapter 18
After the Pandemic (2020–2022), the Regard About Disease and Health Has Changed in France: Has COVID-19 Sparked New Health Practices?

Isabelle Roussel

> The COVID-19 pandemic has shaken the planet and worried consciences, as so many tragedies have failed to do in recent years, the war in the Balkans, those in Iraq and Syria, without talk about forty years of conflicts in Afghanistan and the tragedies of Central Africa and the Congo. by hitting the planet, would the pandemic succeed in uniting people for the first time? (Anne-Marie Moulin 2024)

Abstract The World Health Organization (WHO) raised the highest level of alert on the COVID-19 pandemic in May 2023[1], which does not mean that COVID-19 has disappeared. It is time to draw conclusions. In the aftermath of the pandemic (2020–2022), the way we look at illness and health has changed in France since this episode which caused the death of 116,000 people. On the one hand, it is a question of learning the lessons of this pandemic which has led people to say "that nothing will ever be the same again" and, on the other hand, the spread of this virus has left many after-effects which are necessary to deal with. The hospital crisis has not disappeared, the psychological consequences combined with solastalgia linked to climate change are numerous, mistrust against vaccines and against the scientific approach to medicine has increased while prevention in the framework of global health is increasingly necessary.

Keywords Long COVID · Eco-anxiety · Prevention · Global health · Zoonosis

I. Roussel (✉)
1151 route des Dombes, 01320 Villette-sur-Ain, France
e-mail: appa.irou@gmail.com

R. Akhtar (ed.), *COVID-19: Impacts on Health and Health Systems*,
https://doi.org/10.1007/978-3-031-90255-0_18

18.1 Introduction

The mortality rates presented for several countries (Fig. 18.1) show how much France has suffered from the pandemic. Norway, Australia, and New Zealand held up much better. The pandemic had revealed the fragilities of French society (Roussel 2022) which the weakening of the virus did not consolidate. The World Health Organization (WHO) declare "COVID-19 over as a global health emergency" in May 2023, which does not mean that COVID-19 has disappeared. The virus, even if it is no longer very active and the symptoms of the COVID is confused with those of the flu, has left numerous after-effects like long-term COVID, in the face of which medicine is helpless. It is essential to learn the lessons of this pandemic because, according to biologists, the risk of the emergence of new zoonoses remains high. The healthcare system, despite its fragility, has guided health policies during the pandemic. It was able to resist thanks to the involvement of caregivers but the structural weaknesses of the hospital system persist despite "*Ségur de la santé*" (IRDES 2023) which attempted to improve the material conditions of hospital staff. French society is faced with numerous crises which accentuate the feeling of vulnerability of individuals since climate change and its series of disasters worry individuals overcome by eco-anxiety. The discoveries of ecological sciences have introduced a systemic approach to health which involves a complex process of interweaving between individuals and the environment. Health goes beyond the provision of care and the curative system to move toward prevention in a global systemic approach.

To deal with the looming crises, new practices are being put in place to build an environment favorable to health: transformation of cities, use of alternative medicine,

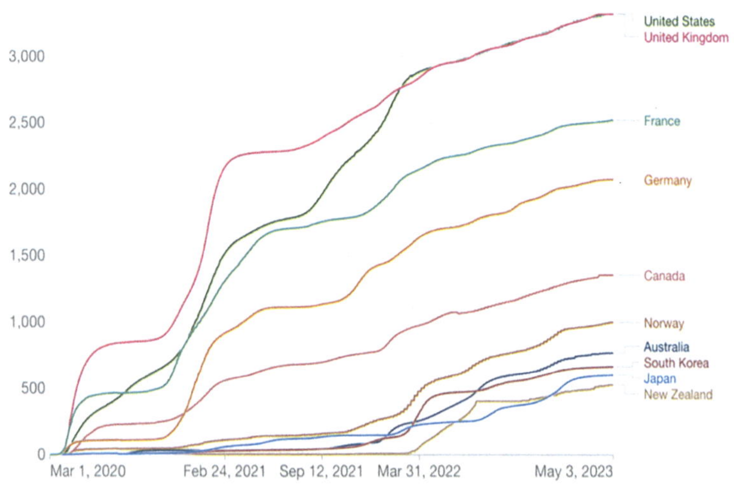

Fig. 18.1 Cumulative confirmed COVID-19 deaths per million people. *Due to varying protocols and challenges in the attribution of the cause of deaths, the number of confirmed deaths may not accurately represent the true number of deaths caused by COVID-19.* WHO COVID-19 dashboard

care taken with food. These new societal orientations are indications of the beginning of profound transformations contributing to the emergence of a new world. Can we then speak of the pandemic as a providential "shock" which, combined with climate change, has made it possible to initiate salutary transformations that must be sustained?

18.2 The Healthcare System in Difficulty

Health in France is less and less associated with the care system which shows numerous failures. The French health system, considered efficient, showed, during the epidemic, its fragility which continues to assert itself even after the disappearance of the pressure linked to the influx of patients affected by COVID-19.

Its fragility was revealed by the pandemic, the management of which consisted, not without difficulty, in saving the hospital. Indeed, the management of health policy during the epidemic and the implementation of various confinements were determined by the hospital's reception capacities and the fear of the failure of resuscitation systems.

However, during the epidemic, caregivers demonstrated quite remarkable capacities for innovation and adaptation. Some consider that, faced with the emergency, the grip of the administrative and financial shackles has loosened to make room for individual initiatives. Garbage bags replaced the insufficient number of gowns while the use of homemade masks made it possible to compensate for the lack of SFP2 masks, "because in this case it was a vast field of improvised acute care which saved the sick, to the detriment of other care suddenly suspended, thus unbalancing the health situation in the country" (Barbara Stiegler[1] 2019).

In the aftermath of the epidemic, the structural weaknesses of the hospital resurfaced and amplified as the aging French population increasingly resorted to care while the number of beds fell in line with the disaffection of staff.

18.2.1 The Shortage of Caregivers

The movement of caregivers away from hospitals predates the pandemic. The rate of increase in the number of hospital employees has declined since 2010. In 2018, the number of hospital employees even declined for the second consecutive year (−0.4%, after −0.3% in 2017). This decline concerns both the public and private sectors. The number of full hospitalization beds (Fig. 18.2) decreased at a fairly regular rate before the health crisis (0.9% per year on average, from 2013 to 2019). The drop was 1.1% in 2021, after 1.2% in 2020.

[1] Stiegler B (2019) AOC Media 24 Janvier 2019.

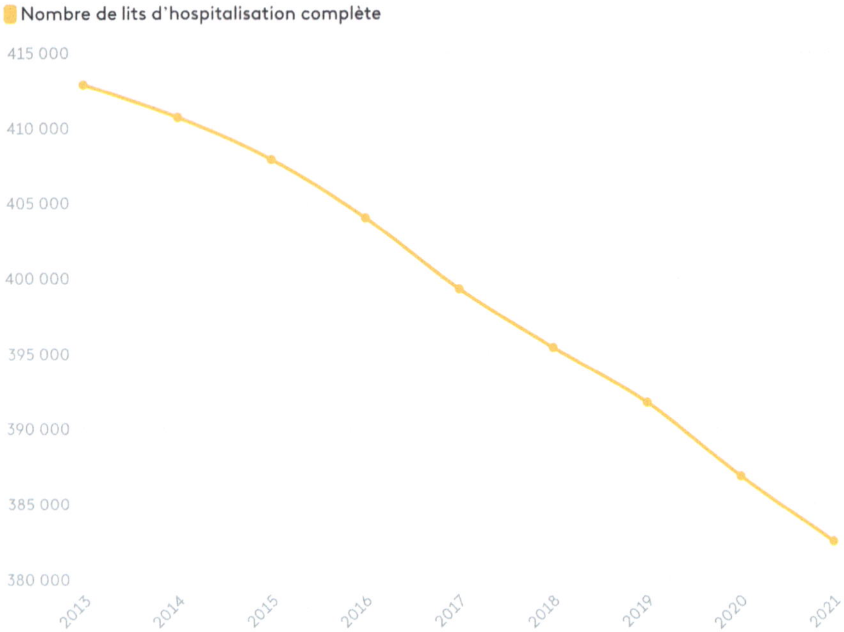

Fig. 18.2 Evolution of the number of full hospitalization beds since the end of 2013 (DREE)

18.2.2 If the Number of Beds is Decreasing, It is Because of the Difficulty in Recruiting Staff

One in ten hospital nurses left their salaried job for an exclusive liberal practice after 10 years of career. Ten years after starting as a hospital nurse, 17% of them are self-employed, as a nurse or in another profession.[2] Even if the shortage of personnel is general in the French economy, these percentages are significantly higher than what we observe in other professions.

This shortage makes working conditions difficult, the deterioration of which can also be explained by other reasons: low remuneration, management guided by reasons of economic profitability, too heavy a hierarchy, etc. Added to these reasons was the trauma dictated by the vaccination obligation imposed on caregivers. Faced with these difficulties, hospital staff report symptoms of depression and anxiety more often than all employed people: in hospitals, 41% of people have symptoms of mild to severe depression, compared to 33% in all employed people. The same is true for anxiety symptoms which are declared by 30% of hospital staff compared to 25% of all employed people.

[2] https://drees.solidarites-sante.gouv.fr/publications-communique-de-presse/etudes-et-resultats/.

Many caregivers are leaving the hospital despite the commitments made by the State during the "*Ségur de la santé*" on July 21, 2020; it aimed to revalorize the careers of caregivers who had been applauded at 8 p.m.

18.3 The Care Offer is also Very Degraded in the Territories

The map (Fig. 18.3) shows how the offer of care is poorly distributed since large cities and the south of France are favored while the provision of care is a real problem in rural areas despite the advantageous conditions offered by rural mayors to practitioners. This shortage is explained by the reduction in the number of general practitioners because only 8000 new practitioners are trained per year, compared to 13,000 in 1970, and retirements have increased sixfold in ten years.

> The COVID-19 pandemic has brought to light some structural weaknesses within the French health system, but it has also provided opportunities for improving its sustainability (IRDES 2023).

18.3.1 Alternatives to Deal with the Shortage of Doctors

The May 2023 law introduces direct access to advanced practice nurses (APNs), physiotherapists and speech therapists. It also broadens the skills of several paramedical professions and reviews the permanence of care.

To compensate for this deterioration in the care offer, teleconsultation, initially marginal, took off with the COVID-19 pandemic. The number of teleconsultations increased from around 80,000 in 2019 to nearly 18 million in 2020.[3] It has since fallen, and remains fairly stable at around 1–1.2 million monthly appointments reimbursed, which represents approximately 4% of all medical consultations. Most appointments are carried out by general practitioners and the patients concerned remain predominantly young and urban. Is this system set to expand or can its technicality contribute to discrediting it?

18.3.2 The French Population, More Vulnerable, is Seeking Psychiatric Care, the Supply of Which is in Decline

According to data from Public Health France, it is students whose studies has been disrupted by the pandemic who are the most affected. 20% of the 2.78 million students

---

[3] https://drees.solidarites-sante.gouv.fr/sites/default/files/2023-06/ER1270.pdf.

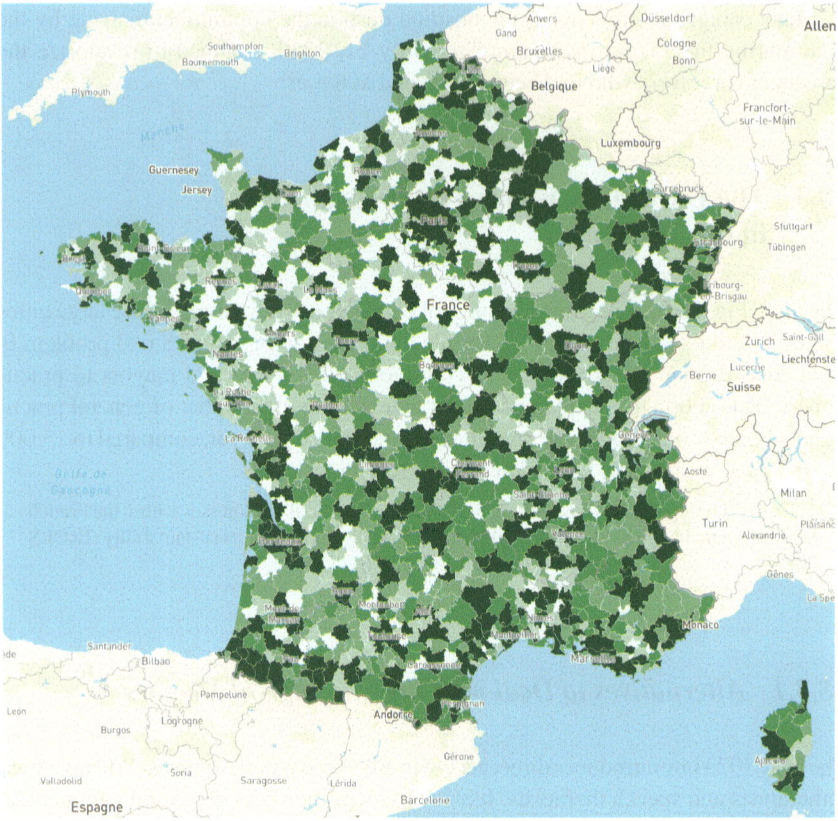

Fig. 18.3 Number of general practitioners per 10,000 inhabitants

enrolled in 2020–2021 in a higher education establishment[4] would be susceptible to depression. However, after the pandemic, the psychological situation of many students remains precarious: competition, uncertainty about the future, isolation. The support systems for student life with regard to the objective of combating student precariousness (MESRI) have therefore been strengthened but remain random and insufficient.

This psychological fragility of the French population is not unrelated to the crises experienced which give rise to eco-anxiety (see below).

Psychiatric structures are particularly insufficient while this sector has often been compressed by the "white plan" to accommodate COVID-19 patients as a priority.

[4] According to the INSEE census.

18.3.3 Growing Distrust of the Hospital Structure

Paradoxically, the healthcare system has gradually become discredited by the population due to its inefficiency. The persistence of long-term COVID-19 which is difficult to treat, the blatant inequalities in the face of the disease, the distrust of science, the importance of the role played by the environment. So many reasons which have contributed to losing confidence in the hospital system while it remains open to all and, compared to other countries, it remains very efficient.

In March 2024, 5 years after the start of the lockdown, the French Hospital Federation publishes its barometer.[5] Although deprogramming has become rarer, the deterioration of care offer remains. Appointment obtaining times have practically doubled in five years. 4 chronically ill patients (or affected by a physical disability) out of 10 share the observation that their medical care has deteriorated since the pandemic.

18.3.4 The Healthcare System and the Technocratic Paradigm

Attention was drawn to the end-of-life situation in EPHADs[6] which were particularly affected by COVID. The increase in life expectancy is a conquest of modernity which everyone welcomes without wanting to look at the conditions offered to people at the end of their lives or to question these institutions in which economic profitability is often more determining than the quality of care. As in the hospital sector, the field of personal assistance suffers from a labor shortage.

The pandemic has raised questions about death and the end of life, thus laying the groundwork for a legislative response on the right to die. The spectacle of COVID-19 patients dying in complete solitude, deprived of visits from their loved ones, has fueled this existential and metaphysical debate which divides society.

Medicine, very dehumanized, concerned with the preservation of biological life, is very often criticized; it is assimilated to the technical coldness of the technocratic paradigm which places little value on ethics and human warmth. But the empathy which constitutes the heart of the nursing profession is hardly compatible with the management criteria imposed by the hospital; criteria that the labor shortage makes all the more strict.

This questioning of the technicality of medicine is associated with a distrust of science.

---

[5] https://www.fhf.fr/sites/default/files/2024-03/202403_FHF_Synth%C3%A8seBarom%C3%A8tre_VDEF_0.pdf.

[6] Accommodation establishment for dependent elderly people "Ehpad, after the scandal, the family dilemma," Le Monde, February 10, 2022, pp. 1, 9–11.

18.3.5 Mistrust of Science

The French were able to discover, thanks to the strong media coverage of caregivers during the pandemic, the diversity of points of view within the medical world: virologists, epidemiologists, and clinicians highlighted the difference in their points of view. The plurality of visions is completely natural but individuals, like Didier Raoult, with radically different opinions, have been able to disrupt public opinion and contribute to discrediting vaccination. This internationally renowned Professor would have found, thanks to a certain pragmatism overriding the determination of scientific evidence, a miracle drug against COVID, hydrochloroquine, the ineffectiveness of which was subsequently certified.

Furthermore, during the pandemic, the rapid media coverage of scientific research results has neglected the importance of the long time required for research. A proliferation of publications, neglecting official validation standards, have been brought to the forefront, exposing the variety of scientific hypotheses that have confused residents, thus contributing to the dissemination of "fake news" and the search for information, solutions outside of traditional medicine.

This distrust of science and medicine explains why many French people are turning to alternative medicine following the movement generated by "antivax" (Jérôme Fourquet[7] 2022). If alternative medicine is successful, it is because it appears to be more respectful of the body and the environment but also because it seems to constitute a substitute or a refuge for certain conspiracy theorists who no longer trust the speeches from health experts or public authorities. Is this a movement dictated by the pandemic or is it a more fundamental trend which joins climate skeptics and divides French society? However, confidence in vaccination seems to have returned because the recent study by Santé Publique France shows that 80% of the French population remains in favor of vaccines despite the distrust aroused by the vaccine against COVID-19. Several studies, 4 years after the pandemic highlight how vaccination has prevented serious forms of the disease and saved 1.4 million people, mainly elderly people in Europe (Grana et al. 2022).

But the new viruses to come risk having different behaviors, hence the importance of research on vaccines, an area in which France has shown its weakness.

To keep health in its rightful place (Laurent 2020) while the healthcare system presents failures, it is necessary to expand the healthcare system to the dimensions of a society that is very destabilized by the prospect of climate change. However, the connections between climate change and health are increasingly better documented and make it possible to link the management of the climate crisis to that of zoonoses and environmental health by relying on another way of conceiving relationships between man and nature at different scales.

[7] Fourquet J, Alternative medicine, one of the components of the anti-sanitary pass movement *L'Express*, 26/07/2022.

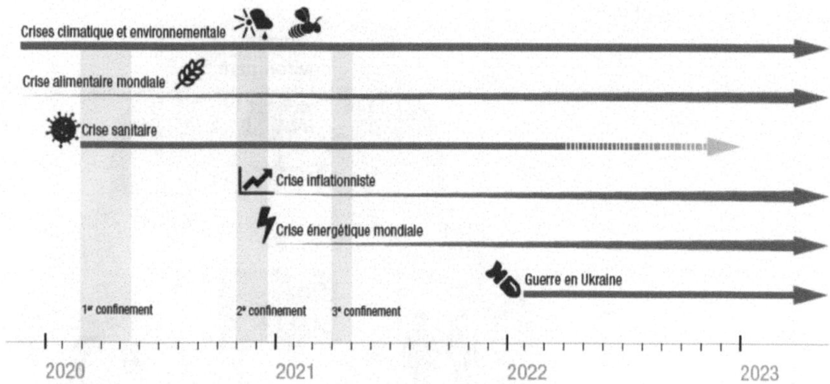

Fig. 18.4 Different crises that the French population has had to face in recent years: a climate and environmental crisis, a global food crisis, the health crisis, the inflationary crisis and the decline in purchasing power and the worrying context of the war on gates of Europe (Paris-region Institute)

18.4 The Expansion of the Health System to the Dimensions of Society

The powerlessness of medicine in the face of a virus has encouraged a different outlook on health to the extent that a healthy lifestyle and a healthy environment were factors that made it possible to avoid the disease or its serious forms. The lockdown and the pandemic have highlighted new environmental factors to consider in defining health and well-being such as the desire for nature, animal welfare, larger housing, the presence of a garden, a more local food supply.

18.4.1 Crises Accumulate and Challenge the Health System

The pandemic is part of a series of different crises affecting French society (Fig. 18.4).[8]

Faced with these crises, the changes in orientation initiated at the time of the pandemic continue, imposing a new outlook on health that is broader and more oriented toward prevention.

These different crises call into question the foundations of consumerist society and question the habitability of the planet. Climate warming and the rapidity with which a new zoonosis can be transmitted form a planetary threat that imposes profound transformations of the way of live to find a better balance with the planet. "*The objective pursued is no longer just to lessen the shock or to protect ourselves from it, but rather to initiate a profound change in consciousness by questioning the origin and*

[8] http://www.comite21.org/docs/ttresilience/220209-ttresilience.pdf.

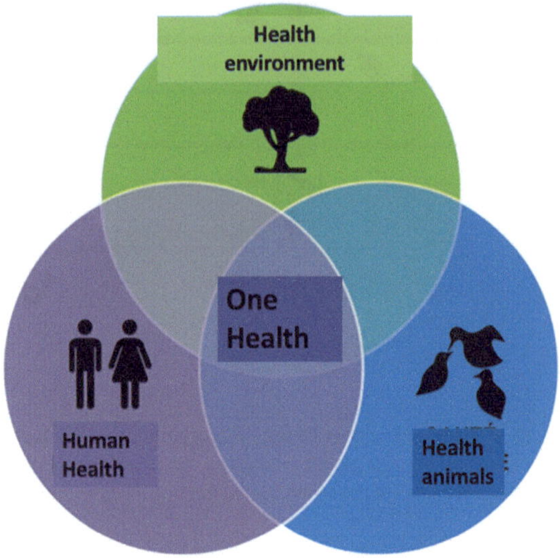

Fig. 18.5 Three poles of the "one health" system

convergence of the crises we face" notes the Committee 21.[9] Faced with these crises, the changes in orientation initiated at the time of the pandemic continue, imposing a new outlook on health that is broader and more oriented toward prevention.

***The "One Health" Concept** Already Mentioned* (Roussel 2022)

This concept integrates the foundations of scientific ecology, namely the importance of relationships and exchanges and the erasure of an anthropocentric vision.

The COVID-19 pandemic leads to reconsider the relationships between men, animals, and other living things. Studying the behavior of reservoir animals and transmission vectors is essential because if viruses remain in the wild part of the world, without contact with humans, they are harmless (Morand 2020). However, galloping urbanization which reduces the share of wild nature while increasing the artificialization of land is a source of dysfunction (Zinsstag 2020) (Fig. 18.5).

In fact, viruses and bacteria are not only enemies, they can constitute objective allies in prevention as shown by recent discoveries on the microbiota (Selosse 2022).

18.4.2 Global Health

Global health is on a planetary scale, as the rapid spread of the virus across the entire earth has shown. Climate change confirms the expansion of the notion of health to the entire planet. For the first time, during the COP (conference of the parties) in Dubai in 2023, a day was dedicated to examining the health consequences of climate change.

[9] http://www.comite21.org/docs/ttresilience/220209-ttresilience.pdf.

Tedros Adhanom Ghebreyesus, Director General of the World Health Organization, underlines the importance of the link between climate policies and the "one health" concept[10]: "*The threats posed by climate change, pollution and loss of biodiversity on health are not hypothetical future risks. These are very real current risks. Health is therefore the first reason which must push us to act on the climate plan.*"

Climate change results in the emergence of new pathologies that only a global approach to health can prevent. With climate change, many animal species are moving, taking with them their parasites and pathogens, changing their geographic distribution[11] (Lenoir 2024[12]), thus requiring close monitoring of the possible emergence of new zoonoses. Thus certain arboviruses (chikungunya, west Nile, Zika) resulting from the migration of pathogenic vectors depending on the modification of their habitat, threaten the French regions. The seasonality of allergenic pollen is also changing.[13]

18.4.3 Policies Favorable to Health Have to Be Available at Different Scales

A global vision is necessary for the climate but it must also be rooted in local life and the territory; it is about trusting civil society and local initiatives without neglecting major international agreements. Tedros Adhanom Ghebreyesus, also emphasizes the urgent need for transformative actions (2024): "*At the national, regional and global levels, we must take transformative action that targets energy and transport systems, food systems and health systems. And we, UN agencies, must do better to move beyond our siled mentalities and work together towards effective, inclusive and sustainable multilateral action.*" According to these words, current health policies go far beyond the medical world, they concern the whole of society which must adapt to the new conditions resulting from the climate crisis (through housing, food, transport, etc.) and therefore fundamentally transform societal paradigms by knowing that human health and overall health are linked.

These necessary transformations can only be achieved through a dialectic between the community and the individual who must, ultimately, be the beneficiary of these policies which, as during the pandemic, must put human health at the heart of decisions.

---

[10] *Le Monde*, 28 mars 2024.

[11] Researchers have observed an average movement towards the poles of 2.7 km per year since 1960.

[12] https://www.college-de-france.fr/sites/default/files/media/document/2024-03/Porcher%202023-2024%20-%20S%C3%A9minaire%2023%20f%C3%A9vrier%20Lenoir.pdf.

[13] Le Monde, 28 mars 2024.

18.4.4 The Dialectic Between the Individual and the Community

The still mysterious mechanisms of contamination and the influence of transmission by a diffuse aerosol, have amplified the question of the relationships between the individual and society and the tendency to withdraw into oneself at a time when responses to crises that across society require collective responses. However, the lockdown linked to COVID-19 appeared in societies where already a certain physical, psychological, political, and moral isolation was increasing, encouraged by the generalization of mobile phones well before.

For some people, due to public health imperatives during the pandemic, the virus has encouraged an individual response to major collective challenges to the detriment of health democracy as described by Tabuteau (2013). According to Moulin (2023), the failure of the health system has forced individuals to manage: "*their personal health by obeying the dictates of medicine or rather the directives relayed by advertising centered on food, sport and the philosophy of healthy living, which would protect against illness... we were therefore on the eve of a new health utopia where the people could now take responsibility for taking care of their health.*" If individual autonomy is essential, it is dangerous that it can be achieved by neglecting the register of public health and avoiding the collective dimension of health. Vaccination is an individual approach intended to protect oneself but also to protect other individuals by allowing access to collective immunity, thus guaranteeing, consequently, the protection of everyone. Individual hygiene, which still remains relevant, is part of the same logic which requires accepting binding collective rules for the greater good of all.

This solidarity works in the same way with regard to the climate; small individual gestures in favor of greater sobriety help to limit carbon emissions; they also make it possible to be more sensitive to regulatory measures and to accept their merits and constraints.

The concern for maintaining individual freedoms is contradictory with the demand for security that the tradition of a strong state in France can offer.

Territories and cities represent essential intermediaries between the standards decreed by the state and the abilities of residents to implement them. It is at this local scale that it is possible to break down the existing walls between the different activities and better control possible perverse effects (injustices and inequalities).

18.4.5 New Territorial or Urban Practices Often in Tension to Improve Public Health

Cities concentrate many negative externalities (pollution, traffic, density, etc.) which they must face; they have united to form networks of international cities which have acquired a certain expertise and can be considered as "laboratories of change."

The urban landscape of many cities has been transformed, green spaces and vegetation have reduced the mineral aspect of the landscape, encouraging the growth of flora and fauna while reducing the high temperatures of summer. Cyclists have invaded urban space for the greater health benefit of city dwellers, even in a country that is not very favorable to cycling like France. (Emilie Schwarz[14] 2024). Low emission zones limit car traffic and polluting emissions.

The fight against animals carrying health risks, rats, bedbugs, mosquitoes, galls, is becoming a major concern for cities which no longer want to use biocides. Certain aspects of the Parisian waste management system have been modified, the trash cans have been replaced by parallelepiped bins inaccessible to rodents (Charles and Roussel 2023). The town-planning uses nature-based techniques. The frequency of heatwaves, the energy crisis and the transmission of viruses have highlighted the importance of housing conditions, small-sized and overcrowding accommodations were factors in mortality during the pandemic. Improving indoor air quality is a major challenge for reducing the risk of contamination in schools, hospitals, and nursing homes in particular.

It is by placing the health of populations and that of the planet at the center of planning logic that the evolution of practices toward prevention and adaptation to global warming takes place. The frequency of climatic disasters in France should draw the attention of doctors to the example of the USA where there is a proliferation of individuals suffering from "post-traumatic stress," called "Katrina brain" due to the trauma caused by hurricane Katrina (Obradovich 2018).

But the deployment of these new planning logics faces numerous obstacles.

18.4.6 Many Obstacles to Overcome

Since the beginning of the twentieth century, France, for cultural reasons, has always experienced difficulties in implementing public health or environmental health policies.

Health in France, with the Pasteurian discoveries and then that of antibiotics, underwent a very technical phase during which the healthcare system took a largely curative turn, thus neglecting the field of prevention which, in 2017, represented only 0 0.02% of the budget allocated to health expenditure. However, the influence of the environment on health has gradually been recognized but with difficulties specific to a strong state culture which, in France, tends to regulate processes, to plan them based on the measurement and quantification, to the detriment of local initiatives. In 2021, the publication of a fourth national health-environment plan (2021–2025) with the subtitle: "*One environment, one health...*" *more broadly affirms the recognition of the relationship between health and environment, drawing lessons of the pandemic: "the COVID-19 health crisis has brought to the surface territorial, social and environmental inequalities in terms of mortality and morbidity in the face of*

[14] https://www.thelancet.com/journals/lanepe/article/PIIS2666-7762(24)00040-1/fulltext.

infectious risk... The environment (housing conditions in particular overcrowding, use of public transport, profession exercised, etc.*) would have contributed to this increase in deaths given the difficulties in respecting physical distancing and the increase in the risk of contamination*".

18.5 Pandemic Has Highlighted Social and Environmental Inequalities

Generational inequalities since the elderly were preferential targets that the solidarity linked to lockdown was able to protect while waiting for the vaccine. But also inequalities linked to professional activity, housing, place of life, gender (Roussel 2023).

The consequences of climate disasters particularly affect the most deprived. Hence, the importance of prevention. In addition, access to healthier living conditions has a financial cost. Thus, a more balanced diet and better insulated housing are not accessible to all budgets, which imposes a policy of redistribution and pooling that is difficult to implement even in a country whose motto is "equality."

"Post COVID spleen" (Dagnaud 2022) is amplified by the concern caused by climate disasters which contribute to fueling a feeling of vulnerability. A WHO note, dated June 2022, emphasized the consequences of climate change on population anxiety: "*this is eco-anxiety, this new form of anxiety which results from paralyzing consciousness. of the destruction of our planet. This concept, initially called solastalgia (a term derived from Latin which means the pain of seeing a once comforting living environment disappear*, (Bohler 2019)". Solastalgia is not a pathology covered by the healthcare system; for older, it can translate into nostalgia for "the good old days" and, for young people, this awareness of their vulnerability can also serve as a reason to act as many young climate activists show. French population, is, according to Lammel (2012), the characteristic of the populations of industrialized countries, often wrongly considered as the least vulnerable due to their efficient healthcare system; however, they find themselves in a situation of cognitive vulnerability because they do not have a holistic approach to the world and are afraid of natural phenomena which they feel are a threat.

18.6 Conclusion

National difficulties combine with a chaotic global geopolitical situation abandoning multilateralism at a time when health and climate issues are global (Lancet 2022). The climate, zoonoses, and trade are globalized and withdrawal, whether at the level of individuals or that of nations, is an obstacle to the good management of crises which affect all of humanity. Yet a new vision of society seems to be emerging at the end of the pandemic, calling into question values that seemed consensual such as consumerism, competition, exploitation of resources and anthropocentrism? The time of confinement has undoubtedly contributed to considering health, life, and death in a different way while a certain distrust of all collective systems, including the health system, has tended to establish itself. The model developed by scientific ecology invites us to rethink the proper functioning of ecosystems which conditions the balance of man on the planet, the prevention of zoonoses and numerous pathologies which have the environment as their origin. The emphasis placed on better knowledge of the environment emphasizes prevention in a timely manner because the consequences mainly affect the most disadvantaged populations. Hence, the harmful influence of disinformation, populism, and anti-science movements because transformations in society require quality information, calm debates, and a high degree of public trust in institutions. This more systemic approach to health requires a change in outlook which must respond to a more integrative vision essential to promoting a new conception of overall health and well-being.

Appendix

- However, there are large geographical differences, with a ninefold variation in standardized suicide rates across local areas, as well as variations according to socioeconomic status (ONS 2022).

https://www.slate.fr/story/266345/generation-z-progressiste-reactionnaire-opinions-jeunesse?utm_source=ownpage&utm_medium=newsletter&utm_campaign=daily_20240403&_ope=eyJndWlkIjoiZmUyOGUyMzc5Y2Y2OTQxYjQxY2JlODUyOTQ3MGYyYzAifQ%3D%3D

References

Bohler S (2019) Le bug humain. Laffont, p 270

Charles L, Roussel I (2023) Le service de santé environnementale parisien: continuité et renouvellement. https://journals.openedition.org/bagf/10335

Dagnaud M (2022) https://www.telos-eu.com/fr/societe/faut-il-sinquieter-du-spleen-post-covid.html

Grana C, Ghosn L et al (2022) Efficacy and safety of COVID-19 vaccines. Cochrane Database Syst Rev 12:CD015477

IRDES (2023) Sustainability and resilience in the French Health System Ministry of Health, 'Ségur de la santé: les conclusions.' https://solidarites-sante.gouv.fr/systeme-desante-et-medico-social/segur-de-la-sante-les-conclusions/

Lammel AM (2012) L'apport de la psychologie cognitive à l'étude de l'adaptation aux changements climatiques: la notion de vulnérabilité cognitive. Vertigo 12. https://id.erudit.org/iderudit/1015103ar

Laurent E (2020) Et si la santé guidait le monde ? Les liens qui libèrent, p 192

Morand S (2020) L'homme, la faune sauvage et la peste. Fayard, p 352
Moulin A-M (2023) La vie avec les virus, PUF, p 286
Obradovich N (2018) Empirical evidence of mental health risks posed by climate change. PNAS 115(43):10953–10958
Roussel I (2022) The pandemic in France reveals the vulnerability of society. In Akhtar R(ed) Coronavirus (COVID-19) outbreaks, vaccination, politics and society. Springer, pp 259–289
Roussel I (2023) La pandémie et le changement climatique revisitent la notion d'habitat favorable à la santé» https://journals.openedition.org/bagf/10189
Selosse MA (2022) Le microbiote à la croisée de l'humain et de la biodiversité. Environnement Risque Sante 21:221–222
Tabuteau D (2013) Démocratie sanitaire. Les nouveaux défis de la politique de la santé. Paris, Odile Jacob
The Lancet Commission on lessons for the future from the COVID-19 pandemic, Lancet (2022). https://www.thelancet.com/journals/lancet/article/PIIS0140-6736(22)01585-9/fulltext
Zinsstag J (2020) One health, une seule santé. Quae, p 575

Chapter 19
Health System Policy and the Pandemic in England, Assessing Its Impact and Shape

Michael Calnan

This chapter comprised revised and updated material some of which was first published in Calnan M and Douglass T (2022) Power, Policy and the Pandemic, Emerald in Chapters 2 and 4.

Abstract The policy response to the pandemic in England was characterised by a mixture of non-pharmaceuticalised and pharmaceuticalised policies with the former generally being a precursor to the latter. However, the adoption of and success of these policies hinged on the strength of the health system (health and social care and public health). Thus, the role played in this policy response by the health system, the extent to which it was prepared for a pandemic, its effectiveness at managing the pandemic and why and what the consequences were for the health of its workforce, the public's health and for patient care will be the focus of this policy analysis. The analysis will draw on the recent evidence given at the current public inquiry in England.

Keyword Pandemic · Health and social care · Public health · England

19.1 Introduction

> "We coped, but only just," said Prof Summers and Dr Suntharalingam in their evidence to the inquiry.... "We would have failed if the pandemic had doubled for even one more week, or if a higher proportion of the NHS workforce had fallen sick. It is crucial to understand how very close we came to a catastrophic failure of the healthcare system" (BBC News Indepth 2024).

M. Calnan (✉)
University of Kent, Canterbury, UK
e-mail: m.w.calnan@kent.ac.uk

R. Akhtar (ed.), *COVID-19: Impacts on Health and Health Systems*,
https://doi.org/10.1007/978-3-031-90255-0_19

The policy response to the COVID-19 pandemic in England was characterised by a combination of non-pharmaceuticalised (or public health), such as lockdowns, and pharmaceuticalised policies, such as the national vaccination programme, with the former tending to be a precursor to the latter. However, it was the former that came in for more critical scrutiny than the latter as England consistently displayed one of the highest excess death rates (Griffin 2020a) when measured against other high-income countries and this was particularly evident after the first wave (Calnan and Douglass 2022). More than 220,000 people have died in the UK with COVID-19 on the death certificate (Dyer 2023). Further analysis (Greener 2021) has shown that this poor performance in relation to the early policy response to the pandemic was more generally associated with countries with liberal political regimes reflecting their reluctance to intervene. However, as the first module of the statutory public inquiry (Hallet 2024) reports in summarising the state of the populations health, the impact of the pandemic might be understood, at least in part, by the policies adopted over the decade which preceded the pandemic:

> In 2020, the UK lacked resilience. Going into the pandemic, there had been a slowdown in health improvement, and health inequalities had widened. High pre-existing levels of heart disease, diabetes, respiratory illness and obesity, and general levels of ill-health and health inequalities, meant that the UK was more vulnerable. Public services, particularly health and social care, were running close to, if not beyond, capacity in normal times (Hallett 2024).

The focus of this chapter, however, is specifically on the health system to describe its experience of and impact on the pandemic and explain why this took the shape it did. However, before this is examined, as the quotation from the public inquiry suggests, it is necessary to understand the position of the health system in the years prior to the pandemic to explain why there was argued to be a lack of resilience at the level of the public's health and in the provision of health and social care. An analysis of the vaccination programme (sometimes described as vaccine 'alone' policy, Paton 2022) rolled out through the NHS and what shaped it has been analysed in detail elsewhere (Calnan and Douglass 2022; Douglass and Calnan 2022) so this will not be a major focus here although its salience in the health systems response to the pandemic was important with some commentators referring to it as a 'game changer' (Paton 2022). The analysis will begin with a consideration of the position of the health and social care system at this time.

19.2 The Health Care, Social Care and Public Health System in England—Pre-Pandemic

Some commentators, as the previous quote from the UK Public Inquiry suggests, argued that the NHS, or at least secondary care, was 'drowning' during the second decade of this century. This was believed to be due to the lack of government investment into this predominantly public tax financed health system. For example, between 2009 and 2013 growth in total healthcare expenditure increased on average by only

1.3% a year (Charlesworth et al. 2016). This was reported to be the lowest rate of growth in NHS spending since 1955. Certainly, this financial crisis had led to a deterioration in the performance of the NHS with key targets, such as waiting times being regularly missed (Kings Fund 2019). However, since then and prior to the pandemic the various conservative governments significantly increased funding although it still remained below the average increases of 3.7% a year since the NHS was founded and is less than the 4% annual increases that policy analysts from the Kings' Fund and elsewhere have argued are necessary to meet rising demand and maintain standards of care (King Fund 2019, Calnan 2020).

The performance of the NHS depends, at least in part, on having a supportive and integrated care system. Social care in England is funded in a different way to the tax-funded NHS in that it is from a mixture of local government and out of pocket payments and the former are not generous (Calnan 2020). Currently, public funding in England is provided only to those in greatest need and the lowest means. For those who do not receive fully funded care, private contributions to social care costs can take different forms although private insurance coverage is limited. The proportion of social care which is publicly funded has been decreasing and the number of people receiving publicly funded social care services fell by 400,000 between 2009/10 and 2015/16 (Bottery et al. 2018). It is more common for people to pay 'out of pocket' to social care providers directly for the full costs of their care. Social care is supplied by a mixture of public, private and third sector providers and reflects a mixed economy approach. Successive governments (Glendenning 2009) encouraged this approach as it was based on a belief in the benefits of marketisation and competition for enhancing quality and efficiency. Successive governments have also, at least officially, emphasised the importance of integration of the health and social care sectors and the mantra of choice and competition has been replaced partially on the grounds it was increasing fragmentation. This is manifest in the NHS with the recent reforms involving the introduction of Integrated Care Systems (Paton 2022). However, the significant financial cuts to local authority funding in the 'austerity' years put severe pressure not only on local government but also on the NHS to provide care and support when social care options are limited. For example, between 2009/10 and 2015/16, spending by councils on social care per adult resident fell by 11% in real terms (Bottery et al. 2018).

What is the position of public health prior to the pandemic which was crucial for addressing this public health problem? Public health can encompass a broad range of areas but could be divided into the public health service which tends to focus on population-based issues, such as vaccination programmes and broader social and public policy which can address social and economic sources of inequalities in health. The public health service at this time was marginalised and in decline in that its responsibilities focussed more on local issues rather than nationwide problems. Coupled with this not only was the nation's health generally poor but there were major social inequalities in health and longevity which were widening. This was hardly surprising given the government's commitment to the neo liberal values of a small state and the promotion of marketisation, competition and individualism which was at odds with the more collective approach of public health (Scally 2024). The public health system's demise dates to the reforms of the coalition government in the 2012 Health and Social Care Act (Calnan 2020; Calnan and Douglass 2022) when it was moved away from the protective haven of the NHS into local government which made it particularly vulnerable to the public expenditure cuts which followed during the period of austerity from 2010. It was also under central control from Public Health England (Paton 2022). The relationship between the public health advisors and the government has also proved problematic in that PHE was not an independent body (Mahase 2020) but under the DHSC (Vize 2020) raising problems about who made decisions and on what basis.

In summary then, in the decade prior to the pandemic, the health system characterised in terms of the health and social care services and the public health was fragmented, lacked investment and was performing increasingly poorly. International comparisons with other high-income countries suggest that the performance of the health system in the UK although strong in the earlier part of the century was, according to a number of mainly process criteria, performing less well over the last decade falling down the performance league tables (Ferris et al. 2024).

19.3 The Health System and the Pandemic

There were conflicting narratives about the position of the NHS in relation to its ability to manage an epidemic of infectious disease (Calnan and Douglass 2022). There was evidence which suggests that the NHS appears to be as prepared as any other healthcare system to meet the challenges and the demands of the COVID-19 pandemic. For example, in 2019 on a GHS healthcare security index measuring preparedness the United Kingdom came second out of 195 countries with an index score of 77.9 out of 100. However, the report concluded that national health security is fundamentally weak around the world, that no country is fully prepared for epidemics or pandemics, and that every country has important gaps to address (GHS 2019). Other evidence, as was previously discussed, suggests that the NHS had not been performing well and has been performing at the limits of its capacity for some time, likely reflecting a period of austerity and the chronic lack of investment over the last decade (Calnan 2020). Evidence from the public inquiry showed that the level of the workforce and hospital equipment was very limited in its capacity to cope with the coming surge in demand (Iacobucci 2023a).

This latter negative narrative was one shared by many (Mahase 2021) and the introduction of the lock downs appeared to be due, at least in part, to a concern that the NHS would be swamped by the demands of the pandemic as the following evidence from Professor Whitty, then the Chief Medical Officer for England, (BBC News Online 2024a) to the public inquiry exemplifies:

> NHS hospitals in England entered the pandemic in early 2020 with a "very low" level of beds in intensive care compared to similar high-income countries. "That's a political choice. It's a system configuration choice, but it is a choice," he told the inquiry." Therefore, you have less in reserve when a major emergency happens, even if it's short of something of the scale

of covid…Sir Chris suggested that countries like the UK had no alternative but to impose lockdown and other social restrictions to avoid a "catastrophic" amount of pressure on the healthcare system. He accepted that "in many individual cases" doctors and nurses found the situation "incredibly difficult" but said without lockdown restrictions "the expectation is it would have got worse. Not a trivial amount worse, but quite substantially worse".

19.4 The Impact of the Pandemic on Healthcare Workers and Patients

Previous government policy in relation to the cutting of the NHS workforce was also imputed as a determinant of the negative impact of the working demands of the pandemic on nursing staff:

> The public inquiry heard evidence from lead nurses from all four UK nations on 17 September, in which the chief nursing officer for England, Ruth May, highlighted the former Conservative government's decision to cut the nursing bursary in 2017 as a major reason why the NHS was so short of staff when the pandemic hit… "it meant that we were 5000 fewer nurses in March 2020 because of that decision, and 700 fewer midwives in March 2020," said May, adding that this equated to around 40 fewer nurses at each hospital. "If we had more nurses there would be less burnout [and] less psychological impact. Removing the bursary, for me, was a catastrophic decision" (Mahase 2024).

The public inquiry received the stories of more than 30,000 healthcare staff, patients and relatives which are to form part of the material entered evidence. The extent to which the experiences of patients and their families will be fully reflected in witness statements has been questioned by groups representing these groups. However, the vulnerability of staff is exemplified by the following quote (BBC News Online 2024b):

> It was horrendous. We were really struggling, having to scrounge around for masks and gloves, "says Mandi Masters, a community midwife from Aylesbury in Buckinghamshire. At that early stage the NHS was, she says, "working in the dark" as the virus spread from China to Italy and then to the UK. Later Mandi caught Covid herself—she is convinced at work—and ended up in hospital on oxygen for three weeks. "My husband took me to A&E but had to leave me there, turn around and walk away," she says. "The news was coming out on how many health professionals were dying of Covid, but I was just too poorly to care at that point," she says. "Looking back, I have to admit, it was extremely frightening. Mandi, 62, has now returned to work part-time, but still struggles to catch her breath after a short walk. Every cold or chest infection "wipes her out" and she "grieves for the person I was before Covid".

Further evidence given to the inquiry by Professor Fong, a consultant anaesthetist, suggested some hospitals were at breaking point (BBC News Online 2024a, b, c).

> It's painful now because it was very clear what was happening to the patients, it was very clear what was happening to the staff. The staff were very injured by just how overwhelmed they were by the whole thing. "In December 2020 as Covid rates were rising again across the UK, he said he was asked to visit an unnamed hospital with a medium-sized intensive care unit..."I'll never forget it," he said. "It was a scene from hell."

"This was a hospital in massive, massive trouble.... there were so few staff that some of the nurses had chosen to either use the patient commodes [or] wear adult diapers because there was literally no one to give them a toilet break," he added. "This was a hospital breaking at the seam's"

The inquiry also received evidence about distressing patient experiences. For example, patients in antenatal and maternity care in hospitals during the early stages of the pandemic, sometimes experienced inadequate communication and difficulties with how hospitals applied COVID restrictions. Mothers were left on their own for lengthy periods and parents felt excluded (Limb and Wise 2024). There were also problems in Intensive Care Units with the use of do not attempt cardiopulmonary resuscitation (DNACPR) orders. Clinicians reported that one of the most difficult parts of ICU care during the pandemic was that relatives were not aware, as they had not allowed to visit, how much patients had deteriorated (Limb and Wise 2024). Some of these orders (DNACPR) had to be communicated online which made the discussions remote and disconnected although some did not seem to have taken place at all:

Heather Hallett, chair of the inquiry, said it had heard from a large number of bereaved families that they were not consulted about a DNACPR notice on their loved ones' records (Limb and Wise 2024).

19.5 Public Health at the Margins

It has been argued that the focus on the ability of the health and social care system to cope with the pandemic reflected the limitations of the public health system and its inability to deal with what was very much seen as a public health problem:

Too little, too late, too flawed...How did a country with an international reputation for public health get it so wrong? Their answer is a sad litany of past and present decisions that have fragmented, decimated, and marginalised public health in the run up to this moment when it is most needed. Their overriding message is that clear leadership from the centre needs to be matched with strong operational capacity at the local level. The UK currently has neither. (Godlee 2020, p. 1)

Survey evidence has suggested that doctors working in public health and health protection roles reported dangerously high levels of mental and physical fatigue as a result of managing COVID-19, coupled with resentment with the sense their expertise had been overlooked by national decision-makers. This evidence showed criticism and scepticism about the government's plans to reorganise PHE and the opportunities for doctors to help shape any changes (Tonkin 2021). Certainly, evidence from interviews with Directors of Public Health showed that they played both a key role in shaping the local response to COVID-19 and in engaging local communities, but they did not feel that they were properly engaged by central government regarding major elements of the overall response to COVID-19, most notably the national testing strategy and the rollout of NHS Test and Trace (Ross et al. 2021). The lack of community testing in the early stages of pandemic led to limited data being available despite the presence of substantial expertise and there was a lack of sharing of data between national and local agencies (The House of Commons Health and Social Care and Science and Technology Committees 2021). Public Health England was a casualty of the pandemic, with the government laying blame for the response to the first wave at its door, and was replaced (at least in part with the new Office of Health Promotion moving into the DHSC), by the UK Health Security Agency (UKHSA) although serious doubts were expressed about its focus:

> Since 2010, successive Westminster governments have dismantled and defunded public health rather than recognising the importance of a healthy population and a robust and effective public health function. Important alterations include the transfer of local directors of public health from the NHS to a much reduced and poorly resourced role within a simultaneously weakened local authority system. The central public health agency, Public Health England, has been abolished and replaced by the UK Health Security Agency. The agency, which is now part of the UK's national infrastructure and security system, has dropped public health from its title. Notably, three of the four chief medical officers in the UK are not public health doctors. (Scally 2021).

One area where public health practitioners did play an important role was at the local level in facilitating the NHS roll out of the vaccine programme.

> Where we get real traction is through very strong leadership from local government and directors of public health, and really understanding your citizens. What the NHS brings is the scale of the offer. But the interface with the citizen and the community is where local government and directors of public health really come in. We could not have done it without working together (Timmins and Baird 2022, p. 6).

19.6 The Role of the Private Sector

Governments over the last two decades have increasingly favoured a public–private partnership and there has been evidence of both increases in marketisation and, at the least, 'creeping' privatisation in the NHS. (Calnan 2023). Thus, the apparent need to turn to the private sector during the pandemic can be explained by the lack of investment in the infrastructure of the NHS and public health combine with the socio-political values of governments over the last 10 years (Lawrence et al. 2020). A centralised approach and a lack of trust in or support for local authorities also underpinned the government turning to the private sector (Shilson-Thomas and Mills 2020) for the building of temporary hospitals, provision of PPE, and the NHS Test and Trace system amongst other aspects during the COVID-19 pandemic (Garside and Neate 2020).

The close relationship with and reliance on the private sector has certainly been central to the government's non-pharmaceutical policy or public health response—though how successful this relationship has been in the policy response is questionable as the following quotation suggests showing governments were not effective at managing their relationships with the private sector:

> The BMJ's investigation into the private sector being underused during the first year of the covid-19 pandemic—despite an expensive arrangement that private hospitals would help the NHS—is yet more evidence that the government, the Department of Health and Social Care, and the NHS central machinery have great difficulty in contracting successfully for large complex packages of services and managing relationships with suppliers. This has happened many times before, including during the pandemic. The Test and Trace system's deal with Deloitte paid senior staff £6000 a day, yet largely failed to meet its targets. Possibly worse was the waste of at least £10bn on personal protective equipment that was incorrectly procured through incompetence (Edwards 2023).

There was, as the last quotation implies, particular criticism about the effectiveness and the value for money of substantial investment into the national NHS Test and Trace system (Iacobucci 2020). Despite its name, the system had been contracted, and sub-contracted, to companies with limited public health experience. The expertise of local authority-funded public health teams was, by contrast, overlooked (Griffin 2020b). For much of the pandemic, evidence shows that the NHS Test and Trace system was only able to detect a maximum of 60% of contacts with only 5–10% of those contacted self-isolating (Griffin 2020b). This has been compounded by the inaccuracies, mistaken false negatives, found in polymerase chain reaction (PCR) testing by some private laboratories (Torjesen 2021). Concerns were also expressed about the substantial fees paid to private consultants in the provision of PPE. Over half of the £18 billion spent on pandemic-related contracts was awarded without competitive tender and the government was not transparent about suppliers and services (Good Law Project 2021). Thus, despite the apparent necessity of and commitment to privatisation in the policy response, poor quality procurement practices combined with a lack of understanding of what the private sector had to offer and how to use it efficiently limited the extent to which the governments could leverage private sector resources effectively (Shilson-Thomas and Mills 2020).

19.7 The Social Care Sector

Much needed reform and investment in social care has been low on the government's policy agenda (Calnan 2020) and continues to be despite the policy rhetoric and this neglect was clearly manifest in the policy response to the pandemic (Foster 2021). The focus in government policy had in the early stages of the pandemic been on hospital care, but the pandemic and the policy response, certainly during the first wave, ineffectively protected care home residents, and had major consequences for social care. Since the beginning of the pandemic, the House of Commons Health and Social Care and Science and Technology Committees (2021) stated that between 16th March and 30th April 2021 there were 41,675 care home residents who died of COVID-19. There were also increased deaths among people receiving care at home, people who work in social care provision and people with learning disabilities and dementia (Comas-Herrera et al. 2020).

This crisis has highlighted the difference between the NHS and Social Care in terms of funding and priority (Foster 2021; House of Commons Health and Social Care and Science and Technology Committees 2021). During the first wave, the decision to discharge hospital patients back to care homes reflected a government desire to protect NHS acute care capacity but had the unintended consequence of spreading infection in care homes. The government claimed that discharge of patients to the community did not have substantial consequences but given the lack of adequate testing capacity during the early stages of the pandemic, it is likely that their analysis is based on substantial under-estimation of the number of cases in care homes during the first wave. Furthermore, staff shortages (both an existing problem as well as because of COVID-19 staff illness), lack of personal protective equipment (PPE) and design of care homes all hampered isolation and infection control (House of Commons Health and Social Care and Science and Technology Committees 2021).

Compulsory vaccination for care home staff was one way the government sought to protect care home residents from COVID-19 along with prioritising residents and staff in the initial vaccination rollout and then in the booster programme. However, COVID-19 vaccines did not guarantee the health of care home residents or staff. Adequate numbers of appropriately trained and paid staff, availability of PPE and appropriate equipment, and risk assessments are also important in protecting the health of care home residents and staff. The high numbers of deaths in care homes during the pandemic reflected other shortcomings in care home provision and infection control and more general pandemic mitigation that could not be remedied solely by vaccine booster programmes or compulsory vaccination of staff (Calnan and Douglass 2022).

19.8 Impact on Untreated Health Problems

> Lynda Ross, from Portadown, was booked in for spinal surgery in early 2020.Her operation was cancelled and, by the time she could see her specialist again, she was told it was too late for treatment. "The consequence of that is I have to live my life in a level of pain that has me on the same medication as someone who is dying from cancer." she says." It feels like the rest of my life has been destroyed." (BBC News Online 2024d).

The non-pharmaceutical policy approach also produced what is termed a 'parallel'epidemic (Calnan and Douglass 2022) due to the prioritisation of hospital care for COVID-19 patients which might have freed up hospital capacity but has led to longer waiting lists particularly for routine surgery, a decline in referrals for cancer surgery, a decline in young people seeking help for mental health distress and a decline in the use of A&E and primary care services. This indicated a build-up of untreated health problems. Approximately two thirds of the patients waiting to start treatment at the end of May 2021 waited up to 18 weeks. This was far below the NHS standard, which stated that 92% of patients should have to wait no more than 18 weeks from referral to treatment (Mahase 2021). This was exemplified by a hospital trust purchasing the site of its local Nightingale Hospital because it wants to continue using it to help tackle the elective surgery backlog (Carding 2021).

Recent data (Stiebahl et al. 2024) shows that the pandemic was still having a negative impact. Before 2020, the NHS in England experienced increased demand alongside declining performance on its main waiting time measures. In many cases, these pressures increased following the COVID-19 pandemic. For example, the waiting list for hospital treatment rose to a record of nearly 7.8 million in September 2023. The waiting list rose consistently between 2012 and 2019 and rose more quickly since early 2021. The 18-week treatment target has not been met since 2016. The number of people visiting A&E in summer 2024 was above pre-pandemic levels. The percentage of patients waiting over 4 h in hospital A&E rose consistently between 2015 and 2020. A new record high was reached in December 2022. The 62-day waiting time standard for cancer has also not been met in recent years. Reduction in waiting lists and an increased capacity for procedures in the NHS were the targets of the most recent labour government's significant increases (4%) in funding for the NHS in the autumn budget of 2024.

19.9 Social Inequalities in Health and Public Health Policy

> The best phrase that i've heard was that we were all in the same storm but not in the same boat, (Iaccobbi 2023b)

This quotation was used to describe the position of black and ethnic minority groups in that these groups were more likely to be over exposed to and under protected against the pandemic compared with white British people. Survey evidence shows that black and minority ethnic people have faced greater barriers in shielding from COVID-19 as a result of the types of employment they held (black and minority ethnic men and women are overrepresented in key worker roles); having to use public transport more; living in overcrowded and multigenerational households more; and not being given appropriate PPE at work (Haque et al. 2020).

> The BMA's chair of council in his evidence to the Public Inquiry said that "bias and discrimination" in the NHS contributed to the pandemic's worse impact on black and Asian healthcare workers, with discrepancies in the way staff were treated and in their experiences.
>
> He said, "People from an ethnic [minority] background are less likely to seek out and be upheld with their risk assessments, they are less likely to be forthright about saying, 'I need to have appropriate respiratory protective equipment,' they are more likely to be posted to the front line and exposed to high-risk cases. The recognition that that was the case emerged across the pandemic, and it has been recognised by the NHS." In the longer term, he said, "We need to make sure we narrow and resolve inequalities. This country cannot go on like this." (Iaccobbi 2023b)

It was reported at the public inquiry that the first 10 doctors in the UK to die from COVID-19 were of ethnic minority origin. JS Bamrah, a member of the Federation of Ethnic Minority Healthcare Organisations, stated.

> It didn't seem as if our voice was being heard, although many of us were trying to shout from the rooftop. we said, 'Look, there's something going on here that shouldn't be happening (Limb and Wise 2024).

This discussion also raised the need to dig deeper and look at the structural and systemic causes of these inequalities, such as the need to address the question of "*Why is it that ethnic minority staff are more likely to find themselves on the frontline and less likely to find themselves in managerial positions?*" (Limb and Wise 2024).

These social inequalities also clearly reflect intersectionality and the additive interaction in the relationship between ethnicity and material circumstances. A similar social pattern was also evident in the inequalities in participation in the vaccination (Douglass and Calnan 2022). The greater risk of serious illness and death experienced by poorer and marginalised socio-economic groups were reflected and exacerbated directly by the COVID-19 pandemic. The more deprived a local authority was, the higher the mortality rate has been during the COVID-19 pandemic (Marmot et al. 2020). COVID-19 and the lockdown policies which accompanied it have also been experienced unequally—with a significant impact, for example, as noted above, on the public's mental health which varied in its intensity by social position. For example, younger people, women and people living in precarious economic circumstances were more likely to experience symptoms of depression because of the COVID-19 pandemic (ONS 2020). Long COVID, the bio mental illness, was also more prevalent amongst people living in deprivation (Duddy 2024).

There were many other social groups who were over exposed to the pandemic for example older people. There was evidence of varying prioritisation and treatment of older people in COVID-19 policy (Alaszewski 2021). The initial approach to the COVID-19 pandemic by the government was defined by attempting to achieve herd immunity which suggested that older people might die letting younger people continue with their social and economic lives and the lack of concern for the longevity of older people also appears to be manifest in the slow response to the second wave. There is additional evidence of older people being the casualties of implicit or informal rationing decisions about the need to protect the NHS and gaining access to hospital intensive care (although this occurred in other countries, such as Italy) (Lewis 2020). However, later policy in the form particularly of the vaccination programme has prioritised older age groups.

19.10 Conclusion

This chapter has considered the impact of COVID-19 on the health system in England. The message that emerges from this analysis was that the health system and each of its different sectors were not prepared for a pandemic of this type. A common rationale for the introduction of the lock down policy was that the healthcare system was struggling to manage the pandemic. This is a message emerging from the earlier sessions of the UK Public Inquiry and even though the more recent session (module

3) on staff and patient experience had yet to conclude the evidence suggests a health service in crisis. One aim of public inquiries is to inform policies to avoid repetition and learn lessons from the successes and failures of previous approaches although an analysis of the literature on lesson learning activities suggests an overall lack of clarity in approach (Powell 2023). One clear and key weakness was a lack of a coherent public health policy to address an essentially public health problem which reflected the neo liberal values and economic and social policies of the governments since 2010. One of the strength's was the relatively efficient roll out of the vaccination 'alone' (not vaccination' plus' as adopted in other countries, Paton 2022) programme through the NHS and facilitated by public health services at the local level (although not without its problems, Majeed et al. 2022, Wise 2025). However, as COVID-19 has reflected and exacerbated existing social inequalities in health, there is a clear need to refocus public health policies further upstream digging deeper into the structural causes of absolute and relative deprivation.

References

Alaszewski A (2021) Covid19 and risk-policy making in a global pandemic, Bristol
BBC News In Depth (2024) Evidence to the inquiry about the health system during Covid 19, Third stage 28/10/24
BBC News Online (2024a) C Whitty's evidence to the Public Inquiry on the state of the NHS, Third stage 26/09/24
BBC News Online (2024b) Experience of a member of staff, Third stage 26/09/24
BBC News Online (2024c) N Fong's evidence on the conditions in Hospitals to the Public Inquiry, Third stage 26/09/24
BBC News Online (2024d) Impact on waiting lists-evidence to the Public Inquiry, Third stage 09/09/24
Bottery S, Varrow M, Thorlby R, Wellings D (2018) A Fork in the road: next steps for social care funding reform. Health Foundation and Kings Fund
Calnan M, Douglass T (2022) Power, policy and the pandemic: a sociological analysis of COVID-19 policy in England. Emerald
Calnan M (2020) Health policy, power and politics sociological insights. Emerald
Calnan M (2023) What are the costs of privatisation in the UK's healthcare system?—economics observatory. University of Bristol
Carding N (2021) Trust buys Nightingale Hospital to help tackle elective backlog, Health Service Journal. Retrieved from https://www.hsj.co.uk/finance-and-efficiency/trust-buys-nightingale-hospital-to-help-tackle-elective-backlog/7030541.article
Charlesworth A, Robert A, Lafond S (2016) NHS finances under the coalition. In: Exworthy M, Mannion R, Powell M (eds) Dismantling the NHS? Policy Press chapter 3, pp 39–64
Comas-Herrera A, Glanz A, Curry N, Deeny S, Hatton C, Hemmings N, Humphries R, Lorenz-Dant K, Oung C, Rajan S, Suarez-Gonzalez A (2020) The COVID-19 Long-Term Care situation in England. LTCcovid.org, International Long-Term Care Policy Network. Retrieved from https://ltccovid.org/wp-content/uploads/2020/11/COVID-19-Long-Term-Care-situation-in-England-19-November-2.pdf
Douglass T, Calnan M (2022) Without risk: a social analysis of the vaccination programme in England. In: Akhtar R (eds) coronavirus (COVID-19) outbreaks, vaccination, politics and society. Springer, Cham. https://doi.org/10.1007/978-3-031-09432-3_18
Duddy C (2024) Coronavirus: long COVID, House of Commons, CBP 214

Dyer C (2023) Covid-19: UK's pandemic inquiry prepares to hear from witnesses. BMJ 381: 1314. https://doi.org/10.1136/bmj.p1314

Edwards N (2023) The role of the private sector during the pandemic. BMJ 380:444. https://doi.org/10.1136/bmj.p444

Ferris T, Dixon J, Blumenthal D (2024) Lessons from England's National Health Service Published November 13, 2024. N Engl J Med 391(19):1852–1859. https://doi.org/10.1056/NEJMhpr2407495

Foster D (2021) Coronavirus: adult social care key issues and sources. House of Commons Library. Retrieved from https://researchbriefings.files.parliament.uk/documents/CBP-9019/CBP-9019.pdf

Garside J, Neate R (2020) UK government using pandemic to transfer NHS duties to private sector. The Guardian. Retrieved from https://www.theguardian.com/business/2020/may/04/uk-government-using-crisis-to-transfer-nhs-duties-to-private-sector

GHS Global Health Security Index (2019) GHS index country profile for United Kingdom. Retrieved from: https://www.ghsindex.org/country/united-kingdom

Glendenning C (2009) Social care: relationships, markets and ethics, Chapter 10. In: Gabe J, Calnan M (eds) The new sociology of the health service. Routledge, pp 201–221

Godlee F (2020) Covid-19: a wake-up call. BMJ 369. https://doi.org/10.1136/bmj.m2021

Good Law Project (2021) Government in breach of law. Retrieved from https://goodlawproject.org/update/government-in-breach-of-the-law/

Greener I (2021) Comparing country risk and response to COVID-19 in the first 6 months across 25 organisation for economic co-operation and development countries using qualitative comparative analysis. J Int Comparat Soc Policy 37:211–225

Griffin S (2020a) Covid-19: England had worst excess mortality in Europe in April. BMJ 370:m3058. https://doi.org/10.1136/bmj.m3058

Griffin S (2020b) Covid-19: NHS England should manage test and trace system, says Independent SAGE. BMJ 371:m4330. https://doi.org/10.1136/bmj.m4330

Hallett B (2024) UK Covid-19 inquiry module 1: the resilience and preparedness of the United Kingdom, HMSO ISBN 978-1-5286-4936-0E03069239 07/24

Haque Z, Becares L, Treloar N (2020) Over-exposed and under-protected. Runnymede Trust. Retrieved from http://www.runnymedetrust.org/uploads/Runnymede%20Covid19%20Survey%20report%20v3.pdf

House of Commons Science and Technology Committee (2021) The UK response to Covid-19: use of scientific evidence. Retrieved from https://assets.publishing.service.gov.uk/government/uploads/system/uploads/attachment_data/file/986148/government-response-to-science-and-technology-committee-report-uk-response-to-covid-19-use-of-scientific-advice.pdf

Iacobucci G (2020) Covid-19: government is criticized for 'scandalous'£10bn spent on test and trace. BMJ 370:m2805

Iacobucci G (2023a) Covid-19 inquiry: experts clash with political leaders over effects of austerity: BMJ 381:1443. https://doi.org/10.1136/bmj.p1443

Iacobucci G (2023b) Covid-19 inquiry: ministers abandoned public health principles during pandemic, says BMA chief. BMJ 383:2320. https://doi.org/10.1136/bmj.p2320

Kings Fund (2019) Report of the NHS ten-year plan (07/01/2019) January 2019

Lawrence F, Gardside J, Pegg D, Conn D, Carrell S, Davies H (2020) How a decade of privatisation and cuts exposed England to the Coronavirus. Guardian. Retrieved from https://www.theguardian.com/world/2020/may/31/how-a-decade-of-privatisation-and-cuts-exposed-england-to-coronavirus

Lewis R (2020) How rationing care in the pandemic is killing elderly people. Healthcare Business International. Retrieved from https://www.healthcarebusinessinternational.com/special-report-how-rationing-care-in-the-pandemic-is-killing-the-elderly/

Limb M, Wise J (2024) Covid inquiry: ethnic minority doctors felt vulnerable and afraid to raise concerns. BMJ 2024(387):2247. https://doi.org/10.1136/bmj.q2247

Mahase E (2020) New Public Health agency must be independent says doctors. BMJ 370:3590

Mahase E (2021) Covid-19: doctors warn of' disconnects' between NHS pressure and lifting of restrictions. BMJ 374:1762. https://doi.org/10.1136/bmj.n1762
Mahase E (2024) Covid inquiry: UKHSA chief is challenged on view that evidence for FFP3 masks is "weak" BMJ 386: 2075. https://doi.org/10.1136/bmj.q2075
Majeed A, Pollock K, Hodes S, Papaluca M (2022) Implementation of covid-19 vaccination in the United Kingdom. BMJ 378:e070344. https://doi.org/10.1136/bmj-2022-070344
Marmot M, Allen J, Goldblatt P, Herd E, Morrison J (2020) Build back fairer: The COVID-19 marmot review: the pandemic, socioeconomic and health inequalities in England. Institute of Health Equity, London
ONS (2020) Coronavirus and depression in adults. Retrieved from: https://www.ons.gov.uk/peoplepopulationandcommunity/wellbeing/articles/coronavirusanddepressioninadultsgreatbritain/januarytomarch2021
Paton C (2022) NHS reform and health politics in the UK, revolution, counter—revolution and Covid Crisis. Palgrave Macmillan
Powell M (2023) Learning lessons from the Covid 19 pandemic. Health Econ Policy Law 18:88–103. https://doi.org/10.1017/S1744133122000160017
Ross S, Fenney D, Thorstensen-Woll C, Buck D (2021) Directors of public health and the Covid-19 PANDEMIC: 'A year like no other' The Kings Fund. Retrieved from https://www.kingsfund.org.uk/sites/default/files/2021-09/A%20Year%20Like%20No%20Other%20online%20version_0.pdf
Scally G (2021) A new public health body for the UK. BMJ 373: 875
Scally G (2024) Covid inquiry: the flaws that led to system failure. BMJ 386:1865. https://doi.org/10.1136/bmj.q1865
Shilson-Thomas A, Mills W (2020) All hands-on deck: public-private partnerships during COVID-19. Retrieved from http://allcatsrgrey.org.uk/wp/download/health_economics/All-hands-on-deck-041220-1.pdf
Stiebahl S, Kirk-Wade E, Harker R (2024) Key NHS statistics. House of Commons Library, Number 07281 25 October
Timmins N, Baird B (2022) The Covid-19 vaccination programme, trials, tribulations and successes. Kings Fund
Tonkin T (2021) Public health doctors 'completely shattered'. British Medical Association. Retrieved from https://www.bma.org.uk/news-and-opinion/public-health-doctors-completely-shattered
Torjesen I (2021) Covid 19: PCR testing suspended at private laboratory after high rate of false negatives. BMJ 375:2535. https://doi.org/10.1136/bmj.n2535
Vize R (2020) Controversial from creation to disbanding, via e-cigarettes and alcohol: an obituary of Public Health England. BMJ 371: 4476. https://doi.org/10.1136/bmj.m447
Wise J (2025) UK Covid inquiry: bereaved families question vaccination strategies. BMJ 388:114. https://doi.org/10.1136/bmj.r114

Chapter 20
Impacts of COVID-19 on Health Care for the Aging Population in the United States

Michele Kekeh and Hedieh Sirjani

Abstract COVID-19 has caused severe adverse outcomes among older adults. According to the Centers for Disease Control and Prevention, the death rate among older adults was 533.5 per 100,000 in 2020. Their risks increased with age and underlying medical conditions. In the United States, the number of people with at least one chronic disease in the older adult population is estimated to increase from 71.52 million in 2020 to 142.66 by 2050. Within that timeframe, this population is expected to grow from 137.25 to 221.13 million. The rapidly growing aging population is still at high risk of adverse health outcomes from any future pandemic. This chapter explores how healthcare access and quality were altered for older adults, highlights emerging challenges and opportunities, and discusses innovative approaches and policies to improve healthcare service delivery for the aging population. Additionally, the study investigates requirements from caregivers as they adapt to new protocols to ensure the well-being of the older adults for whom they provide care.

Keywords COVID-19 · Aging population · Healthcare access · Healthcare quality · Chronic diseases · Caregivers

20.1 Introduction

Older adults commonly deal with multiple risk factors for diseases and are prone to a variety of concurrent health issues. Multimorbidity, the coexistence of two or more chronic diseases (Aïdoud et al. 2023), can increase the risk of complications and vulnerability to infections and harsh environments. Moreover, coexisting health issues can lead to polypharmacy, taking multiple medications to treat different

M. Kekeh (✉)
Joint School of Public Health at ODU and NSU, Old Dominion University, Norfolk, VA, USA
e-mail: mkekeh@odu.edu

H. Sirjani
Nursing and Allied Health, College of Science, Engineering, and Technology, Norfolk State University, Norfolk, VA, USA

R. Akhtar (ed.), *COVID-19: Impacts on Health and Health Systems*,
https://doi.org/10.1007/978-3-031-90255-0_20

diseases at once (National Institute on Aging 2021). In multimorbidity, patients may receive too many medications that do not always contribute to their health. Unnecessary polypharmacy could lead to unwanted conditions and physical or cognitive health problems, further increasing the patient's vulnerability and reducing quality of life.

Aging by itself is the most potent risk factor (Bektas et al. 2018) and can increase the vulnerability of people 65 years or older. Frailty in older adults, due to a decline in multiple physiological systems, predisposes them to more diseases (Yanase et al. 2018). The most prevalent chronic diseases among older adults are type 2 diabetes, arthritis, heart disease, and dementia (CDC 2022). COVID-19 impacted the lives of older adults, both directly and indirectly. The direct impact was exposure to the virus; the indirect impact was facing the healthcare system's challenges and enduring routine life during the pandemic. The present chapter examines healthcare access and quality for older adults as a result of COVID-19, highlights emerging challenges and opportunities, and discusses innovative approaches and policies to improve healthcare service delivery for the aging population. Additionally, the study investigates requirements from caregivers as they adapt to new protocols to ensure the well-being of the older adults for whom they provide care.

20.2 Vulnerability and Health Challenges of Older Adults

The critical impact of lifestyle should not be overlooked. Some lifestyles and habits increase the risk of chronic disease, especially in older adults. Physical inactivity intensifies the incidence of chronic disease in older adults (Dogra et al. 2022). Smoking or exposure to air pollutants due to poor environmental health or job conditions could cause chronic obstructive pulmonary disease (COPD) in older adults, who are more likely to be diagnosed with it (CDC 2023; López-Campos et al. 2016).

A poor lifestyle and lack of a healthy diet can contribute to type 2 diabetes. Insulin resistance and type 2 diabetes are more prevalent among older populations (Assar et al. 2019). Patients who have diabetes can experience multiple complications, such as neuropathy, cardiovascular disease, kidney disease, and dementia or Alzheimer's, which are already more common among people aged 65 and over. Eating habits may also increase the potential for Alzheimer's disease. Cardiovascular disease and heart issues are more prevalent among people with poor eating habits and who are sedentary. It is critical to note that older adults with heart disease typically suffer from at least one other health issue, adding to their day-to-day challenges (Forman et al. 2017).

Multimorbidity is further complicated when one chronic disease is a risk factor for another, negatively impacting patient outcomes. Meanwhile, the burden of addressing chronic issues can reduce the patient's ability to manage day-to-day responsibilities and stay active. The problem is enhanced when one suffers from multiple issues and concurrent complications requiring a caregiver. Managing disease symptoms and daily responsibilities may be more challenging for older adults without caregivers.

The mortality rate (death rate) among older adults during COVID-19 has been higher than for other age groups, with men having a higher rate than women (Tejada-Vera and Kramarow 2022). The mortality rate is even higher among those 85 years and older (Tejada-Vera and Kramarow 2022). Older adults with multimorbidity are more vulnerable to death due to COVID-19 because they suffer from multiple health issues and complications. Furthermore, the demand for intensive care in older adults suffering from COVID-19 is higher than in other age groups.

Even though the COVID-19 mortality rate has been in the headlines, the indirect impacts of the pandemic have also received much attention. Food insecurity was a public health issue before COVID-19, and the number of households that could not meet their family needs rose even more during the pandemic (Neff 2020). Social distancing policies made food less accessible to older adults because those policies limited interventions that could help them access food (Nicklett et al. 2021). Food insecurity could intensify health problems in older adults who need healthy food to manage chronic diseases, such as diabetes. An unhealthy diet further increases the risk of blood sugar spikes that could be life threatening for diabetic patients.

Older adults suffering from cognitive decline or physical impairments are more likely to be dependent on caregivers, someone to help them with daily responsibilities, including preparing food and obtaining health services or medication. Social isolation during the pandemic made caregiving to vulnerable populations more challenging.

Many older adults in nursing homes are hospitalized for prolonged periods due to multimorbidity complications. Hence, they are more exposed to viral and bacterial infections than the general population. A higher probability of contamination from airborne particles and droplets containing the virus has increased vulnerability to COVID-19.

20.3 Healthcare Access

As the older adult population grows, the need for medical care is expected to expand. During the COVID-19 pandemic, the health of this age group deteriorated, and access to care became limited. The demand for health care among older adults dropped significantly (by 22%), even though the overall demand for available health services intensified (Vigezzi et al. 2022). The negative impact on healthcare availability has been more prominent among males than females (Vigezzi et al. 2022).

Comorbid conditions are the common risk factor for healthcare and caregiver access limitations. A lack of access to medications results in poor health for older adults with disabilities (Ghimire et al. 2022). Missed appointments were more frequent among patients with cognitive disorders, while people with physical challenges tend to cancel their doctor appointments more often (Vigezzi et al. 2022).

Telehealth could resolve accessibility issues to some extent, especially for older adults who face the most challenges in meeting their healthcare needs (Smolić et al. 2022). Issues that do not require immediate intervention or hospitalization could be addressed remotely. However, the hardship of providing care for older adults, who needed to be seen in person, was not alleviated easily during the pandemic.

Limited supplies of medications, a shortage of available beds and intensive care units, healthcare provider turnover, and a lack of hospital staff are vital issues for all age groups. Nevertheless, since the pandemic primarily impacted older adults, they felt the shortcomings more intensely than others. Older adults suffering from multimorbidity and polypharmacy were usually at higher risk of medication shortages or missed appointments. The dilemma was more prominent due to the fear of exposure to the COVID-19 virus in hospitals or clinics, hence a drop in visitation (Núñez et al. 2021).

Factors impacting healthcare access, such as distance to a healthcare facility, inadequate transportation, poor communication with the provider, and a lack of preparedness for serving a diverse population in need, contribute to healthcare inaccessibility. Older adults deal with these factors as much, if not more intensely. The existing disparities among older adults increased the problem of not receiving timely health services. Since the growing older adult population is more noticeable in rural areas, these areas commonly face a lack of access to health care, creating a disparity in health care between rural and urban areas (James et al. 2018).

Social determinants are significant factors contributing to health and healthcare disparities. Lack of access to care and fewer effective outcomes are common among older adult minorities and lower-income populations (Noren Hooten et al. 2022).

In the United States, most older adults aged 65 years or older can receive health insurance coverage through Medicare. However, Medicare plans may require high deductibles for routine care and may not cover some medications and long-term care. According to the Administration for Community Living (2022), in 2020, the average out-of-pocket cost of healthcare services for older adults was $7000, about a 40-percent increase over 2010. The low-income older adult population struggled the most with healthcare costs during the pandemic.

20.4 Innovative Approaches and Policies

The COVID-19 pandemic has profoundly impacted older adults, highlighting the urgent need for innovative approaches and robust policies to enhance access to healthcare services for this population in a more holistic and multidisciplinary way. Older adults are particularly vulnerable to adverse health outcomes, as demonstrated by the pandemic. The deficiencies in the healthcare system relating to this population revealed during this crisis further underscore the need for solutions to better serve older adults and improve their health outcomes, safety, quality of life, and survival in anticipation of future health crises.

According to Fulmer et al. (2021), there are currently improvements in science and technology that improve preventive and medical care for older adults. However, there is a need for policies that bridge the gap between public health, health care, and the economy. Those policies must emphasize the social determinants of health and preventive actions to reduce and prevent chronic diseases among older adults and provide care that is personalized to each patient.

In 2021, the National Academy of Medicine identified six strategies to improve the health of older adults:

- Creating and training a qualified workforce,
- Making healthy aging one of the priorities of public health,
- Creating agencies at the federal, state, and local levels responsible for implementing policies and programs related to the older adult population,
- Focusing on addressing their social determinants of health,
- Promoting policies for the integration of telehealth into existing practice models,
- Adopting innovative approaches that allow older adults to decide if they want to age in place or to reside in institutional facilities (Fulmer et al. 2021).

These strategies are not just a top-down approach. They require a collaborative effort and coordinated actions from all stakeholders for successful implementation, making each policymaker an integral part of the process.

In February 2024, the U.S. Department of Health and Human Services (DHSS) released updated regulations for the Older American Act (OAA), passed in 1965, and amended in March 2020 (Administration for Community Living [ACL] 2024). The updated regulations (45 CFR Parts 1321, 1322, 1323, and 1324) address gaps in services identified in the previous law and incorporate lessons learned during COVID-19. Furthermore, the law seeks to improve service delivery and benefits for OAA participants, especially those with social and economic needs.

The OAA is a crucial piece of legislation that insures that older adults in the U.S. have access to the necessary resources and services for aging in place (ACL 2024). The OAA includes regulations designed to address issues identified in the original act to ensure that older adults in the U.S. continue to have adequate access to health care. These provisions include clarifying the requirements for each state, specifying a broad range of people who are eligible for the services, outlining conflicts of interest and how to improve them, including provisions for caregivers, establishing clear expectations for legal assistance, promoting the autonomy of older adults, updating the definitions and requirements for nutrition programs, and encouraging innovations to improve nutritional assistance for older adults.

Through the OAA, older adults have access to meals delivered to them at home to ensure that they have access to adequate nutrition, as many of them may be unable to prepare meals for themselves. The act also provides family caregivers with access to a wide range of resources, such as respite care, to alleviate caregiving burdens. These provisions are not just a token gesture, but a testament to the law's commitment to support caregivers in their crucial roles. The provisions also include assistance for preventing illnesses through screening, vaccinations, and health education. Additional assistance includes transportation services, especially for those who no longer drive, legal assistance, and services for elder abuse prevention. These services are designed to collectively enhance the well-being of older adults and allow them to age with dignity and actively participate in their communities of choice. The act also promotes access by participants to high-quality and culturally competent medical and mental health services (ACL 2024).

To bolster many of the programs covered under the OAA, the Biden-Harris Administration allocated $1.4 billion in 2021 through the American Rescue Plan (White House 2021). The funding provided enhanced programs for nutrition services, such as home meal delivery, home and community-based services (support for practical daily living, transportation, support for essential services, such as medical appointments, and psychological and social support to reduce social isolation), assistance for caregiving (respite care, in-home support, and caregiver education training), measures to address fall prevention, manage chronic diseases and reduce depression, safeguards for residents in nursing homes and long-term facilities (promote advocacy for nursing home resident and promote health, safety, and welfare of residents), and disease prevention.

In addition to extending the OAA in 2022, the Biden-Harris Administration recognized that the COVID-19 pandemic severely impacted older adults (White House 2022). In response, the administration developed policies in collaboration with the disability community to promote in-person education support for students living with vulnerable older adults, expanded the disability information access line available through the ACL, and allocated resources to prevent infection.

According to a report by the Brookings Institution, older adults who live in nursing homes and long-term facilities should be part of the approach to providing equity in the healthcare system due to funding challenges. Due to the burdens on Medicare and Medicaid, only limited payment is offered, causing disparities in services between the less fortunate, who can only afford Medicare and Medicaid, and the most fortunate, who can afford private care. The report highlights the challenges that older adults living in nursing homes and long-term facilities face and promotes advocacy in addressing these challenges (Yaraghi 2023).

The COVID-19 pandemic has severely affected older adults, especially those living in nursing homes and long-term care facilities certified by Medicare or Medicaid. According to data reported by the Center for Medicare and Medicaid Services (CMS) (CMS 2022), as of January 2022, 853,910 older adults have been infected by COVID-19 in nursing homes, and 145,270 have died from the disease. The pandemic revealed significant challenges in nursing homes regarding infection and prevention. It was a common practice to have two residents in one room, increasing the risk of infection (Bakerjian 2022). In 2016, the CMS mandated hiring one or more infection specialists for each facility. The specialist would work part time to manage infection control and prevention as a condition for participating in Medicaid and Medicare (GovInfo.gov 2016). As a result of COVID-19, the requirements have increased, and facilities are required to employ full-time infection specialists. This update also includes the need for transparency efforts and mandates that nursing home facilities inform patient family members of the number of COVID-19 cases in their facilities.

Research has shown racial disparities among nursing home residents, with minority groups having unequal access to alternative long-term services (Carlson and Selassie 2022). Despite these disparities, many current policies are color blind and tend to disfavor minority groups. Some policy recommendations to address race-based disparities include eliminating partial Medicaid certification of nursing homes. This certificate allows nursing homes to cap the number of rooms allocated to Medicaid recipients and allows these facilities to service the preferred payment plan they desire. Other recommendations include improving the staffing for nursing home facilities, especially for facilities with a higher percentage of minority patients (Carlson and Selassie 2022).

Additionally, the Federal and State Nursing Homes and Assisted Living regulations provided a regulatory framework for managing the operations and oversight of nursing home facilities. These include certification of nursing homes in compliance with federal Medicare and Medicaid requirements (SeniorLiving.org 2024). Additionally, many states have enacted regulations in addition to federal guidelines that underscore the willingness of some state agencies to provide consistent oversight and protect the health of older adults, especially the most vulnerable living in assisted living facilities (U.S. Government Accountability Office [GAO] 2024). The ombudsman program is an example of state programs that advocate for nursing homes to ensure that safety guidelines are in place to promote the welfare of older adults (GAO 2024). These states require that facilities undergo regular inspections to meet the required standards. However, access to these reports is challenging, making it difficult for families to be well informed in choosing care facilities that maintain high standards and safety.

The Naturally Occurring Retirement Community (NORC) conducted a survey in 2021 among a nationally representative sample of older adults and their caregivers to identify the concerns and public health strategies available to support their well-being during the COVID-19 pandemic. The primary concerns expressed by the participants were mental health challenges exacerbated by social isolation. To address these mental health challenges, the National Council on Aging (NCOA) recommended the implementation of initiatives that provide support to older adults to manage conditions, such as anxiety, depression, and any psychological aftermath of isolation (2024). In addition to these broad needs, older adults and their caregivers living in rural areas highlighted the lack of access to health care, medications, and medical supplies as a significant concern. An environmental review of public health interventions implemented nationally by state and local organizations found that most programs focused on educational interventions, direct services, health care, policy, and system change. Most of these interventions were online, which may be challenging for older adults who do not have access to the Internet or those who are technologically challenged (NORC 2021).

One of the significant recommendations from the research done by the NORC in 2021 was to increase older adults' access to high-speed Internet. *Telehealth expansion* is a crucial innovation that can help improve access to health care for older adults, especially those with mobility issues or those who live in remote areas (Cao et al. 2023). This method of healthcare delivery can promote communication with providers. Adjustments are needed to improve the efficient use of telehealth to provide services to older adults, including addressing Internet issues and patient privacy control. While public health and healthcare resources may be available online, research indicates that over one-quarter of older adults cannot access them. Using systems, such as videoconferencing, to connect with healthcare services may be discouraging for this population. It is crucial to develop and implement interventions that educate older adults on using technology and expand their technology literacy. Providers must provide technological support to educate their patients on how to set up the system and ensure that they know how to set it up and can use it properly.

In 2023, Facchinetti et al. conducted a systematic review to assess the use of smart home technology in managing chronic diseases in older adults. The health outcomes assessed include vital signs, medication management, mobility, falls, and quality of life. The authors found that the use of technologies, such as remote monitoring and smart home devices, offers significant advantages in the management of chronic diseases, including monitoring of vital signs, medication adherence, managing disease symptoms more effectively, improving the support for activities for daily living, and ensuring excellent safety in the home environment. These technologies can alert healthcare providers and family members about potential health issues before they become severe crises (Facchinetti et al. 2023).

According to Facchinetti et al. (2023), the benefits of using technology to improve health care for older adults extend beyond the pandemic. Smart technologies and remote monitoring systems provide timely data collected in real time on patient vitals, help providers better manage the health of older adults, and reduce the frequency of hospital visits. These technologies help enhance the management of health conditions in older adults, medication adherence, fall prevention, and overall quality of life. Smart home technologies can also support older adults with cognitive impairments and improve continuity of care and safety.

COVID-19 has caught many healthcare systems unprepared for a pandemic of this scale and speed of contamination, highlighting the need to rebuild and enhance the healthcare system and services to better care for the most vulnerable, including older adults. Dr. Tedros Adhanom Ghebreyesus, director of the World Health Organization (WHO), noted that to build a safer world, we must endeavor together from the ground up, starting with a robust primary healthcare system with skilled workers and empowering the communities to take charge of their health (WHO 2024). In 2021, the WHO implemented seven policies designed to strengthen and build resiliency in the healthcare system. These policies leverage experiences from COVID-19 to enhance readiness, commit resources to build public health capacity, develop comprehensive care services accessible to all populations, foster a collaborative approach to care, encourage innovations and research, and focus on reducing disparities (WHO 2024). They can also significantly enhance healthcare services for older adults by ensuring that services remain uninterrupted for older adults, even during emergencies.

The ability to manage emergencies will ensure that healthcare services for older adults, such as chronic disease management, will be maintained. Additionally, by investing in and researching senior care, health services for older adults could be more robust and tailored to their ages. The U.S. Centers for Disease Control is working across the U.S. to strengthen the public health system, provide funding for research on social determinants, and provide technical support to strengthen the public health workforce, including providing resources and support to caregivers (CDC n.d.).

20.5 Implications for Caregivers

Many of the policies and guidelines discussed above were intended to improve the health and well-being of older adults during the pandemic and beyond. A number of policies also called for caregivers to help maintain and improve the health of this population. Caregivers are paid and unpaid sources that help older adults provide skilled care, companion services, homemaker services, and personal care services. According to the CDC (2021a), 53 million caregivers in 2020 care for older adults in the U.S. This number is estimated to reach 73 million by 2030. While caregiving, in general, may be rewarding, caregivers faced many health challenges during the pandemic, including financial strain, stress, depression, and anxiety, many of which stem from the need to balance care for older adults, whether it is a paid position or helping an older adult family member and managing their own needs and health (NORC 2020). In response to the pandemic and its detrimental impacts on older adults, the CDC introduced training to support caregivers in providing safe and effective care to older adults during the pandemic and beyond. This support includes training modules on infection control prevention protocols, stress management, mental health support, and technology use (CDC 2021b). These resources are made available on mainstream social media to allow easy access by all caregivers.

Caregivers themselves have health issues. According to the CDC, almost one in five caregivers reported having fair to poor health, and two in five reported having at least two chronic diseases (CDC 2021a). During the pandemic, individuals who had the most significant adverse outcomes included those with existing health conditions, such as chronic diseases. This risk makes caregivers vulnerable to the pandemic. Many policies have been enacted to provide guidelines and recommendations for caregivers caring for older adults; many policies have been designed to protect caregivers. These policies help caregivers understand the protocols to use in their daily activities to reduce the chances of infection among older adults. Additionally, adhering to the protocols helps with consistent care across different settings.

Racial and ethnic disparities exist among caregivers in the U.S. Among the 53 million caregivers, 14% are African American, 17% are Hispanic, and 58% are women (CDC 2021c). Due to the disproportionate level of chronic diseases among older adults, especially those of minority groups, having caregivers with diverse cultural backgrounds can help meet the needs of older adults. Most caregivers, especially family caregivers, are not formally trained and learn as they go.

In 2023, Medicare Part B included a new policy that provides training services for caregivers who care for Medicare patients. Under the policy, the training must help improve the health and treatment goals of the patient and be recommended by their provider if the caregiver's involvement is crucial for the success of the treatment (Medicare.gov 2023). The training covers assistance with daily tasks, safety, effective communication, and ways to better understand the patient's health condition, promoting social support and personalized care. This initiative was designed to support caregivers in providing the best and most culturally sensitive care possible for older adults under their care.

The CDC recommends that healthcare workers and caregivers maintain strict hygiene to prevent infection. This protection includes frequent hand washing, use of alcohol-based sanitizers, wearing gloves correctly when caring for older adults, and any protective gear when necessary (CDC 2024). Additionally, the CDC recommends regular cleaning and disinfection of the patient environment and medical devices. Furthermore, caregivers must follow evidence-based handling and injection procedures to prevent infection. Caregivers must be attentive to the older adults under their care, monitor the prevention practices effectively, and contact providers on a timely basis to receive feedback for any concerns to improve care delivery. These practices are designed to minimize the risk of infection among older adults and improve their health and safety.

Older adults and their caregivers still face significant challenges beyond COVID-19. Some of the key issues include the ongoing need to manage their chronic health conditions and the increased incidence of conditions, such as dementia and mobility impairments. Many caregivers, especially family caregivers without specific training, struggle with burnout and a lack of resources, training, and adequate finances. Additionally, older adults living in rural areas continue to face additional challenges regarding access to healthcare services due to the scarcity of medical services in these regions and limited access to telehealth. These challenges underscore the need for healthcare providers, policymakers, and researchers to continue working collaboratively to develop comprehensive approaches, integrated policies, and support systems to address the needs of older adults and their caregivers. These requirements are essential to ensure that caregivers provide safe, effective, and compassionate care to older adults under their supervision during and beyond the pandemic.

20.6 Conclusion

The COVID-19 pandemic has been significantly challenging for the aging population. The vulnerability of this population to diseases is the predisposing factor to adverse health outcomes. Multimorbidity and frailty among older adults commonly cause polypharmacy and dependency on caregivers. Due to physical vulnerability among older adults, exposure to COVID-19 often leads to severe symptoms and fatal complications. During the pandemic, providing routine needs, including healthy food, medication, and receiving health care, was problematic for the aging population. Social distancing added more issues to the existing disparities experienced by this group. Given the critical role of caregivers in older patients' lives, it is crucial to provide them with the necessary training and support. This can significantly enhance the patients' outcomes. The established policies to assist caregivers, the CDC recommendations, and Medicare resources and training are valuable tools that can be utilized to achieve this goal.

The experience gained from the COVID-19 pandemic could guide preparedness efforts for future pandemics and upcoming health challenges. The necessity for creating robust infrastructure in the healthcare system, including adequate and accessible medical supplies and hospital beds, equipped nursing homes, telehealth and access to the required technology, attention to mental health, training providers and caregivers, and comprehensive health insurance coverage, was underscored during the pandemic.

To avoid the intense impacts seen with COVID-19 in any future pandemics, it is imperative that we foster global collaboration. By considering the unique challenges and needs of each region, all countries can play a role in building a safer future for all populations, including older adults. This shared responsibility can instill a sense of unity and purpose in our efforts.

References

Administration for Community Living (2022) 2021 Profile of Older Americans. https://acl.gov/sites/default/files/Profile%20of%20OA/2021%20Profile%20of%20OA/2021ProfileOlderAmericans_508.pdf

Administration for Community Living (2024) Older Americans Act: Grants to State and Community Programs on Aging; Grants to Indian Tribes and Native Hawaiian Grantees for Supportive, Nutrition, and Caregiver Services; Grants for Supportive and Nutritional Services to Older Hawaiian Natives; and Allotments for Vulnerable Elder Rights Protection Activities. Department of Health and Human Services. 45 CFR Parts 1321, 1322, 1323, and 1324. RIN: 0985-AA17

Aïdoud A, Gana W, Poitau F, Debacq C, Leroy V, Nkodo JA, Poupin P, Angoulvant D, Fougère B (2023) High prevalence of geriatric conditions among older adults with cardiovascular disease. J Am Heart Assoc 11, e026850. https://www.ahajournals.org/, https://doi.org/10.1161/JAHA.122.026850

Assar ME, Laosa O, Rodríguez Mañas L (2019) Diabetes and frailty. Curr Opin Clin Nutr Metab Care 22(1):52–57. https://doi.org/10.1097/MCO.0000000000000535

Bakerjian D (2022) Coronavirus disease 2019 (COVID-19) and safety of older adults residing in nursing homes. Agency for healthcare research and quality (AHRQ), PSNet. https://psnet.ahrq.gov/primer/coronavirus-disease-2019-covid-19-and-safety-older-adults

Bektas A, Schurman SH, Sen R, Ferrucci L (2018) Aging, Inflammation and the environment. Exp Gerontol 105:10–18. https://doi.org/10.1016/j.exger.2017.12.015

Cao Y, Chen D, Smith M (2023) Use telehealth as needed: telehealth substitutes in-person primary care and associates with the changes in unplanned events and follow-up visits. BMC Health Serv Res 23(1):426. https://doi.org/10.1186/s12913-023-09445-0

Carlson E, Selassie G (2022) Racial disparities in nursing facilities—and how to address them. Justice in Aging. https://justiceinaging.org/wp-content/uploads/2022/09/Racial-Disparities-in-Nursing-Facilities.pdf

Centers for Disease Control and Prevention (2021a) Supporting caregivers. https://www.cdc.gov/aging/publications/features/supporting-caregivers.htm

Centers for Disease Control and Prevention (2021b) Training for healthcare professionals. https://www.cdc.gov/coronavirus/2019-ncov/hcp/training.html

Centers for Disease Control and Prevention (2021c) COVID-19: A public health emergency for older adults and their caregivers. https://www.cdc.gov/aging/publications/features/covid-19-help-older-adults-caregivers/index.html

Centers for Disease Control and Prevention (2022) Promoting health for older adults. https://www.cdc.gov/chronicdisease/resources/publications/factsheets/promoting-health-for-older-adults.htm

Centers for Disease Control and Prevention (2023) Basics about COPD. https://www.cdc.gov/copd/basics-about.html#ref2

Centers for Disease Control and Prevention (2024) Infection control basics. https://www.cdc.gov/infection-control/about/index.html

Centers for Disease Control and Prevention (n.d) Public health professionals gateway. https://www.cdc.gov/public-health-gateway/php/index.html

Center for Medicare and Medicaid Services (2022) COVID-19 Nursing home data. https://data.cms.gov/stories/s/COVID-19-Nursing-Home-Data/bkwz-xpvg/

Dogra S, Dunstan DW, Sugiyama T, Stathi A, Gardiner PA, Owen N (2022) Active aging and public health: evidence, implications, and opportunities. Annu Rev Public Health 43:439–459. https://doi.org/10.1146/annurev-publhealth-052620-091107

Facchinetti G, Petrucci G, Albanesi B, De Marinis MG, Piredda M (2023) Can smart home technologies help older adults manage their chronic condition? A systematic literature review. Int J Environ Res Public Health 20(2):1205. https://doi.org/10.3390/ijerph20021205

Forman DE, Arena R, Boxer R, Dolansky MA, Eng JJ, Fleg JL, Haykowsky M, Jahangir A, Kaminsky LA, Kitzman DW, Lewis EF, Myers J, Reeves GR, Shen WK (2017) Prioritizing functional capacity as a principal end point for therapies oriented to older adults with cardiovascular disease: a scientific statement for healthcare professionals from the American heart association. Circulation 135, e894–e918. https://www.ahajournals.org/doi/https://doi.org/10.1161/CIR.0000000000000483

Fulmer T, Reuben DB, Auerbach J, Fick DM, Galambos C, Johnson KS (2021) Actualizing better health and health care for older adults. Health Affairs 40(2). https://www.healthaffairs.org/doi/https://doi.org/10.1377/hlthaff.2020.01470

Ghimire S, Shrestha A, Yadav UN, Mistry SK, Chapadia B, Yadav OP, Ali AM, Rawal LB, Yadav P, Mehata S, Harris M (2022) Older adults with pre-existing noncommunicable conditions and their healthcare access amid COVID-19 pandemic: a cross-sectional study in Eastern Nepal. BMJ Open 12(2):e056342. https://doi.org/10.1136/bmjopen-2021-056342

GovInfo (2016) 81 FR 68688—Medicare and Medicaid programs; reform of requirements for long-term care facilities. https://www.govinfo.gov/app/details/FR-2016-10-04/2016-23503

James W, Cossman J, Wolf J (2018) Persistence of death in the United States: the remarkably different mortality patterns between America's Heartland and Dixieland. Demographic Research 39:897–910. https://www.demographic-research.org/articles/volume/39/33

López-Campos JL, Tan W, Soriano JB (2016) Global Burden of COPD. Respirology (Carlton, Vic.), 21(1):14–23. https://doi.org/10.1111/resp.12660

Medicare.gov (2023) Caregiver training services. https://www.medicare.gov/coverage/caregiver-training-services

National Institute on Aging (2021) The dangers of polypharmacy and the case for deprescribing in older adults. https://www.nia.nih.gov/news/dangers-polypharmacy-and-case-deprescribing-older-adults

National Council on Aging (2024) Anxiety, depression, and sleep: the pandemic's toll on older adults' mental health. https://www.ncoa.org/article/anxiety-depression-and-sleep-the-pandemics-toll-on-older-adult-mental-health

Naturally Occurring Retirement Community (2020) Your health and well-being matter. https://www.norc.org/content/dam/norc-org/pdfs/OACCaregiverOnePager.pdf

Naturally Occurring Retirement Community (2021) NORC needs assessment and environmental scan report: Maintaining physical and mental well-being of older adults and their caregivers during public health emergencies. Centers for Disease Control and Prevention

Neff LM (2020) Hidden hunger: food insecurity in the age of coronavirus. Am J Clin Nutr 112(5):1160–1161. https://doi.org/10.1093/ajcn/nqaa279

Nicklett EJ, Peng S, Brown RT (2021) Food insecurity and social distancing among middle-aged and older adults during COVID-19. J Gerontol: Ser B 76(7):e287–e296. https://doi.org/10.1093/geronb/gbab003

Noren Hooten N, Pacheco NL, Smith JT, Evans MK (2022) The accelerated aging phenotype: the role of race and social determinants of health on aging. Ageing Res Rev 73:101536. https://doi.org/10.1016/j.arr.2021.101536

Núñez A, Sreeganga SD, Ramaprasad A (2021) Access to healthcare during COVID-19. Int J Environ Res Public Health 18(6):2980. https://doi.org/10.3390/ijerph18062980

SeniorLiving (2024) Federal and state nursing home & assisted living regulations. https://www.seniorliving.org/nursing-homes/state-federal-regulations/

Smolić Š, Blaževski N, Fabijančić M (2022) Remote healthcare during the COVID-19 pandemic: findings for older adults in 27 European countries and Israel. Front Public Health 10:921379. https://doi.org/10.3389/fpubh.2022.921379

Tejada-Vera B, Kramarow EA (2022) COVID-19 mortality in adults aged 65 and over: United States, 2020. NCHF Data Brief, 446. https://www.cdc.gov/nchs/products/index.htm

U.S. Government Accountability Office (2024) Requirements for protecting elders in nursing Homes. https://www.gao.gov/search?keyword=requirements+for+protecting+elders+in+nursing+homes+

Vigezzi GP, Bertuccio P, Amerio A, Bosetti C, Gori D, Cavalieri d'Oro L, Iacoviello L, Stuckler D, Zucchi A, Gallus S, Odone A, Lost in Lombardia Project Investigators (2022) Older adults' access to care during the COVID-19 pandemic: results from the lockdown and lifestyles (LOST) in lombardia project. Int J Environ Res Public Health 19(18):11271. https://doi.org/10.3390/ijerph191811271

White House (2021) FACT SHEET: biden-harris administration delivers funds to support the health of older Americans. https://www.whitehouse.gov/briefing-room/statements-releases/2021/05/03/fact-sheet-biden-harris-administration-delivers-funds-to-support-the-health-of-older-americans/

White House (2022) FACT SHEET: administration announces new actions to address the needs of people with disabilities and older adults in response to and recovery from COVID-19. https://www.whitehouse.gov/briefing-room/statements-releases/2022/02/24/fact-sheet-administration-announces-new-actions-to-address-the-needs-of-people-with-disabilities-and-older-adults-in-response-to-and-recovery-from-covid-19/

World Health Organization (2024) WHO's 7 policy recommendations on building resilient health systems. https://www.who.int/news/item/19-10-2021-who-s-7-policy-recommendations-on-building-resilient-health-systems

Yanase T, Yanagita I, Muta K, Nawata H (2018) Frailty in elderly diabetes patients. Endocr J 65(1):1–11. https://doi.org/10.1507/endocrj.EJ17-0390

Yaraghi N (2023) Transforming health care for older adults: innovations, equity, and economics. Brookings Institution. https://www.brookings.edu/articles/transforming-healthcare-for-older-adults/

Chapter 21
The Impact of COVID-19 on Mental Health and Substance Use in the United States

Olaniyi Olayinka and Terence Tumenta

Abstract This book chapter examines the impact of the COVID-19 pandemic on mental health and substance use in the United States. The global health crisis triggered a cascade of psychological and behavioral effects, reshaping healthcare services and delivery. Drawing on a review of studies, clinical reports, and epidemiological data, the chapter highlights the connection between the pandemic and changes in mental health disorders and substance misuse patterns and care. The chapter concludes with a discussion of the consequences for healthcare systems, including the expansion of telemedicine that increased access to mental health and substance use treatment. By offering a nuanced understanding of the intricate relationship between the COVID-19 pandemic, mental health, and substance use, this chapter aims to contribute to our understanding of the enduring impact of the crisis on the well-being of individuals and communities in the United States.

Keywords COVID-19 · Mental health · Substance use · Public health · Vulnerable populations · Healthcare systems

21.1 Introduction

The COVID-19 pandemic has posed significant challenges to public health and triggered a cascade of psychological and behavioral effects, reshaping the landscape of mental health and substance use in the United States. The immediate aftermath of the outbreak revealed significant impacts on access to mental health and substance use services (MHSUS) and the mode and quality of these critical services (Bojdani et al. 2020). A cross-sectional survey of approximately 200 million emergency department

O. Olayinka (✉)
University of Texas Health Science Center at Houston, Houston, USA
e-mail: olaniyi.o.olayinka@uth.tmc.edu

T. Tumenta
Yale University School of Medicine, New Haven, US
e-mail: terence.tumenta@yale.edu

R. Akhtar (ed.), *COVID-19: Impacts on Health and Health Systems*,
https://doi.org/10.1007/978-3-031-90255-0_21

(ED) visits found significantly higher median visit rates for psychiatric problems, drug overdoses, domestic violence, and child abuse in the year following COVID-19 compared to the year preceding the outbreak (Holland et al. 2021). Similarly, text messages for crisis intervention to a mental health support platform increased from March to July 2020 compared to the same period in 2019 (Runkle et al. 2021).

After COVID-19 was declared a national emergency on January 31, 2020, protocols were developed to reduce the virus's spread in hospitals and residential drug treatment programs. Access to MHSUS was particularly impacted as in-person medical services were discouraged in favor of telemedicine (Hollander and Carr 2020). Kantamneni (2020) highlighted that the pandemic disproportionately affected marginalized populations, who often keep society running (Kantamneni 2020). This chapter examines the impact of the pandemic on mental health disorders, substance misuse patterns, and care delivery systems. Drawing on empirical studies, clinical reports, and epidemiological data, we highlight the direct and indirect consequences of the virus and its disproportionate impact on vulnerable populations.

21.2 COVID-19 and Mental Health

The psychological impact of COVID-19 on Americans is well established. Anxiety, depression, and post-traumatic stress syndrome were particularly prevalent during the pandemic compared to the months leading up to the outbreak in December 2019. The negative consequences of COVID-19 converged to impact mental health directly or indirectly. Those who contracted the virus experienced physical symptoms that resulted in anxiety, depression, and fatigue. Additionally, those who lost loved ones and friends to COVID-19 reported depression and trauma-associated symptoms. Other indirect psychological consequences stemmed from isolation due to social distancing and financial stress from underemployment and job losses.

Several studies have examined the predictors of psychological distress among Americans during the pandemic. For example, a 2020 online survey of 2787 adult Americans showed that women, non-binary individuals, and participants younger than 45 had a significantly higher prevalence of anxiety, depression, and post-traumatic stress (Prout et al. 2020). Killgore et al. found that those who lost their jobs due to COVID-19 restrictions had a higher prevalence of anxiety disorders (30–46%), depression (27–32%), insomnia (25%), suicidal ideation (18%), and acute/post-traumatic stress (15–18%) compared to those who did not (Killgore et al. 2021). They estimated that mental health problems were 1.5–1.7 times higher among individuals who reported job loss due to COVID-19 restrictions.

Anxiety related to COVID-19 was particularly high after the outbreak. The emergence of COVID-19 and its associated uncertainties, including fears of infection, social isolation, economic instability, and disruptions to daily routines, exacerbated anxiety symptoms and heightened vulnerability in susceptible populations. An online survey of over 50,000 adult Americans between April 2020 and May 2022 showed that approximately 44% of low socioeconomic status participants and 35% of higher socioeconomic status participants reported high levels of COVID-19 health worry (Cozen et al. 2024). A similar analysis of the mental health of over 1400 participants during the early periods of the outbreak showed high levels of anxiety, especially among young, female, and low-income participants (Shuster et al. 2021). A survey of a national sample of American adults in March 2020 found average levels of worry and fear were high, with scores of 3.4 out of 5 and 7 out of 10, respectively (Fitzpatrick et al. 2020). The physiological stress response played a pivotal role in the exacerbation of anxiety disorders during the COVID-19 pandemic. The activation of the hypothalamic–pituitary–adrenal (HPA) axis and the release of stress hormones, such as cortisol, in response to perceived threats contributed to heightened anxiety levels (Lupien et al. 2009). Moreover, dysregulation of the immune system and increased inflammation, as observed in COVID-19 patients, further amplified anxiety symptoms through neuroinflammatory pathways (Mazza et al. 2020). Neuroimaging studies have demonstrated alterations in brain regions implicated in anxiety processing, including the amygdala and prefrontal cortex, in individuals exposed to chronic stressors, such as the pandemic (Holmes et al. 2020). The psychological impact of COVID-19 on anxiety disorders was multifaceted, encompassing heightened fear of illness and death, uncertainty about the future, and social isolation. Fear of contagion and perceived threats to personal and familial safety contributed to anticipatory anxiety and hypervigilance among individuals susceptible to anxiety disorders (Taylor et al. 2020). Prolonged exposure to pandemic-related stressors, coupled with limited access to coping mechanisms and social support networks, exacerbated pre-existing anxiety symptoms and precipitated the onset of new cases (Gao et al. 2020). The pervasiveness of misinformation and conflicting media narratives surrounding COVID-19 further fueled anxiety and exacerbated feelings of helplessness and distress (Mertens et al. 2020). The societal response to the COVID-19 pandemic, including public health measures, such as lockdowns, social distancing, and quarantine protocols, had profound implications for individuals with anxiety disorders. While these measures were essential for containing viral transmission, they also entailed significant psychosocial consequences, including increased social isolation, economic hardship, and disruptions to healthcare services (Brooks et al. 2020). Vulnerable populations, including frontline healthcare workers, individuals with pre-existing mental health conditions, and marginalized communities, were disproportionately affected, exacerbating existing health disparities (Huang and Zhao 2020). Stigmatization and discrimination against individuals perceived as high-risk or infected with COVID-19 further contributed to psychological distress and exacerbated anxiety symptoms (Bagcchi 2020).

The prevalence of depression also increased during the pandemic. Prolonged social isolation has been linked to an increased risk of mood disorders. Research indicates that social connections are crucial for maintaining mental health, and their disruption due to isolation can lead to feelings of loneliness, depression, and anxiety (Loades et al. 2020). Additionally, individuals already vulnerable to mood disorders may experience heightened symptoms in the absence of social support systems, exacerbating their condition. The economic repercussions of the COVID-19 pandemic further exacerbated the burden on mental health. Individuals facing job loss or reduced income experienced hopelessness and despair, known triggers for depression. Moreover, uncertainty about the future economic landscape amplified anxiety and distress, perpetuating a cycle of negative mood states. The disruption of daily routines and lifestyle habits due to the pandemic significantly impacted mood disorders. Many individuals rely on structured routines and coping mechanisms to manage their mental health, and the upheaval caused by COVID-19 destabilized these strategies. Factors, such as disrupted sleep patterns, changes in exercise routines, and limited access to recreational activities, contributed to stress and agitation, further exacerbating mood disorders (Holmes et al. 2020). Additionally, the inability to engage in pleasurable activities or hobbies due to lockdown measures led to a loss of motivation and interest, common symptoms of depression. A survey of a nationally representative sample of 1441 US adults found that moderate-severe and severe depressive symptoms during COVID-19 were reported in 7.9% and 5.1% of participants, respectively, compared to 2.1% and 0.7% in a similar survey in 2017/18 (Ettman et al. 2020). The risk of anxiety and depression was notably higher among younger Americans, those with low socioeconomic status, individuals exposed to multiple stressors, residents of areas with higher COVID-19 infection rates, those who were separated or widowed, users of cannabis and alcohol, and those with a history of depressive symptoms (Ettman et al. 2020; Holingue et al. 2020). The disruption to daily life, coupled with fears of infection and uncertainty about the future, heightened stress, anxiety, and depression across the country. This was exacerbated by the pandemic's negative impact on social connections. A study by Rosenberg et al. highlighted that individuals with in-person social connections had a lower risk of depression and loneliness (Rosenberg et al. 2021).

The COVID-19 pandemic exposed individuals to a myriad of traumatic stressors, ranging from fear of infection and illness to the loss of loved ones and economic instability. Research suggests that exposure to such stressors can increase the risk of developing trauma-related symptoms, including PTSD and ASD (Holmes et al. 2020). The illness-induced psychological distress and grief associated with loss added additional layers of complexity to the mental health landscape. Ashby et al.'s survey of 527 workers and 218 undergraduates in the US found a positive correlation between COVID-19 traumatic stress, a history of cumulative trauma, and post-traumatic stress disorder symptoms (Ashby et al. 2022). An important outcome of the COVID-19 outbreak worth highlighting is its impact on frontline healthcare workers. Of over 2500 healthcare workers involved in the care of COVID-19 patients at Mount Sinai in spring 2020, more than half (and up to 80%) reported significant distress (Norman et al. 2021). The workers' distress was related to concerns about family, infection, and work and correlated highly with the severity of COVID-19-related PTSD symptoms.

21.3 COVID-19 and Substance Use

Parallel to the increase in mental health challenges, the COVID-19 pandemic witnessed a rise in substance use and behavioral addictions across the United States (Avena et al. 2021; Lee and Singh 2023; Wang et al. 2021). Factors contributing to this trend included heightened stress levels, coping mechanisms, increased leisure time, and disruptions to access to traditional treatment modalities. Unfortunately, the increased substance use was associated with rises in substance misuse, overdose, and use disorders, particularly in 2020 and 2021. Drug overdose deaths increased by 30% between 2019 (70,459 deaths) and 2020 (91,536 deaths) (Lee and Singh 2023). Additionally, the risk of severe outcomes among those with COVID-19 was higher among individuals with substance use disorders (Kumar and Nayar 2021; Wang et al. 2021). An analysis of over 12,000 Americans diagnosed with COVID-19 by June 15, 2020, found that the odds of COVID-19 among those with a substance use disorder (SUD) diagnosis in the past year were approximately eight times higher than those without an SUD diagnosis (Wang et al. 2021). This association was strongest for individuals with opioid use disorder (AOR = 10.244 [9.107–11.524]) and tobacco use disorder (AOR = 8.222 [7.925–8.530]). Additionally, changes in drug supply chains and availability impacted the types and potency of opioids circulating within communities, altering consumption patterns and overdose risks (Haley and Saitz 2020; Slavova et al. 2020). The pandemic posed significant challenges to individuals seeking treatment for opioid use disorder (OUD) and those in recovery. Disruptions to in-person healthcare services, including medication-assisted treatment (MAT) programs and counseling sessions, hindered access to critical supports for individuals with OUD.

The relationship between COVID-19 and alcohol use is bidirectional, with the pandemic exacerbating existing mental health challenges and alcohol consumption serving as both a coping mechanism and a risk factor for poor mental health outcomes. Moreover, alcohol misuse can exacerbate mental health symptoms, creating a cycle of negative reinforcement that poses significant challenges for individuals seeking to maintain sobriety during these uncertain times.

The COVID-19 pandemic also unleashed a cascade of challenges, reshaping the landscape of substance use and addiction worldwide. Stimulants, including cocaine, amphetamines, and prescription medications like Adderall, garnered increased attention amidst the pandemic due to shifts in consumption patterns and associated health risks (Dubey et al. 2020).

Social isolation and boredom stemming from lockdown measures led some individuals to increase their cannabis consumption as a means of alleviating distress and monotony (Håkansson et al. 2020). Economic hardships, including job loss and financial strain, also impacted cannabis use patterns, with some individuals turning to cannabis as a more affordable alternative to alcohol or other substances (Avery et al. 2020).

The COVID-19 pandemic created a fertile ground for the emergence and proliferation of novel substances, including synthetic drugs, designer drugs, and non-traditional psychoactive compounds. The emergence of novel substances during the

COVID-19 pandemic presented unique challenges and risks to public health. Limited knowledge of the composition, potency, and effects of these substances posed and continue to pose challenges for harm reduction efforts and medical interventions (EMCDDA 2020).

The pandemic amplified existing health disparities and exacerbated pre-existing mental health conditions and substance use among vulnerable populations (Avena et al. 2021; Wang et al. 2021). Racial and ethnic minorities, individuals with disabilities, and those living in congregate settings bore a disproportionate burden of the pandemic's psychological toll. The confluence of factors that led to spikes in substance use (alcohol, cannabis, tobacco), opioid overdoses, and other substance-related harms strained an already burdened healthcare system.

21.4 COVID-19 and the Healthcare Landscape

The delivery of mental health and substance use services was significantly impacted, with many facing challenges in accessing care due to overwhelmed healthcare systems and disruptions to traditional treatment modalities (Filip et al. 2022). The enduring impact of the COVID-19 pandemic on mental health and substance use led to a reevaluation of healthcare delivery and policy frameworks with far-reaching consequences. For example, the rapid expansion of telemedicine services during the pandemic increased access to mental health and substance use treatment, offering a potential avenue for addressing longstanding disparities in care access (Kichloo et al. 2020; Lau et al. 2020). New York City is a prime example of a major city that experienced significant growth in virtual services in 2020, with more than 30,000 mental health services delivered via telehealth (Lau et al. 2020). However, challenges, such as digital inequities and regulatory barriers, must be addressed to ensure equitable access to telehealth services (Kichloo et al. 2020).

21.5 Conclusion

In conclusion, the COVID-19 pandemic significantly exacerbated mental health disorders, triggering new symptoms and worsening existing conditions due to social isolation, economic stress, and disrupted routines. Addressing these mental health challenges requires a comprehensive approach that includes social support, financial relief, accessible mental health services, and public health initiatives to reduce stigma and build resilience. Integrating mental health into the broader COVID-19 response can mitigate the long-term effects on individuals' well-being and foster a more resilient society. Recognizing the unique challenges faced by those with trauma and stressor-related disorders, it is crucial to ensure access to trauma-informed care and tailored interventions. Additionally, understanding the interplay between COVID-19 and substance use will aid in developing strategies to reduce substance-related harm.

Continued research and evidence-based approaches are essential to support mental health and build stronger communities.

References

Ashby JS, Rice KG, Kira IA, Davari J (2022) The relationship of COVID-19 traumatic stress, cumulative trauma, and race to posttraumatic stress disorder symptoms. J Community Psychol 50(6):2597–2610

Avena NM, Simkus J, Lewandowski A, Gold MS, Potenza MN (2021) Substance use disorders and behavioral addictions during the COVID-19 pandemic and COVID-19-related restrictions. Front Psych 12:653674

Avery AR, Tsang S, Seto EYW, Duncan GE (2020) Stress, anxiety, and change in alcohol use during the COVID-19 pandemic: findings among adult twin pairs. Front Psych 11:571084

Bagcchi S (2020) Stigma during the COVID-19 pandemic. Lancet Infect Dis 20(7):782

Bojdani E, Rajagopalan A, Chen A, Gearin P, Olcott W, Shankar V, Delisi LE et al (2020) COVID-19 pandemic: impact on psychiatric care in the United States. Psychiatry Res 289:113069. https://doi.org/10.1016/j.psychres.2020.113069

Brooks SK, Webster RK, Smith LE, Woodland L, Wessely S, Greenberg N, Rubin GJ (2020) The psychological impact of quarantine and how to reduce it: rapid review of the evidence. The Lancet 395(10227):912–920

Cozen AE, Carton T, Hamad R, Kornak J, Faulkner Modrow M, Peyser ND, O'Brien EC et al (2024) Factors associated with anxiety during the first two years of the COVID-19 pandemic in the United States: an analysis of the COVID-19 Citizen Science study. Plos One 19(2), e0297922

Dubey MJ, Ghosh R, Chatterjee S, Biswas P (2020) Impact of COVID-19 on mental health and lockdown-induced lifestyle changes in Indian population: an online survey. Asian J Psychiatr 54:102240

EMCDDA (2020) EMCDDA trendspotter briefing: Impact of COVID-19 on patterns of drug use and drug-related harms in Europe. European Monitoring Centre for Drugs and Drug Addiction. Retrieved from https://www.emcdda.europa.eu/publications/ad-hoc-publication/impact-covid-19-patterns-drug-use-and-drug-related-harms-europe_en

Ettman CK, Abdalla SM, Cohen GH, Sampson L, Vivier PM, Galea S (2020) Prevalence of depression symptoms in US adults before and during the COVID-19 pandemic. JAMA Netw Open 3(9):e2019686. https://doi.org/10.1001/jamanetworkopen.2020.19686

Filip R, Gheorghita Puscaselu R, Anchidin-Norocel L, Dimian M, Savage WK (2022) Global challenges to public health care systems during the COVID-19 pandemic: a review of pandemic measures and problems. J Personalized Med 12(8):1295

Fitzpatrick KM, Drawve G, Harris C (2020) Facing new fears during the COVID-19 pandemic: the State of America's mental health. J Anxiety Disord 75:102291

Gao J, Zheng P, Jia Y, Chen H, Mao Y, Chen S, Dai J et al (2020) Mental health problems and social media exposure during COVID-19 outbreak. PLoS ONE 15(4):e0231924

Håkansson A, Fernández-Aranda F, Menchón JM, Potenza MN, Jiménez-Murcia S (2020) Gambling during the COVID-19 crisis—a cause for concern. J Addict Med 14(4):e10–e12

Haley DF, Saitz R (2020) The opioid epidemic during the COVID-19 pandemic. JAMA 324(16):1615–1617

Holingue C, Badillo-Goicoechea E, Riehm KE, Veldhuis CB, Thrul J, Johnson RM, Kalb LG (2020) Mental distress during the COVID-19 pandemic among US adults without a pre-existing mental health condition: findings from American trend panel survey. Preventive Med 139, 106231

Holland KM, Jones C, Vivolo-Kantor AM, Idaikkadar N, Zwald M, Hoots B, Chen MS et al (2021) Trends in US emergency department visits for mental health, overdose, and violence outcomes before and during the COVID-19 pandemic. JAMA Psychiatry 78(4):372–379

Hollander JE, Carr BG (2020) Virtually Perfect? Telemedicine for Covid-19. N Engl J Med 382(18):1679–1681. https://doi.org/10.1056/nejmp2003539

Holmes EA, O'Connor RC, Perry VH, Tracey I, Wessely S, Arseneault L, Bullmore E (2020) Multidisciplinary research priorities for the COVID-19 pandemic: a call for action for mental health science. The Lancet Psychiatry 7(6):547–560

Huang Y, Zhao N (2020) Generalized anxiety disorder, depressive symptoms, and sleep quality during COVID-19 outbreak in China: a web-based cross-sectional survey. Psychiatry Res 288:112954

Kantamneni N (2020) The impact of the COVID-19 pandemic on marginalized populations in the United States: a research agenda. J Vocat Behav 119:103439. https://doi.org/10.1016/j.jvb.2020.103439

Kichloo A, Albosta M, Dettloff K, Wani F, El-Amir Z, Singh J, Chugh S et al (2020) Telemedicine, the current COVID-19 pandemic and the future: a narrative review and perspectives moving forward in the USA. Family Med Commun Health 8(3), e000530. https://doi.org/10.1136/fmch-2020-000530

Killgore WDS, Cloonan SA, Taylor EC, Dailey NS (2021) Mental health during the first weeks of the COVID-19 pandemic in the United States. Front Psychiatry 12. https://doi.org/10.3389/fpsyt.2021.561898

Kumar A, Nayar KR (2021) COVID 19 and its mental health consequences, vol 30, pp 1–2. Taylor & Francis

Lau J, Knudsen J, Jackson H, Wallach AB, Bouton M, Natsui S, Avalone L et al (2020) Staying connected in the COVID-19 pandemic: telehealth at the largest safety-net system in the United States: a description of NYC Health+ Hospitals telehealth response to the COVID-19 pandemic. Health Affairs 39(8):1437–1442

Lee H, Singh GK (2023) Estimating the impact of the COVID-19 pandemic on rising trends in drug overdose mortality in the United States, 2018–2021. Ann Epidemiol 77:85–89

Loades ME, Chatburn E, Higson-Sweeney N, Reynolds S, Shafran R, Brigden A, Crawley E (2020) Rapid systematic review: the impact of social isolation and loneliness on the mental health of children and adolescents in the context of COVID-19. J Am Acad Child Adolescent Psychiatry 59(11):1218–1239

Lupien SJ, McEwen BS, Gunnar MR, Heim C (2009) Effects of stress throughout the lifespan on the brain, behaviour and cognition. Nat Rev Neurosci 10(6):434–445

Mazza MG, De Lorenzo R, Conte C, Poletti S, Vai B, Bollettini I, Benedetti F et al (2020) Anxiety and depression in COVID-19 survivors: role of inflammatory and clinical predictors. Brain Behav Immun 89:594–600

Mertens G, Gerritsen L, Duijndam S, Salemink E, Engelhard IM (2020) Fear of the coronavirus (COVID-19): predictors in an online study conducted in March 2020. J Anxiety Disord 74:102258

Norman SB, Feingold JH, Kaye-Kauderer H, Kaplan CA, Hurtado A, Kachadourian L, Southwick SM (2021) Moral distress in frontline healthcare workers in the initial epicenter of the COVID-19 pandemic in the United States: relationship to PTSD symptoms, burnout, and psychosocial functioning. Depression and Anxiety 38(10):1007–1017

Prout TA, Zilcha-Mano S, Aafjes-Van Doorn K, Békés V, Christman-Cohen I, Whistler K, Di Giuseppe M (2020) Identifying predictors of psychological distress during COVID-19: a machine learning approach. Front Psychol 11. https://doi.org/10.3389/fpsyg.2020.586202

Rosenberg M, Luetke M, Hensel D, Kianersi S, Fu T-C, Herbenick D (2021) Depression and loneliness during April 2020 COVID-19 restrictions in the United States, and their associations with frequency of social and sexual connections. Soc Psychiatry Psychiatr Epidemiol 56:1221–1232

Runkle JD, Sugg MM, Yadav S, Harden S, Weiser J, Michael K (2021) Real-time mental health crisis response in the United States to COVID-19. Crisis

Shuster A, O'Brien M, Luo Y, Berner LA, Perl O, Heflin M, Fiore VG et al (2021) Emotional adaptation during a crisis: decline in anxiety and depression after the initial weeks of COVID-19 in the United States. Translational Psychiatry 11(1):435

Slavova S, Rock P, Bush HM, Quesinberry D, Walsh SL (2020) Signal of increased opioid overdose during COVID-19 from emergency medical services data. Drug Alcohol Depend 214:108176

Taylor S, Landry CA, Paluszek MM, Fergus TA, McKay D, Asmundson GJ (2020) Development and initial validation of the COVID Stress Scales. J Anxiety Disord 72:102232

Wang QQ, Kaelber DC, Xu R, Volkow ND (2021) COVID-19 risk and outcomes in patients with substance use disorders: analyses from electronic health records in the United States. Mol Psychiatry 26(1):30–39

Chapter 22
Disease and Environmental Factors: A Profile of Brazil Post-COVID Scenario

Rhavena Barbosa dos Santos, Júlia Alves Menezes, and Rose Ane Pereira de Freitas

Abstract Environmental transformations, climate change, deforestation, and urbanization have significant impacts on human health. This chapter aims to address the complex relationship between Olayinka these changes and the emergence of diseases, as occurred with the COVID-19 pandemic, in addition to understanding the relationships between environmental changes and the spread of infectious diseases, considering the Brazilian context. The pandemic highlighted human vulnerability to emerging diseases; in addition, the uncontrolled exploitation of ecosystems can facilitate the transmission of viruses from animals to humans. It is crucial to consider the interconnection between human health and the environment and encourage sustainable practices to prevent future pandemics and protect global health.

Keywords Brazil · Climate change · Public health

22.1 Brazil in Transition: Environmental Characteristics and Climate Change

Climate change is not a recent event; it has been occurring over decades due to phenomena beyond the natural variability of the climate, where anthropogenic actions have been the major cause of the changes experienced.

Unsustainable patterns of energy production and land use have contributed to this situation. According to information from the latest report of the Intergovernmental Panel on Climate Change (IPCC 2023), the observed increases in greenhouse gas

R. B. dos Santos (✉)
Department of Medicine and Nursing, Federal University of Viçosa, Viçosa, Brazil
e-mail: rhavena.santos@ufv.br

J. A. Menezes
Laboratory for Analysis and Development of Sustainability Indicators, National Institute for Space Research, São José dos Campos, Brazil

R. A. P. de Freitas
Department of Meteorology, Federal University of Pelotas, Pelotas, Brazil

R. Akhtar (ed.), *COVID-19: Impacts on Health and Health Systems*,
https://doi.org/10.1007/978-3-031-90255-0_22

(GHG) concentrations since 1750 are certainly from human activities during this period. Data show that anthropogenic GHG emissions in 2019 were 54% higher than in 1990, mainly from the burning of fossil fuels, industrial processes, and methane production.

Another important factor related to climate change concerns land use and land cover, which alter interactions between the surface and the atmosphere, the flows and distribution of available energy, and, therefore, the water balance. In this sense, Brazil has experienced significant changes with substantial impacts on its biomes, namely the Amazon, Atlantic Forest, Caatinga, Cerrado, Pantanal, and Pampas (IBGE 2019) (Fig. 22.1).

Furthermore, the country has experienced extreme heat exposure concurrently with a period marked by forest fires and deforestation, especially in the Amazon, a region that, besides its vast biodiversity, also has great ethnological diversity (Alves de Oliveira et al. 2021). The region is an important source of heat for the atmosphere, responsible for about 15% of the freshwater reaching the oceans, and has been synergistically influenced by different factors (e.g., deforestation, land use changes, decreased evapotranspiration, and increased dry season) that can increase exposure to risks and thereby human vulnerability to climate change (Nobre et al. 2016).

The Brazilian Pantanal, in turn, is characterized by interannual variability in rainfall, often influenced by phenomena such as El Niño and La Niña, which affect the region's floods and droughts. In recent decades, meat production and consequently pasture areas have become significant in the local productive sector. Thus, the conversion of areas for agricultural use and upstream water use, combined with climate change, are considered threats that increase both environmental and human vulnerability to climate change (Marengo et al. 2015; Caballero et al. 2021).

The Atlantic Forest is also threatened by continuous habitat loss due to deforestation for agriculture, urbanization, and infrastructure, increasing forest fragmentation, which harms habitat connectivity and species migration. This biome has been suffering from land use and occupation, with only fragments of intact forests and reforestation initiatives remaining in Brazil (Caballero et al. 2022). A study by Santos et al. (2019) on human vulnerability to climate change indicates that by 2071, considering the RCP 8.5 scenario, areas in southeastern Brazil, where the Atlantic Forest once predominated, could see a significant increase in consecutive dry days, a decrease in average annual precipitation, and a rise in average temperature. These characteristics, combined with the environmental degradation scenario, contribute to the onset of a desertification process in the region.

It is evident that human actions have warmed the land, ocean, and atmosphere and that ecosystem and human vulnerability are mutually dependent, exposing the population to conditions such as food insecurity, water stress, and the spread of diseases. High-confidence data from the IPCC indicate that extreme heat has increased mortality and morbidity rates related to vector-borne, water-borne, and food-borne infectious diseases, as well as non-communicable diseases (mental health) related to the occurrence of extreme events (IPCC 2023).

The reality in Brazil is no different. In the country, the vulnerability of the exposed population is characterized by extreme social inequality and strong dependence on

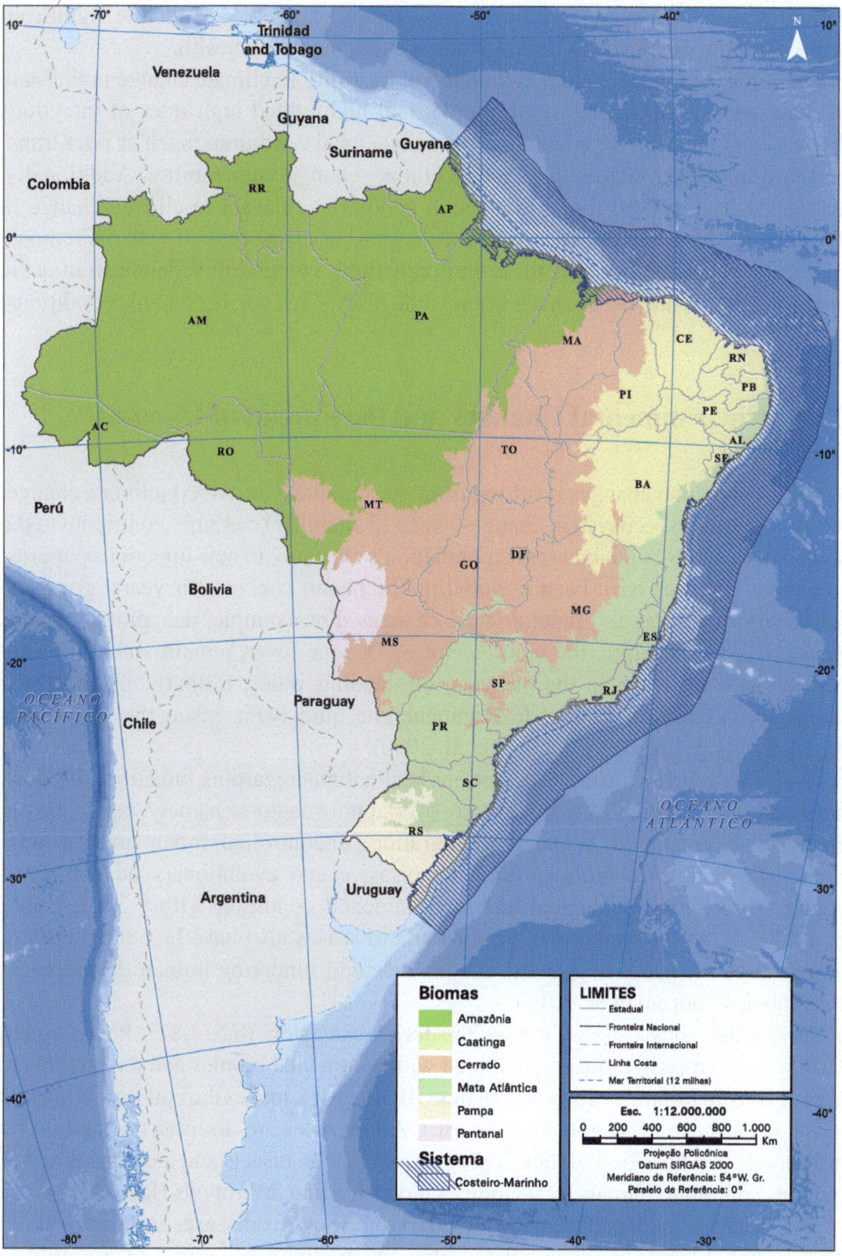

Fig. 22.1 Brazilian States and Biomes. https://educa.ibge.gov.br/jovens/conheca-o-brasil/territorio/18307-biomas-brasileiros.html

natural resources, which, combined with climate changes, corroborate significant impacts on the population across various sectors, including health.

Menezes et al. (2018) studied human vulnerability to climate change in the state of Amazonas (AM), in the north region, and highlighted high rates of infectious diseases and the need for adaptations to unique local conditions, such as river transport, are important factors that elevate climate change vulnerability. Additionally, Vommaro et al. (2020) found that health conditions relevant to climate change in the state of Maranhão (MA), in northeast region, when associated with the region's socioeconomic conditions, increase the population's vulnerability, demonstrating the intrinsic relationship between environmental, health, and socioeconomic conditions.

22.2 Environmental Changes and the Spread of Diseases

The environmental changes resulting from the situations described (climate change, habitat alterations, ecosystem changes, water availability) certainly contribute to the spread of diseases and increase the population's exposure to new infectious patterns. Infectious diseases have been a global public health concern for years, given the sensitivity of vectors to climatological changes. For example, dehydrated eggs of *Aedes aegypti* mosquito, the main vector of Dengue fever, remain viable for long periods and can hatch in the presence of standing water; furthermore, increased temperatures can reduce larval development time, quickly increasing the adult vector population (Epstein 2000; Githeko et al. 2000).

Worldwide, arboviruses pose a serious health threat regarding infectious diseases, particularly in tropical regions due to overlapping factors, namely rapid climate changes, deforestation, population migration, uncontrolled urban area occupation, poor sanitary conditions, and the invasive and evolutionary adaptation of vectors to different ecological and environmental conditions (Rust 2012; Lopes et al. 2020). At the same time, vector-borne diseases also have broader socioeconomic impacts, increasing health inequalities and hindering human development (Campbell-Lendrum et al. 2015).

Given the above, Brazil's vast territorial extension (8,514,877 km^2), which mostly has a tropical climate, combined with continental biomes and the previously mentioned environmental characteristics (IBGE 2004), makes the country conducive to the occurrence of vectors and diseases. Arboviruses are diseases transmitted by arthropods that have part of their reproductive cycle in insects and are transmitted to humans and other animals by the bite of blood-feeding arthropods (Rust 2012).

For decades, studies have shown the risk of arbovirus diseases in Brazil, due to factors such as large densely populated cities and the high proliferation of mosquitoes, especially *Culex* sp. and *Aedes aegypti*, which are highly anthropophilic, and also ticks (Figueiredo 2007; Rust 2012; Donalísio et al. 2017).

Among the most important arboviruses in Brazil, dengue stands out, being considered an endemic disease in the country. It is transmitted by the female *Aedes aegypti* mosquito. So far, four serotypes of the dengue virus (DENV) are known: DENV-1,

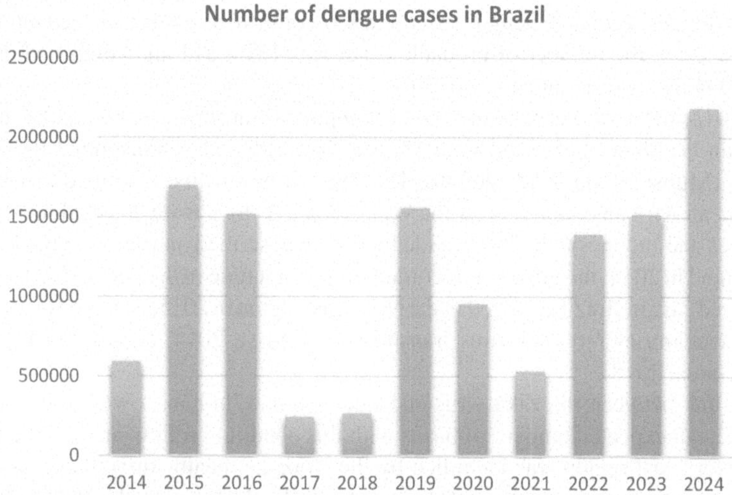

Fig. 22.2 Registered dengue notifications in notifiable diseases information system (SINAN-SUS), from 2014 to 2024—Brazil. http://tabnet.datasus.gov.br/cgi/tabcgi.exe?sinannet/cnv/denguebbr.def

DENV-2, DENV-3, and DENV-4. Evidence indicates that the virus arrived in Brazil on ships from Africa with slaves, and the first epidemic occurred between 1981 and 1982 in Boa Vista, the capital of the state of Roraima, caused by serotypes 1 and 4. After four years, the epidemic reached Rio de Janeiro and the capitals of the northeastern states, with epidemic periods associated with the introduction of new serotypes in previously unaffected areas (Ministério da Saúde 2024a).

A study on the burden of dengue in Brazil between 2000-2015 shows that, despite the low mortality rate, the disease caused a considerable loss of healthy life years in the country by affecting a large number of people of all age groups and causing deaths, particularly in children (Araújo et al. 2015). The following graph depicts the historical series of dengue notifications in the country (Fig. 22.2).

The disease exhibits a seasonal pattern in the country, with the highest number of cases occurring between October and May of the following year, corresponding to spring and summer, as well as the rainiest periods. Urbanization, uncontrolled population growth, inadequate sanitation, and climatic factors maintain favorable conditions for the presence of the vector, impacting the transmission dynamics of these disease (Ministério da Saúde 2024b). Additionally, it also represents a significant economic and social impact on the country, leading to the loss of potential years of life—an estimated 1,391.68 potential years of life lost between 2001 and 2005 and U$D 1.212 million spent (90% CI 904–1526) after adjustment for underreporting for the 2012–13 epidemic (Araújo et al. 2015; Martelli et al. 2015; Suaya et al. 2009).

Data from the National Health Council indicate that in 2024, Brazil will face the worst dengue epidemic in recent years, due to both the factors discussed throughout this text and the occurrence of El Niño, as well as the reappearance of dengue serotypes DENV-3 and DENV-4 (Conselho Nacional de Saúde 2024). By March

2024, 2,406,582 probable dengue cases and 897 confirmed deaths had been reported, while in 2023, the number of probable cases was 1,658,814 and confirmed deaths were 1094 (Ministério da Saúde 2024b).

In 2014, the virus responsible for Chikungunya transmission was added to the Brazilian health scenario, and in 2015, the Zika virus, also transmitted by *Aedes aegypti* (Ministério da Saúde 2024b). The Zika virus was first identified in 1947 in Uganda, an African country. Over the years, isolated cases were detected in African countries and Indonesia. In 2007, epidemics were described on islands in the Pacific Ocean, and in 2014, the virus was recorded for the first time in the Americas. The first confirmed case in Brazil occurred in the Northeast region in 2015, and in 2016, dozens of other countries reported virus transmission (Musso 2014; Nunes and Pimenta 2016; Garcia 2018).

The link between the Zika virus and microcephaly in babies was confirmed in 2016. The unexpected number of births with this condition following the Zika fever in the Northeast region was identified by the Brazilian health surveillance system, allowing for the rapid detection of the event and the declaration of a Public Health Emergency of National Importance. The Brazilian Ministry of Health was among the pioneers in raising the probable causal association between microcephaly and congenital Zika virus infection (Garcia 2018), accomplishment made possible by the live birth notification system (SINASC-SUS) implemented in Brazil since 1990 to collect birth data nationwide.

The disease spread quickly throughout Brazil, highlighting national difficulties in vector control, failures in maternal and childcare, and emphasizing inequalities, as the endemic disease primarily affects less developed regions and poor families (Garcia 2018).

It is noteworthy that, before the emergence of these two diseases, the Oropouche virus (OROV) was the second most frequent arbovirus in the country, second only to the dengue virus (Wesselmann et al. 2024). OROV is a significant arbovirus for the health scenario in Latin America and has gained prominence in the past year. The virus was first isolated in Brazil in 1960, and since then, isolated cases and outbreaks have been reported mainly in the Amazon region. Panama, Argentina, Bolivia, Venezuela, Ecuador, and Peru have also reported cases of the disease (Ministério da Saúde 2024c).

The main vector of OROV, in both the wild and urban cycles, is the mosquito *Culicoides paraensis*, commonly known as the biting midge. The symptoms of the disease include headache, muscle pain, joint pain, nausea, and diarrhea (Ministério da Saúde 2024c). The OROV genome allows for rearrangements, with three known recombinants, and the viruses Madre de Dios and Iquitos have already been associated with human diseases (Wesselmann et al. 2024). In Brazil, notification of the disease is required immediately due to its epidemic potential and high mutation capacity. Most cases are concentrated in the northern region of the country; however, cases have been detected in Paraná and Rio de Janeiro, all likely infected in the northern region (Ministério da Saúde 2024c).

Data from the Pan American Health Organization indicate that between epidemiological weeks 1 and 8, the virus had already been detected in 2104 samples in Brazil

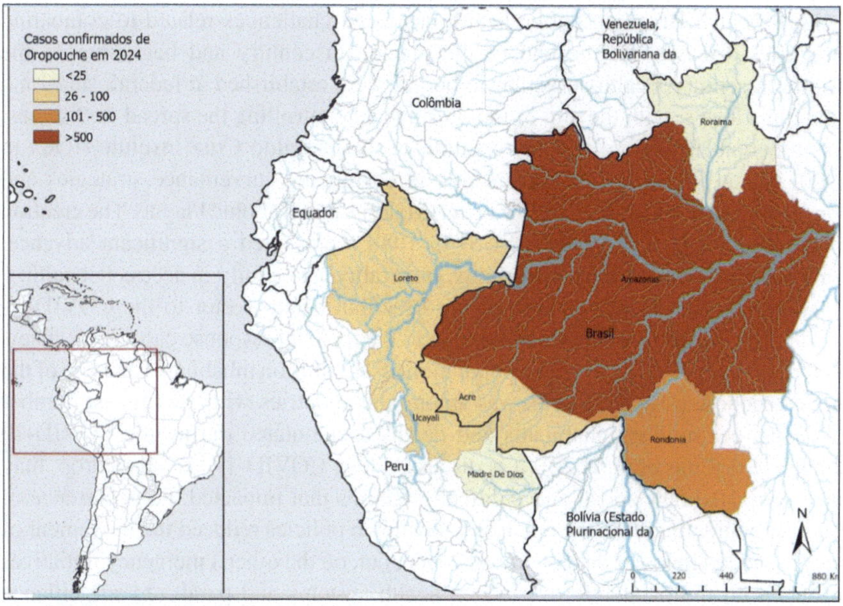

Fig. 22.3 Distribution of confirmed oropouche cases in Brazil and Peru, 2024. https://www.paho.org/pt/documentos/atualizacao-epidemiologica-oropouche-na-regiao-das-americas-6-marco-2024

(1821 in Amazonas, 172 in Rondônia, 51 in Acre, and 12 in Roraima)—Figure 22.3, while in Peru, 146 cases of OROV were detected during the same period (Organização Pan-Americana de Saúde 2024).

Geographic space, population movement, socioeconomic characteristics, and global warming are considered determinants for the dynamics of these diseases. Along with globalization, Latin America, in general, and Brazil, in particular, have experienced the emergence (and re-emergence) of various arboviruses (Wesselmann et al. 2024), which underscores the importance of endemic arboviruses in the country for the global health landscape.

22.3 Health sector Responses: Challenges Posed by the COVID-19 Epidemic

There are many arboviruses circulating in Brazil, such as Mayaro, Venezuelan Equine Encephalitis, Eastern Equine Encephalitis, Rocio, Dengue, Chikungunya, and Zika, just to name a few examples (Lima-Câmara 2016; Confalonieri et al. 2017). Considering the perspective of globalization and the rapprochement between people, the entry of new arboviruses into the country represents a major challenge for health professionals, researchers and governments.

Historically, Brazilian public health has faced challenges related to combating infectious diseases. At the end of the nineteenth century and beginning of the twentieth century, health surveillance bodies were established at federal, state, and municipal levels, with the aim of monitoring and controlling the spread of diseases. An important milestone was the creation of the Oswaldo Cruz Institute (IOC) in 1900, which played a fundamental role in developing surveillance strategies and combating diseases such as Yellow Fever, Malaria, and Bubonic Plague. The creation of the Unified Health System (SUS) in 1988 represented a significant advance, consolidating principles of universality, integrality, and equity in access to health.

However, the recent response of the Brazilian health sector to the COVID-19 pandemic exposes weaknesses in relation to the national response capacity to a new pattern of diseases. Brazil, with a population of 212 million inhabitants and one of the largest economies in the world, was among the countries with the highest number of deaths per million inhabitants and deaths accumulated during the COVID-19 pandemic (Freitas et al. 2022). It is discussed how COVID-19, as a backdrop, may have generated new social and political scenarios that impacted the occurrence of arbovirus epidemics. On the one hand, lockdown policies reduced the movement of people and led to lower disease transmission, but, on the other, emergency initiatives to deal with the pandemic overloaded health systems and paralyzed surveillance actions for other diseases. There was also a shortage of financial and human resources to deal with other multiple arboviruses, such as Dengue and Yellow Fever, which were epidemic at the same time as COVID-19 in Brazil (Cimerman et al. 2020; Wilder-Smith et al. 2020; Ma et al. 2022).

This period highlighted the fragility of national coordination, which hampered the health sector's response by not establishing national standards for controlling the pandemic (the country had four health ministers during this period). The political scenario was not favorable to evidence-based medicine, with managers refuting the World Health Organization's recommendations regarding the use of masks and the need for vaccination.

Among social policies, the government granted financial assistance to reduce the impact of poverty in the country, but specific actions for populations with high social vulnerability, such as indigenous people and the homeless population, were scarce.

The health sector's response to the pandemic period can be seen through three main strategies: (1) assistance actions in hospitals and, later, in primary care; (2) health surveillance actions, with emphasis on epidemiological surveillance; and (3) national production of tests and, subsequently, vaccines for COVID-19 by two public scientific institutions, Fiocruz and Butantã, functionally linked to the Ministry of Health and the government of the state of São Paulo, respectively. In addition, documents were released to assist in regulating processes and responsiveness (Freitas et al. 2022).

To improve the organization of this response, in 2021 the Surveillance, Alert and Response Network for Public Health Emergencies of the Unified Health System (Rede Vigiar-SUS) was established, which can be considered an important milestone for tackling other potential events of public health of national and international concern (Brasil 2022).

Another important Brazilian strategy in combating infectious diseases in general is the Brazilian National Immunization Program (PNI), recognized worldwide. The program offers 31 vaccines designed to protect the individual from birth to adulthood (Brasil 2024). Through routinely developed campaigns and advertisements, the Ministry of Health proposes to offer equitable access to these vaccines throughout the country, contributing to the eradication of diseases such as Polio, Measles, Rubella, Tetanus, and Pertussis, which are no longer a health problem public.

Even with a turbulent performance during the pandemic, mainly related to political issues, the PNI and SUS are perhaps the greatest highlights of Brazilian public health to deal with the perspectives presented in this chapter.

Finally, it is worth highlighting that in December 2023 the country made progress in the fight against Dengue, with the inclusion of the free vaccine in the national vaccination calendar, an important tool for classifying the disease as vaccine preventable in a country where it has such significant impacts (Ministério da Saúde 2024a).

22.4 Final considerations

Brazil, with its vast territorial dimensions and distinct environmental characteristics, is prone to the proliferation of vectors and, consequently, of arboviruses, a fact that, enhanced by climate change, represents a significant challenge for public health.

Tackling arboviruses and other emerging diseases requires the involvement of various sectors of society, in addition to public health. The implementation of multisectoral policies and interventions is essential to effectively deal with these challenges and guarantee the health and well-being of the Brazilian population. Prevention and preparation of the health sector for these situations are essential in a national context prone to arboviruses, in a country marked by huge inequalities.

References

Alves de Oliveira BF, Bottino MJ, Nobre P et al. (2021) Deforestation and climate change are projected to increase heat stress risk in the Brazilian Amazon. Commun Earth Environ 2:207 https://doi.org/10.1038/s43247-021-00275-8. Accessed in 23 Feb 2024

Araújo VEM de, Bezerra JMT, Amâncio FF, Passos VM de A, Carneiro M (2015) Aumento da carga de dengue no Brasil e unidades federadas, 2000 e 2015: análise do Global Burden of Disease Study. Revista Brasileira De Epidemiologia 20:205–216. 2017. https://doi.org/10.1590/1980-5497201700050017. Accessed in 05 May 2024

Brasil (2022) Ministério da Saúde. Secretaria de Vigilância em Saúde. Departamento de Saúde Ambiental, do Trabalhador e Vigilância das Emergências em Saúde Pública. Rede Nacional de Vigilância, Alerta e Resposta às Emergências em Saúde Pública no âmbito do Sistema Único de Saúde—Rede VigiAR-SUS : Resultados das ações nos anos 2020 e 2021 [recurso eletrônico]/ Ministério da Saúde, Secretaria de Vigilância em Saúde, Departamento de Saúde Ambiental, do Trabalhador e Vigilância das Emergências em Saúde Pública. Brasília: Ministério da Saúde.

Available in: http://bvsms.saude.gov.br/bvs/publicacoes/rede_nacional_vigil. Accessed in 05 April 2024

Brasil (2024) Ministério da Saúde. Vacinação. Available in: https://www.gov.br/saude/pt-br/vacinacao. Accessed in 19 April 2024

Campbell-Lendrum D, Manga L, Bagayoko M, Sommerfeld J (2015) Climate change and vector-borne diseases: what are the implications for public health research and policy?. Philosophical Trans Royal Soc B: Biol Sci 370(1665):20130552. Available in: https://www.ncbi.nlm.nih.gov/pmc/articles/PMC4342958/. Accessed in 05 May 2024

Caballero CB, Ruhoff A, Biggs T (2022) Land use and land cover changes and their impacts on surface-atmosphere interactions in Brazil: a systematic review. Sci Total Environ 808. Available in: https://doi.org/10.1016/j.scitotenv.2021.152134. Accessed in 23 Feb 2024

Cimerman S, Chebabo A, Cunha CA da, Rodríguez-Morales AJ (2020) Deep impact of COVID-19 in the healthcare of Latin America: the case of Brazil. Brazilian J Infect Diseases 24(2):93–95. Available in: https://doi.org/10.1016/j.bjid.2020.04.005. Accessed in 05 May 2024

Confalonieri U, Menezes JA, Margonari C (2017) Environmental change and the emergence of infectious diseases: a regional perspective from South America. In: Examining the role of environmental change on emerging infectious diseases and pandemics. IGI Global, pp 109-137

Conselho Nacional de Saúde (2024) Ministério da Saúde. CNS lança campanha Conselhos contra dengue. Available in: https://conselho.saude.gov.br/ultimas-noticias-cns/3325-cns-lanca-campanha-conselhos-contra-a-dengue#:~:text=Estat%C3%ADsticas%20indicam%20que%20em%202024,em%20investiga%C3%A7%C3%A3o%20e%20122%20confirmados. Accessed in 31 March 2024

Donalisio MR, Freitas AR e Zuben APBV (2017) Arboviruses emerging in Brazil: challenges for clinic and implications for public health. Revista de Saúde Pública. 51. https://doi.org/10.1590/S1518-8787.2017051006889. Accessed in 23 March 2024

Epstein PR (2000) Is global warming harmful to health? Sci Am 283(2):50–57. Available in: https://pubmed.ncbi.nlm.nih.gov/10914399/. Accessed in 05 April 2024

Figueiredo LTM (2007) Emergent arboviruses in Brazil. Rev Soc Bras Med Trop 40(2):224–229. https://doi.org/10.1590/S0037-86822007000200016. Accessed in 23 March 2024

Freitas CM, Pereira AMM, e Machado CV (2022) A Resposta do Brasil à Pandemia da Covid-19 em um contexto de crises e desigualdades. In: Políticas e sistemas de saúde em tempos de pandemia: nove países, muitas lições [online]. Rio de Janeiro, RJ: Observatório Covid-19 Fiocruz; Editora Fiocruz, 342. https://doi.org/10.7476/9786557081594. Accessed in 05 April 2024

Garcia, Leila Posenato (2018) Epidemia do vírus Zika e microcefalia no Brasil: emergência, evolução e enfrentamento. No. 2368. Texto para Discussão, Instituto de Pesquisa Econômica Aplicada. Available in: https://www.econstor.eu/handle/10419/177584 Accessed in 31 March 2024

Githeko AK, Lindsay SW, Confalonieri UE, Patz JA (2000) Climate change and vector-borne diseases: a regional analysis. Bull World Health Organ 78(9):1136–1147. Available in: https://www.researchgate.net/publication/12305249_Climate_Change_and_Vector-Borne_Diseases_A_Regional_Analysis. Accessed in 05 April 2024

IBGE (2004) Instituto brasileiro de Geografia e Estatística. Brasil em síntese. Available in: https://brasilemsintese.ibge.gov.br/territorio.html. Accessed in 23 March 2024

IBGE (2019) Instituo Brasileiro de Geografia e Estatística. Geociências. Biomas. Available in: https://agenciadenoticias.ibge.gov.br/agencia-sala-de-imprensa/2013-agencia-de-noticias/releases/25798-ibge-lanca-mapa-inedito-de-biomas-e-sistema-costeiro-marinho. Accessed in 23 Feb 2024

Lima-Camara TN (2016) Emerging arboviruses and public health challenges in Brazil. Revista De Saúde Pública 50:36. https://doi.org/10.1590/S1518-8787.2016050006791 Accessed in: 23 March 2024

Lopes N, Nozawa C, Linhares REC (2020) Características gerais e epidemiologia dos arbovírus emergentes no Brasil. Revista Pan-Amazônica e Saúde, 5(3):10. https://doi.org/10.5123/S2176-62232014000300007. Accessed in 23 March 2024

Ma J, Guo Y, Gao J, Tang H, Xu K, Liu Q, Xu L (2022) Climate change drives the transmission and spread of vector-borne diseases: an ecological perspective. Biology 11(11):1628. Available in:https://pesquisa.bvsalud.org/global-literature-on-novel-coronavirus-2019-ncov/resource/pt/covidwho-2099326. Accessed in: 05 May 2024

Marengo JA, Oliveira GS, Alves LM (2015) Climate change scenarios in the Pantanal. In: Bergier I, Assine M (eds) Dynamics of the Pantanal Wetland in South America. The Handbook of Environmental Chemistry, vol 37. Springer, Cham. https://doi.org/10.1007/698_2015_357. Accessed in 23 Feb 2024

Martelli CMT, Siqueira Jr. JB, Parente MPPD, Zara A, Oliveira CS, Braga C et al (2015) Economic impact of dengue: multicenter study across four Brazilian regions. PLoS Neglected Tropical Dis 9(9):e0004042. https://doi.org/10.1371/journal.pntd.0004042. Accessed in 05 May 2024

Menezes JA, Confalonieri U, Madureira AP, Duval IB, Santos RB, Margonari C (2018) Mapping human vulnerability to climate change in the Brazilian Amazon: the construction of a municipal vulnerability index. PLoS ONE 13(2):e0190808. https://doi.org/10.1371/journal.pone.0190808. Accessed in 23 March 2024

Ministério da Saúde (2024a) Dengue. Available in: https://www.gov.br/saude/pt-br/assuntos/saude-de-a-a-z/d/dengue. Accessed in 23 March 2024

Ministério da Saúde (2024b) Painel de Monitoramento das arbovioreses—Dengue. Available in: https://www.gov.br/saude/pt-br/assuntos/saude-de-a-a-z/a/aedes-aegypti/monitoramento-das-arboviroses. Accessed in 31 March 2024

Ministério da Saúde (2024c) Febre do Oroupuche. Available in: https://www.gov.br/saude/pt-br/assuntos/saude-de-a-a-z/f/febre-do-oropouche Accessed in 31 March 2024

Musso D (2014) Rapid spread of emerging Zika virus in the Pacific area. Clinical Microbiology Infections 20:595–596. Available in: https://www.clinicalmicrobiologyandinfection.com/article/S1198-743X(14)65391-X/fulltext. Accessed in 05 April 2024

Nobre CA et al. (2016) Land-use and climate change risks in the Amazon and the need of a novel sustainable development paradigm. Proc Natl Acad Sci USA 113:10759. https://doi.org/10.1073/pnas.1605516113. Accessed in 23 Feb 2024

Nunes J, Pimenta DN (2016) A epidemia de Zika e os limites da saúde global. Lua Nova: Revista De Cultura E Política (98):21–46. https://doi.org/10.1590/0102-6445021-046/98. Accessed in 05 April 2024

Organização Pan-Americana de saúde (2024) Atualização epidemiológica—Oropouche na Região das Américas - 6 de março de 2024. Available in: https://www.paho.org/pt/documentos/atualizacao-epidemiologica-oropouche-na-regiao-das-americas-6-marco-2024. Accessed in 31 March 2024

IPCC (2023) Painel Intergovernamental sobre Mudança do Clima. Mudança do Clima 2023: relatório síntese. Traduzido para o português pelo Governo do Brasil. 19 a 21. Available in: https://www.gov.br/mcti/pt-br/acompanhe-o-mcti/sirene/publicacoes/relatorios-do-ipcc. Accessed in 23 Feb 2024

Rust RS (2012) Human arboviral encephalitis. Semin Pediatr Neurol 19(3):130–151. https://doi.org/10.1016/j.spen.2012.03.002. Accessed in 23 March 2024

Santos RB. dos, Menezes JA, Confalonieri U, Madureira AP, Duval I de B, Garcia PP, Margonari C (2019) Construção e aplicação de um índice de vulnerabilidade humana à mudança do clima para o contexto brasileiro: a experiência do estado do Espírito Santo1. Saúde E Sociedade 28(1):299–321. https://doi.org/10.1590/S0104-12902019180431. Accessed in 23 March 2024

Suaya JA, Shepard DS, Siqueira JB, Martelli CT, Lum LC, Tan LH, et al (2009) Cost of dengue cases in eight countries in the Americas and Asia: a prospective study. Am J Trop Med Hyg 80(5):846–855. Available in: https://www.researchgate.net/publication/24394157_Cost_of_Dengue_Cases_in_Eight_Countries_in_the_Americas_and_Asia_A_Prospective_Study. Accessed in 05 May 2024

Vommaro F, Menezes JA, Barata MML (2020) Contributions of municipal vulnerability map of the population of the state of Maranhão (Brazil) to the sustainable development goals. Sci Total

Environ 1, 706:134629. https://doi.org/10.1016/j.scitotenv.2019.134629. Epub 2019 Nov 22. PMID: 31864137. Accessed in 23 March 2024

Wesselmann KM et al (2024) Emergence of Oropouche fever in Latin America: a narrative review. The Lancet Infectious Diseases, 2024. Available in: https://www.sciencedirect.com/science/article/pii/S1473309923007405?casa_token=XAUB_upLmKoAAAAA:HMJc26WPr5X0JXhn5i8JkAMKpVL8ibpzUkzK2keGggjq2rMqqXEH0ZxJnGGpCQ59seORbKP-91o. Accessed in 31 March 2024

Wilder-Smith A, Tissera H, Ooi EE, Coloma J, Scott TW, Gubler DJ (2020) Preventing dengue epidemics during the COVID-19 pandemic. Am J Trop Med Hyg 103:570–571. Available in: https://www.ncbi.nlm.nih.gov/pmc/articles/PMC7410414/. Accessed in 05 May 2024

Chapter 23
Argentina in the Post-COVID Era: Social, Health, and Economic Impact of the Pandemic

Daniel Oscar Lipp

Abstract The impact of the pandemic in the country was widespread, affecting all areas of the economy, in addition, of course, to the numerous direct consequences on health. This work describes what happened in the socioeconomic field particularly affected by the crisis. The growing levels of poverty, lower levels of well-being, growing public spending, informality, and unemployment show how extensive and deep our economic and social deterioration has been as a result of the pandemic.

Keywords Pandemic · COVID-19 · Economic activity · Gross domestic product · Unemployment · Preventive and mandatory social isolation · Teleworking · Telemedicine

23.1 Introduction

The pandemic impacted Argentine society with a virulence never imagined. It is most direct, and dramatic impact was on health. Mortality rates were significantly high. By mid-2020, according to official statistics, 45,568 people had died in the country, although there are estimates that the excess deaths were even higher. Of the six countries with the highest number of confirmed deaths from COVID-19 at the regional level, three are Latin American: Brazil, Chile, and Argentina. At the same time, the health crisis triggered by the pandemic and the public health measures implemented to stop the spread of the virus had a far-reaching economic impact on the country due to the paralysis of tourism, the suspension of non-essential activities and the decrease in remittances and fall in international trade. As a consequence of all this, Argentina fell significantly in the last decade (Rearte 2021, p. 3).

The COVID-19 pandemic in the country therefore not only generated serious consequences on the health of the population, directly exposed to the virus, but other

D. O. Lipp (✉)
Universidad de Buenos Aires, Buenos Aires, Argentina
e-mail: daniellipp@arnet.com.ar

Universidad Católica de Salta, Salta, Argentina

R. Akhtar (ed.), *COVID-19: Impacts on Health and Health Systems*,
https://doi.org/10.1007/978-3-031-90255-0_23

impacts that are not directly related to the disease caused by the virus must also be considered. I am referring to the economic and social aspects that the pandemic brought about upon its arrival in Argentina. For this reason, I will first limit myself to developing aspects that the pandemic generated within the health system in the country and then I will address the effects that are not directly related to health services, but rather their consequences are generated in other areas or services.

23.2 Materials and Methods

The methodology consisted of a search and selection of specialized bibliography, followed by its analytical treatment. The data were collected from the National Institute of Statistics and Censuses (INDEC), the nation's main body for statistical information and socioeconomic indicators. Consulting specialized journals on the subject, institutional documents on websites and other secondary sources was certainly not omitted. With this information collected, it was systematized, processed, analyzed, and evaluated. The tables and graphs allow, in a simple and quick way, the comparison of data in the present study.

23.3 Results

One of the most immediate impacts that the pandemic produced in the world, and particularly in Argentina, was the interruption or delay in the care of different health services. For example, a study by the Economic Commission for Latin America and the Caribbean (ECLAC) and the Pan American Health Organization (PAHO) on the region of the Americas (28 countries) concluded that 28% of the countries had partial interruptions in oncological treatments and that, in addition, in 17% of them the care of cardiovascular emergencies was partially affected. Similar situations were observed in Argentina. A survey of 31 healthcare centers showed, comparing 2020 with 2019, that emergency consultations decreased by 74%; admissions for angina pectoris and acute coronary syndrome, 62%, and those due to cerebrovascular accidents and transient ischemic attacks were reduced by 46%. Another report from the country showed a 16% decrease in hospitalizations for acute myocardial infarction (Perman et al. 2020, p. 186). In other specialties, services have also been partially or completely interrupted. Consultations in clinical dermatology decreased by 85%, and within this percentage, there are patients both in the diagnosis stage (laboratory studies, images, etc.) and already diagnosed, in gynecology consultations decreased by 71% and those in mastology, 61%. At the Austral University Hospital (HUA), the number of mammograms and breast ultrasounds was reduced by 83% in April. While in the gynecology area of the *"Halitus Medical Institute"* consultations decreased by almost 90% (Clarin 2020, p. 1).

The arrival of the pandemic in Argentina also had other impacts with serious social consequences. These had no direct relationship with health services, but rather strictly affected the mood of the population. Social isolation and strict confinement generated uncertainty, fear, worry, anguish, and anxiety. An electronic survey carried out at the beginning of the quarantine showed, indeed, a very depressed mood in people, in addition to an increase in the consumption of alcohol and other substances in young people and unemployed people. The occurrence of self-induced injuries, or due to accidents or violence related to the consumption of alcoholic beverages generated enormous damage in the country that will surely increase in the future without strict control by the authorities. However, an area of concern since the beginning of the pandemic was responding to mental health-related healthcare needs. This has had a considerable impact because added to the difficulties inherent in maintaining and guaranteeing attention to their usual demand, the considerable increase in cases of anguish, anxiety, feelings of fear, and panic that the pandemic would bring.

Another negative impact of social isolation is the increase in episodes of domestic violence. One report estimates that the projected number of children exposed to domestic violence due to COVID-19 could be closer to 52–84 million worldwide. On the other hand, a working document from the United Nations (UN) projects an excess of 15 million cases of gender-related violence. As for the situation of older adults, it is no less dramatic. It can also have very important consequences in terms of loneliness (26%) or strict isolation (29%) (Perman et al. 2020, p. 187). Incalculable efforts are being made in Argentina to try to alleviate this serious situation, although the impact is probably even greater than expected.

23.4 Post-COVID-19 Syndrome

The pandemic had a great impact in terms of the collateral damage it caused. We have already listed, for example, the care that was no longer provided to many patients with chronic diseases, the stoppage of preventive and vaccination programs for patients with organic disorders, just to mention a few. However, COVID has generated other physical and mental consequences that have been recently investigated, and this is what infectious disease specialists define as post-COVID syndrome or long COVID.

Post-COVID-19 syndrome is a new medical category that began to spread in 2020 to describe the long-term effects of the disease caused by the SARS-CoV-2 virus. This appears weeks or months after recovery from active COVID and even after vaccination. The World Health Organization (WHO) has defined post-COVID syndrome as a set of multisystemic, subacute, or chronic symptoms, which can represent a significant decrease in the functional capacity of patients (Halabe et al. 2022, p. 8).

The Pan American Health Organization (PAHO) summarized which organs can be affected by this condition (Fig. 23.1). This includes neuropsychiatric, hematological, cardiovascular, pulmonary, and other organ and system manifestations. To make its interpretation more precise, the National Institute for Health and Care Excellence

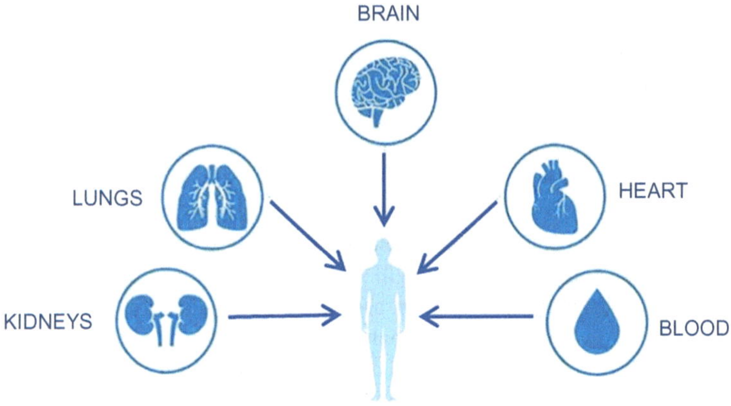

Fig. 23.1 Organs affected by post-covid syndrome. *Source* Pan American Health Organization (PAHO)

(NICE) in the United Kingdom uses the term long COVID to define symptoms that continue beyond the first month or appear after the initial infection. It should be noted that several of the scientific publications related to the post-COVID-19 syndrome integrate both researchers and specialists, as well as patients and families, which results in the important and necessary link between science and people in the knowledge of the problems of health.

23.5 Social and Economic Impacts

Most countries, due to the COVID-19 pandemic, have imposed a series of restrictions on activities that generated, of course, a devastating socioeconomic effect. Quarantines, confinements, no trips, cancelation of events, and a total closure of certain establishments were just to prevent the spread of the virus. This caused a collapse in the market, leading many companies to their definitive closure, and leaving, of course, millions of workers without jobs. Argentina was one of those countries to prioritize the health system, establishing a strict confinement. This measure, although it was necessary to contain the virus and avoid health collapse, will trigger a serious economic crisis and tremendous social conflicts (Gozza and Fontana 2020, p. 10).

The economic-social impact that this crisis has had are varied and global. Argentina has a lot to do in this regard. At the beginning of 2020, the Argentine economy had accumulated eight years of falling GDP per capita, inflation was above 50% annually and the poverty rate had increased to more than a third of the population (United Nations 2021, p. 11). At the same time, the public debt, which had reached the highest level in Latin America, was in the process of restructuring. The coronavirus disease pandemic broke out in this situation of economic fragility, which, of course, contributed to deepening pre-existing problems.

Firstly, the pandemic aggravated the poverty that was already experienced in Argentina decades ago, a product of the recession and poor management of the economy. In reality, the crisis unleashed by the pandemic affected the entire society, but it had a much greater impact on households that were already in a vulnerable situation and/or received low income, as well as in some labor sectors, such as construction, commerce, hotels and restaurants, and domestic work. As a result of this crisis, the national government initiated a series of actions to benefit such sectors to mitigate the impact. One of the measures carried out were monetary transfers to the most vulnerable families with adolescent children, to holders of social programs and to the elderly. At the same time, in addition to this, and with the objective of contributing to the well-being of the sectors whose income was reduced to a minimum, the national government ordered the creation of an Emergency Family Income (IFE) consisting of an exceptional payment of $ 10,000 (Argentina Gob. AR. 2021, p. 1). These initiatives were complemented by other types of social policies, such as programs to strengthen access to remote education, the provision of basic goods and services, and investment in infrastructure and health prevention. Such measures made it possible to mitigate the consequences of the disease and the collapse of the health system.

However, despite the efforts seen above, the outlook remained critical throughout the pandemic. Already in the first half of 2020, poverty reached 40% of the population, according to INDEC data. This means that between 2019 and 2020, 2.5 million people entered poverty. It should be noted that, without the mitigation measures proposed by the national government, the poverty rate could have increased by 2.6% points more, reaching 43.5% of the population (Pagina 12 2024, p. 2).

The pandemic also had a strong impact on the Argentine labor market, which, of course, was already in a critical situation before the health emergency (Fig. 23.2). Thus, in the last quarter of 2019, a slight increase in the unemployment rate was recorded in the country, but its evolution in the previous two years, the rate remained stable and still declining. The most recent data on unemployment records that in the fourth quarter of 2019 around 1.2 million people were actively looking for work, with an unemployment rate of 8.9%. In the months of March and April 2020, falls of approximately 3% annually were recorded in private employment registered in companies with more than 10 workers. This is the largest drop in labor activity since 2002 in Argentina. Although in the country, the majority of labor institutions are designed to fundamentally protect formal workers, these crises, like the pandemic, reveal a situation of lack of protection for a large percentage of workers who are active in informality and those who work on their own.

Regarding economic activity in 2020, GDP contracted by 9.9%, the largest annual drop since the 2001/2002 crisis (Fig. 23.3). This decrease is explained by the negative impact that the pandemic had on both supply and demand, due to restrictions on circulation, high uncertainty and lower international trade flows, which, therefore, were reflected in the retraction of private consumption, investment, and exports. The important package of measures implemented by the national government only partially offset the negative impact of the pandemic (United Nations 2021, p. 12).

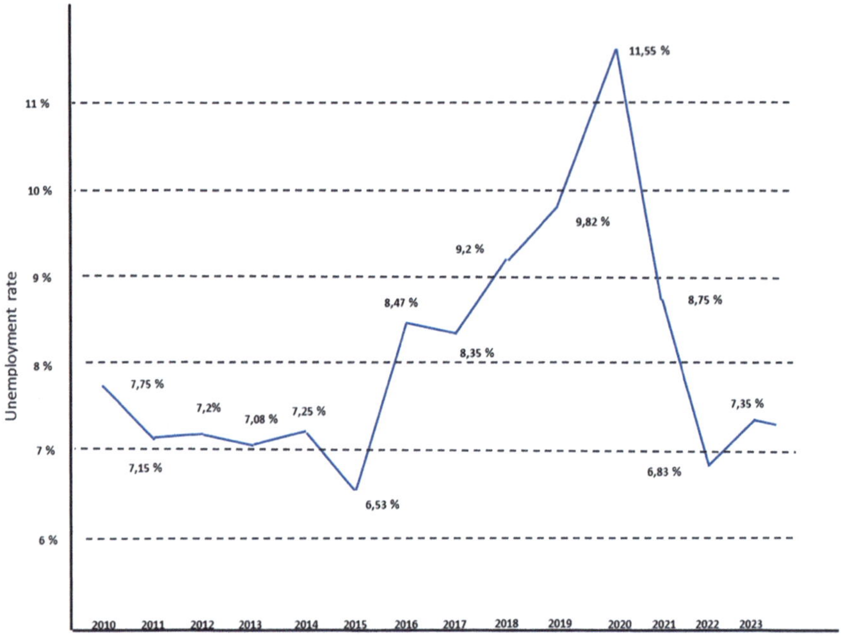

Fig. 23.2 Unemployment rate in Argentina between 2010 and 2023. *Source* National institute of statistics and censuses (INDEC)

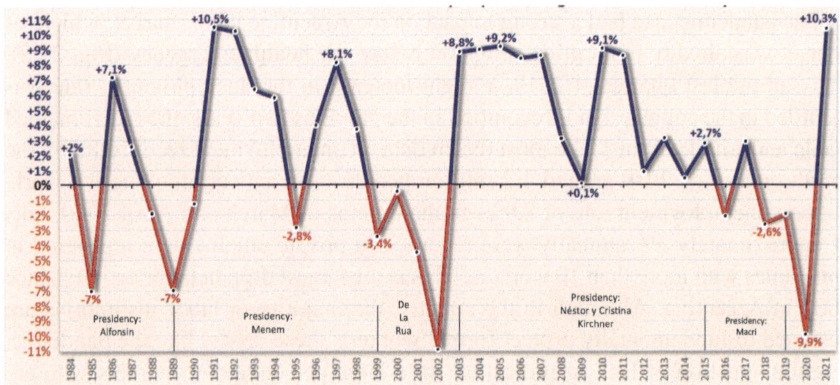

Fig. 23.3 Annual variation of gross domestic product (gdp) of Argentina between 1984 and 2021. *Source* Instituto Nacional de Estadísticas y Censos (INDEC)

Regarding, however, the fall in economic activity at the sectoral level was general, with the exception of residential public services (electricity, gas, and water), which, due to confinement (quarantine) and disinfection measures, increased, but very slightly compared to the previous year. Activities linked to financial intermediation also increased, which, toward the end of the year, recovered, closing with a growth of 2.1% compared to the previous year.

Regarding the economic sectors most affected by the crisis, they were those linked to tourism and cultural and recreational activities, such as hotels and restaurants, and other community, social and personal service activities, which fell by 49% and 39%, respectively. The negative impact was also quite intense in the construction sector (−23%), and in private domestic service households (−19%), which concentrate a high percentage of urban jobs, high rates of informality, low salaries, and high levels of poverty (Table 23.1) (NOTIAR.COM. 2021, p. 3).

23.6 New Work Modalities: Teleworking

The pandemic not only had economic impacts but also employment impacts. Workers who kept their jobs had no other option to adapt to new modalities, such as teleworking, also called *Home Office*. Although this form of employment had been occurring before the appearance of COVID-19, it was a process that was later accelerated to comply with the measures dictated by the government regarding the isolation and mandatory distancing of people in the context of the pandemic.

Thus, on the worker's side, they had to adapt and become familiar with the new computer tools that they may not have been used to using before the pandemic and adapt their home as a workplace. Although these changes may have been friendlier for adolescents or young adults, this is not the case for older people, for whom the adaptation process may have brought greater difficulties. And, on the other hand, the companies because they had to provide every worker who needs them with the basic resources necessary to be able to continue carrying out their work tasks and thus be able to keep their job. This type of work was regulated in Argentina by law no. 27,555 and law no. 20,744.

These laws established a regulatory framework that protects the rights of employees working remotely, establishing rights and obligations for both employers and workers. However, not all companies were sure to adopt it. Not being in a workplace under the control of supervisors, many employers saw the *Home Office* as something negative, where they did not have control over the worker and whether they worked more or less time than being in the office. In any case, many Argentine companies adopted this modality, especially those in the administrative, financial, technological, and electronic commerce sectors. This modality of work was also commonly used in the education sector, where virtual classes began to be taught through various digital platforms, such as *Zoom, Google Meet, Skype, and Discord*, just to name a few. Figure 23.4 shows a wide variety of telematic equipment and services used by teleworkers, 18% use mobile telephony, 16% Internet access, 14% desktop equipment, another 14% fax, 13% landline telephony, 9% other types of equipment, such as radios and GPS, 7% portable equipment, 7% video conferencing services, and 2% PDA.

Table 23.1 Selected indicators of activity level, jobs, feminization, and incidence of poverty in employed people (inter-annual variation in percentages)

	2020					1ST Semester 2020	
	GDP (%)	Total (%)	Registered employees (%)	Unregistered employees (%)	Non-salariated (%)	They live in poor homes (%)	Feminization rate (%)
Financial intermediation	2	−3	−2	−20	−4	7	45
Electricity, gas, and water	1	−1	−1	−2	6	28	17
Teaching	−5	−2	−1	−17	−13	13	74
Trade	−5	−8	−3	−12	−10	33	39
Real estate, business, and rental activities	−6	−7	−1	−11	−15	19	40
Public administration and defense	−7	1	1	0	0	15	45
Primary activities	−8	−3	−2	−13	3	20	22
Manufacturing industry	−8	−1	−3	1	−1	36	33
Social and health services	−10	−3	1	−2	−15	12	71
Transportation, storage, and communications	−17	−13	−5	−25	−20	20	12
Private homes with domestic service	−19	−19	−2	−26	–	52	96
Construction	−23	−15	−23	−16	−10	49	3
Other community, social, and personal service activities	−39	−11	−3	−19	−15	34	47
Hotels and restaurants	−49	−19	−11	−28	−22	35	38
Grand total	−9.9	−7.5	−3	−16	−10	28	44

Sources Instituto Nacional de Estadísticas y Censos (INDEC)

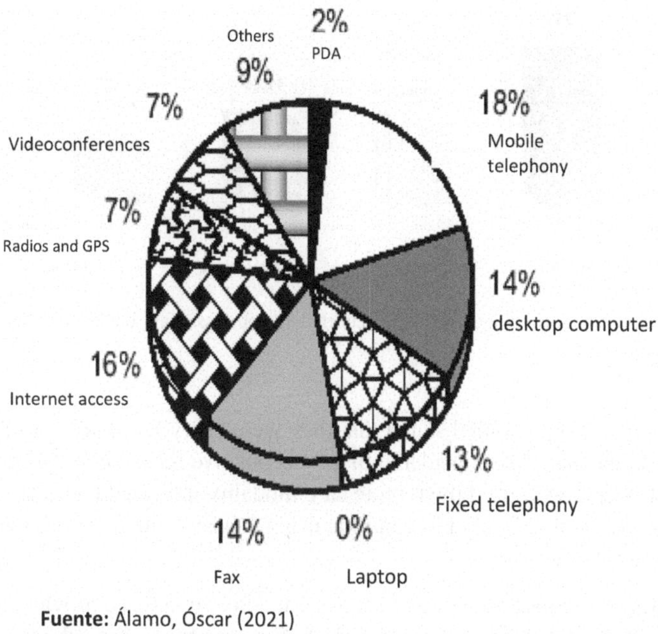

Fig. 23.4 Equipment and services used in telework

However, these changes in work modalities brought imbalances in people's lives, affecting their psychophysical state, family coexistence, and rest. Regarding manufacturing activity, they did not have these problems for the most part since the virtual work modality was less. But there are others: the fear of personal contagion and that of their colleagues, the fear of bringing the virus home, very restrictive security protocols that hinder tasks, among others (Albrieu 2020, p. 45).

23.7 Telemedicine

Health care also had to familiarize itself with new IT tools after the pandemic began. Although the use of these new technologies was a frequent modality in Argentine medicine, before the appearance of COVID-19, it was urgent to implement it in its entirety during the crisis. The arrival of the pandemic in the country definitively established this modality and more and more patients are widely verifying its benefits. In this sense, at the end of 2020, the Ministry of Health of the Nation released a survey on the subject with very eloquent data (UNIDIVERSIDAD 2022, p. 8).

1. During 2020, 72% of all patients accessed the health system through their cell phone.

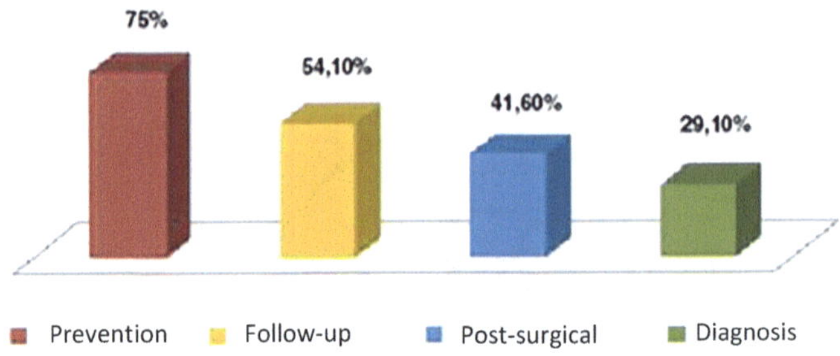

Fig. 23.5 Main functions of telemedicine reported by professionals during pandemic. *Source* Pariset al. (2023)

2. 82% of medical personnel stated that they were "very familiar" with the use of telemedicine and 75% considered that "it is positive for the population."
3. 77% of people who used the remote care modality indicated that it is "good" or "very good" and 73% plan to continue using it post-pandemic with the purpose of saving time or money.

Telemedicine, whose name involves these medical practices, provides excellent benefits to doctors and patients. Figure 23.5, for example, details the functions of telemedicine in medical consultations made by patients. The most valuable function was prevention (75%) over follow-up, postsurgical and diagnostic functions (Parise et al. 2023, p. 10). On the other hand, telemedicine helps eliminate territorial barriers and improves equal access to services that would otherwise be impossible or unavailable in remote communities. It is about transmitting voice, data, images, and information instead of physically moving patients or health personnel, which, of course, would involve travel and time costs. It should be noted that another important aspect of telemedicine is the use of telecommunications between professionals when providing a medical service.

Argentina is a country that decades ago became a pioneer in the use of telemedicine. The renowned pediatric hospital *"Professor Dr. Juan P. Garrahan"* has been, for decades, the only medical institution that had its own telemedicine program in Argentina. This medical center organized a very extensive telemedicine network that goes from Buenos Aires to the rest of the country's provinces. In its beginnings, it had the purpose of preventing its patients from traveling to receive medical care. In this way, patients from distant geographical areas were able to access highly complex treatments and subsequent follow-up, without leaving their homes.

23.8 Conclusions

The COVID-19 pandemic had a great impact on the world, not only did it affect the health of millions of human beings, generating multiple consequences, but it also led to the death of a large number of people infected by the virus both in the short term and in the long term. Argentina was no exception; here, the disease spread at great speed and with serious direct consequences in morbidity and mortality. The economic and social crisis generated by the pandemic was another of the aspects, perhaps the most devastating, that was unleashed due to the expansion of the virus. By limiting production and consumption activities, it caused a market collapse in Argentina, leading many companies to bankruptcy and permanent closure, and, therefore, leaving millions of workers without a livelihood. In addition, the zero growth of the economy, the low domestic savings, and the very low levels of domestic and foreign direct investment caused the increasing levels of poverty that the country had during the entire time that the COVID pandemic lasted. In addition to this, Argentina's inflationary history determined that during the crisis our currency lost two of the three functions attributed to money: means of value accumulation and unit of account, reserving exclusively the transactional function.

However, in response to the COVID-19 crisis, new forms of work were generated by implementing information and communication technologies (ICTs). Although these were available before the pandemic as a work modality, the development achieved during the crisis. It was spectacular, at least in the different sectors of the economy that made use of it. Telemedicine, a variety of virtual work modality, increased enormously during the pandemic in Argentina, improving medical coverage and mitigating the effects of the crisis.

References

Albrieu R (2020) Evaluando las oportunidades y los límites del teletrabajo en Argentina en tiempos del COVID-19. CIPPEC, Buenos Aires. https://www.cippec.org/publicacion/evaluando-las-oportunidades-y-los-limites-del-teletrabajo-en-argentina-en-tiempos-del-covid-19/

ARGENTINA GOB. AR (2021) Ministerio de Economía. Medidas Económicas COVID-19. "Ingreso familiar de emergencia" Buenos Aires, Argentina. https://www.argentina.gob.ar/economia/medidas-economicas-COVID19/ingresofamiliardeemergencia

CLARIN J (2020) Efectos colaterales del coronavirus: la desatención de otras enfermedades podría dejar al menos 6.000 muertos. Argentina, Buenos Aires. https://headtopics.com/ar/efectos-colaterales-del-coronavirus-la-desatenci-n-de-otras-enfermedades-podr-a-dejar-al-menos-6-00-13773043

Gozza MA, Fontana MR (2020) Impacto de una pandemia en la economía argentina. Universidad Nacional de San Martín, UNSAM, Trabajo de investigación, Buenos Aires, Argentina

Halabe J, Robledo Z, Fajardo G (2022). Síndrome post-COVID-19. Certezas e interrogantes. Editorial Médica Panamericana. México

NACIONES UNIDAS (2021) Análisis Conjunto del Sistema de Naciones Unidas 2021: los efectos de la pandemia por covid-19 en la argentina. Documento de Actualización del Análisis Común

de País 2020–2021. https://argentina.un.org/es/145708-an%C3%A1lisis-conjunto-del-sistema-de-naciones-unidas-2021-los-efectos-de-la-pandemia-por-covid
NOTIAR.COM (2021) Duro informe de la ONU sobre el impacto negativo del Covid en la Argentina. Argentina, Buenos Aires. https://notiar.com.ar/index.php/politica/114393-duro-informe-de-la-onu-sobre-el-impacto-negativo-del-covid-en-la-argentina-por-martin-dinatale
PAGINA 12, (2024) La pandemia podría haber hecho más daño. Buenos Aires, Argentina. https://www.pagina12.com.ar/296197-la-pandemia-podria-haber-hecho-mas-dano
Parise AC, Coria H, López V, Santini Araujo G (2023) Pie diabético y pandemia de la COVID-19: adaptación al uso de la Telemedicina por los especialistas de pie y tobillo de los Servicios de Traumatología de la Argentina. Revista de la Asociación Argentina de Ortopedia y Traumatología. https://raaot.org.ar/index.php/AAOTMAG/article/view/1623/5098
Perman G, Puga C, Ricci I, Terrasa SA (2020) Daños colaterales de la pandemia por COVID-19: ¿consecuencias inevitables?. Rev Hosp Ital B Aires (2004), 40(4):185–190
Rearte A (2021) Exceso de mortalidad por todas las causas en contexto de pandemia de COVID-19. Argentina, 2020. Rev argent salud pública vol. 13 supl. 1 Buenos Aires abr. http://www.scielo.org.ar/scielo.php?script=sci_arttext&pid=S1853-810X2021000200018
UNIDIVERSIDAD (2022) Ministerio de Salud de la Nación, Buenos Aires. Argentina. https://www.unidiversidad.com.ar/telemedicina-72-de-pacientes-uso-el-celular-y-52-consulto-por-covid

Chapter 24
Conclusion and Suggestions

Rais Akhtar

Country studies included in the book encompasses both from developed and developing countries, reveals the pandemic has pointed to the significance of collaboration at local, regional, and international levels to successfully manage unforeseeable events such as COVID-19.

The impact the corona crisis was a short-term devastating effect causing at the same time severe suffering all over the world. Long-term improvements can only be achieved with targeted climate and environmental policies that sustainably and permanently change production structures, infrastructures and consumption and mobility patterns. The goal is not to return to pre-industrial societies, but to invest in resilient economies and communities that can prosper without damaging their foundations. Economic stimulus/investment packages that are crucial to get economies back on track in region such as Pacific Ocean countries, need to be combined with sustainability goals. Climate protection, emission reductions, and resource conservation are crucial measures to continue. Whether overcoming COVID-19 encouraged structural changes toward climate protection and sustainability is still to be seen. It might be that people (incl. politicians) only respond and take decisive action, according to John Connell, until the water is up to one's neck. Therefore, researchers have rightly urged countries to invest in strategies to improve the resilience of their health systems for future emergencies, as new data sheds light on the extent of disruption to essential services like cancer screening, maternal health, and vaccination during the COVID-19 pandemic (Geddes 2022), as well as cardiovascular and vector-borne diseases.

As the world warms, researchers assert, mosquitoes will move beyond their current habitat, shifting the burden of diseases like malaria, dengue fever, chikungunya, and West Nile virus to new regions.

R. Akhtar (✉)
Formerly Professor of Geography, University of Kashmir, Srinagar, Jammu and Kashmir, India
e-mail: raisakhtar@gmail.com

R. Akhtar (ed.), *COVID-19: Impacts on Health and Health Systems*,
https://doi.org/10.1007/978-3-031-90255-0_24

Researchers forecast different scenarios depending on the extent of climate change (Jordan 2019). In an analyses of pattern of dengue incidence before, during, and after, the coronavirus disease 2019 in Taiwan by Mei-Hui highlights that, dengue transmission is influenced by multiple factors, such as temperature, rainfall, population immunity, urbanization, population mobility, and health systems with multiple scales. The year 2023 saw Taiwan experience its second most severe dengue outbreak, with 277 imported cases and approximately 26,000 indigenous cases. An increased risk of exposure to the imported cases is not adequate to explain an outbreak of such a large scale. Similarly, climatic factors alone cannot explain the low ingenious incidence of dengue in 2019 or the second most severe outbreak in 2023. The changes in the pattern of dengue outbreaks in Taiwan reflect the complex interplay among social, political, and environmental factors on multiple levels. This had been brought to our attention to a great extent much before the outbreak of Covid-19, that we must not neglect outbreak of tropical diseases (Fürst et al. 2017). It should be noted that climate change has caused vectors like ticks and mosquitoes to expand their range of altitudes and latitudes. For example, the Asian tiger mosquito, Aedes albopictus, has spread to many European countries, including the Balkans, France, and the Mediterranean basin. Lyme disease, also known as Lyme borreliosis, is a prevalent tick-borne disease in Europe. Local transmission of malaria remains possible in Europe. With reference to USA, Michele Kekeh and Hedieh Sirjani assert that the experience gained from the Covid-19 pandemic could guide preparedness efforts for future pandemics and upcoming health challenges. The necessity for creating robust infrastructure in the health care system, including adequate and accessible medical technology, attention to mental health,training providers and caregivers, and comprehensive health insurance coverage, was underscored during pandemic.

The study on Japan by Ai Tashiro emphasizes importance of global health governance which is a framework that addresses global health issues using various methods involving states and non-state actors. With the progression of globalization and the transformation of infectious diseases into global threats, health governance has become more complex and intertwined with politics. Unfortunately, cooperation is currently hindered by inter-state conflicts and nationalistic tendencies during the COVID-19 pandemic.

The lack of global solidarity in Asia, problems with global health regulations, and the absence of a global response coordination mechanism are the main issues that require active involvement from the international community. Global health governance is a political arena comprising diverse actors, including major powers, smaller states, and non-state actors. If these actors can build strong solidarity in developing COVID-19 vaccines and treatments, ensuring fair distribution, supporting pandemic resolution, and reinforcing response frameworks, Japan's future in health governance in the post-COVID-19 era looks promising. The urgent task for global health governance is to establish a sustainable health cooperation framework that can withstand political changes while utilizing the individual roles of diverse actors.

Following the COVID-19 pandemic, Indonesia strengthened its healthcare system to better combat infectious diseases, including dengue fever, tuberculosis (TB), malaria, and conditions related to air pollution, while also improving worker health. Recognizing its vulnerability, Indonesia invested in health surveillance and disease monitoring systems to provide earlier detection and faster response times for outbreaks, including mosquito-borne diseases like dengue and malaria. Expanded distribution of insecticide-treated nets, environmental management strategies, and public awareness campaigns have helped curb vector populations and reduce transmission.

The study on Thailand highlights the necessity of integrating environmental considerations into policy formulation processes. Furthermore, the COVID-19 pandemic served as a revealing lens, shedding light on both the strengths and weaknesses of Thailand's healthcare system, as well as those of nations worldwide. In response to identified weaknesses, the Thai government implemented measures to fortify the healthcare system, including the recruitment of village health volunteers, migrant volunteers, and the implementation of tele-healthcare services. Additionally, intriguing shifts observed in the disease patterns of NCDs are likely influenced by a combination of factors, including environmental influences, public health interventions, lifestyle modifications, and healthcare accessibility. These findings underscore the importance of holistic approaches to address emerging health challenges and enhance overall population health outcomes.

Finally, the way forward in the process of health and development post–Covid-19 as revealed by Hem and Jhala in their chapter on India and Canada.

The authors assert that nearly five years have passed since the beginning of the COVID-19 pandemic. Over this time period, the virus has successfully mutated, several times, and often managed to evade host's immune defenses. It has also challenged vaccine makers to find an adequate response and often overwhelmed health systems around the world (both in developed and developing countries).

Scientists have urged countries to invest in strategies to improve the resilience of their health systems for future emergencies, as new information reveals the extent of disruption to essential services like cancer screening, maternal health, and climate change induced tropical diseases during the COVID-19 pandemic.

Finally, as Geddes exhorted that "Health system resilience must become a central component of national health plans. Given the likelihood of future pandemics and other major shocks, there is an urgent need to design more resilient health systems capable of addressing a crisis while maintaining essential functions."(Geddes 2022).

References

Fürst T, Salari P, Llamas LM, Steinmann P, Christopher Fitzpatrick FT (2017) Global health policy and neglected tropical diseases: Then, now, and in and in the years to come. PLoS Negl Trop Dis. https://doi.org/10.1371/journal.pntd.0005759

Geddes L (2022) How the COVID-19 pandemic has affected healthcare around the world (New insights into the disruption caused by COVID-19 could help strengthen health systems ahead of future pandemics) Vaccineswork.gavi.org

Jordan R (2019) How does climate change affect disease? Stanford Woods Institute for the Environment, News and events, March 15

Index

R. Akhtar (ed.), *COVID-19: Impacts on Health and Health Systems*,
https://doi.org/10.1007/978-3-031-90255-0

Q

R

S

The manufacturer's authorised representative in the EU is Springer Nature Customer Service Centre GmbH, Europaplatz 3, 69115 Heidelberg, Germany. If you have any concerns regarding our products, please contact ProductSafety@springernature.com

Printed and bound by CPI Group (UK) Ltd, Croydon, CR0 4YY

30/09/2026

02225564-0014